Algal Biorefineries in Circular Economy for Climate Change Resilience

This book explores the multifaceted role of algae in addressing sustainable development, circular economy practices, and climate change mitigation issues. The integration of flue gas for algal biofuel production and its contribution to sustainable energy solutions is included along with role of algae-based biofertilizers in reduction of carbon footprints.

Other topics discussed include algae for biomass production and carbon sequestration, algae-based carbon credits, production of bioplastics for climate-resilient packaging, and their integration into waste management.

Key features—

- Covers algal biorefineries in circular economy and climate change mitigation
- Discusses the role of biogas and microplastics
- Provides insights into waste management strategies
- Reviews the role of biorefineries in the achievement of net zero emission targets
- Explains government incentives, regulations, and the economic feasibility of implementing pertinent technologies

This book is aimed at graduate students and researchers in biotechnology, environmental sciences, and climate change.

Climate Change Resilience through Biorefinery Technologies
Series Editor- Bikash Kumar and Kiran Bala

Algal Biorefineries in Circular Economy for Climate Change Resilience
Edited by Fayaz A. Malla and Suhaib A. Bandh

Algal Biorefineries in Circular Economy for Climate Change Resilience

Edited by
Fayaz A. Malla and Suhaib A. Bandh

CRC Press
Taylor & Francis Group
Boca Raton London New York

CRC Press is an imprint of the
Taylor & Francis Group, an **informa** business

Designed cover image: Getty Images

First edition published 2026
by CRC Press
2385 NW Executive Center Drive, Suite 320, Boca Raton FL 33431

and by CRC Press
4 Park Square, Milton Park, Abingdon, Oxon, OX14 4RN

CRC Press is an imprint of Taylor & Francis Group, LLC

ISBN: 9781032842851 (hbk)
ISBN: 9781041166771 (pbk)
ISBN: 9781003685814 (ebk)

DOI: 10.1201/9781003685814

Typeset in Times
by Newgen Publishing UK

Contents

Preface

There must be an immediate shift from linear resource models to systems that are both sustainable and regenerative if we are to combat the growing threats of climate change, resource shortages, and environmental degradation in this century. Combining algae biorefineries with the principles of the circular economy is a very attractive strategy. Investigating new biological platforms that are adaptable to climate change and efficient with resources is becoming more important as the world's population and industrial demands continue to rise.

Algae, both micro and macro, have a lot of untapped potential as sustainable bioresources that can fix environmental problems and turn trash into treasure, as this book explores. Key participants in the development of green biotechnologies are algal systems because of their adaptability, high carbon capture capacity, and rapid growth. Algae provide a comprehensive answer to resource circularity and climate change mitigation through their use in a wide range of applications, including bioenergy generation, wastewater treatment, biofertilizer manufacturing, bioplastics, nutraceuticals, and pharmaceuticals.

Integrating algal biorefineries into models of the circular economy is one of the main focuses of this book. Systems based on algae close resource loops and lessen environmental footprints by valorizing waste streams like CO_2 emissions, industrial effluents, and agricultural runoffs. The book delves into the ways in which carbon sequestration, nutrient recycling, and the creation of agricultural and industrial systems that can withstand climate change can be enhanced through the use of circular algal biorefineries.

In addition, we talk about how cutting-edge technologies like synthetic biology, nanobiotechnology, multi-omics, and life cycle assessment (LCA) might help improve algae operations. System integration is highlighted in the chapters as a crucial component to maximize economic and environmental returns. This involves connecting the production of algal biomass with downstream processing, renewable energy systems, and waste management infrastructures.

The role of algae in reducing vulnerability to climate change is a prominent theme, as is their ability to remediate wastewater, provide sustainable fuels, and add value to many products. A future that is both economically and environmentally sustainable can be achieved through the complementary efforts of biological innovation and the principles of the circular economy.

Academics, practitioners, politicians, and researchers all looking to learn more about how to use algae technology for sustainable development will find this book to be an invaluable resource. It hopes to inspire a new era of innovation in which biotechnology and circularity come together to make the world more resilient to climate change by showcasing state-of-the-art research, real-world applications, and policy implications.

Acknowledgments

First of all, we would like to convey our gratitude toward CRC Press for providing us the opportunity to submit this book entitled *Algal Biorefineries in Circular Economy for Climate Change Resilience* under the esteemed book series.

This book is a result of invaluable support from researchers and academicians who have shared their expertise and offered insightful feedback throughout the review and revision process. We extend our sincere appreciation to all the contributing authors, whose dedication, commitment, and scientific acumen have been instrumental in bringing this work to fruition. We are equally indebted to the reviewers, who provided rigorous peer-review support and helped us refine the chapters to meet high academic standards. Their collective contributions reflect the dynamic and interdisciplinary nature of algal biorefineries in the context of circular economy models and climate resilience. We hope this book inspires future research and technological advancements that promote sustainability, resource efficiency, and environmental stewardship in the face of global climate challenges.

Drs. Fayaz Ahmad Malla and Suhaib Ahmad Bandh would like to express their deepest appreciation to their schools for providing invaluable academic and infrastructure support during the book's preparation. The editorial perspectives and understanding of the urgent need for sustainable bio-based solutions in climate-resilient systems have been greatly influenced by their research and teaching surroundings.

Dr. Fayaz Ahmad Malla expresses his sincere thanks to his mentors, colleagues, and research collaborators for their continuous encouragement and academic cooperation. He is deeply grateful to his family, especially his parents, for their prayers and unwavering support throughout his professional journey. Dr. Suhaib Ahmad Bandh is extremely grateful to his teachers and students for fostering an environment that encouraged teamwork, curiosity, and new ideas. For their unwavering support, encouragement, and faith in his academic endeavors, he is eternally grateful to his loved ones, especially his parents, siblings, and close friends.

We are deeply thankful to the Almighty for bestowing upon us the strength, wisdom, and perseverance to successfully complete this academic endeavor. This work is dedicated to all those striving to build a sustainable future through science, innovation, and circular thinking.

Fayaz Ahmad Malla
Suhaib Ahmad Bandh

About the Editors

Fayaz A. Malla finished his PhD in Environmental Science on "Biogas methane enrichment using selective chemical scavengers and microalgae" from the Indian Agricultural Research Institute in New Delhi. He has a strong interest in waste management, pollution, and renewable energy as his PhD was focused on the same issue. Moreover, in the university he has also dealt with water conservation and management, water auditing and water use efficiency in agriculture. He has published many research articles in reputed, referred national and international journals. He has also presented many research papers at national and international conferences. Previously, he worked as Research Associate with Water Resources Division in The Energy and Research Institute (TERI) and currently working as Assistant Professor in Higher Education Department Government of Jammu and Kashmir. Dr. Malla has recently edited two books with Elsevier Press on *Valorization of Microalgal Biomass and Wastewater Treatment* and *Environmental Sustainability of Biofuels*. He has also edited one book published by Springer Nature Singapore on *Biofuels in Circular Economy*.

Suhaib A. Bandh is Assistant Professor of Environmental Science at Govt. Degree College DH Pora, teaching the courses of Environmental Science and Disaster Management. He is a member of the Academy of Plant Sciences, India, and National Environmental Science Academy. He is a recipient of many awards. He has participated in several national and international conferences held by different reputed scientific bodies in India and abroad. He has a few monographs and several scientific publications to his credit, published in some highly reputed and impactful journals, which attest to his scientific insight, fine experimental skills, and outstanding writing skills. Dr. Bandh has recently edited and authored around 18 books. Currently Dr. Bandh is currently working on several book Project like *Climate Change and Microbial Diversity: Impact and Vulnerability* with AAP/CRC Press, a Taylor & Francis Group. • *Climate Change and Microbes: Construction of Sustainable Future* with AAP/CRC Press, a Taylor & Francis Group. • *Climate Change: The Social and Scientific Construct* with Springer Nature • *Freshwater Biodiversity* with AAP/CRC Press, a Taylor & Francis Group. • Dr. Bandh holds a bachelor's degree from Sri Pratap College, Srinagar, a master's degree in Environmental Science from the University of Kashmir, an M.Phil. degree from Awadhesh Pratap Singh University, Rewa (M.P.), and a PhD in Environmental Science with a specialization in freshwater fungi from University of Kashmir, Srinagar.

Contributors

Ashfaq Ahmad
Department of Biological Sciences, College of Medicine and Health Sciences
Khalifa University
Abu Dhabi, United Arab Emirates
ASPIRE Research Institute for Food Security in the Drylands (ARIFSID)
United Arab Emirates University
Al Ain, United Arab Emirates

Shailendra Kumar Arya
Biotechnology, UIET
Panjab University
Chandigarh, India

Syed Salman Ashraf
Department of Biological Sciences, College of Medicine and Health Sciences
Khalifa University
Abu Dhabi, United Arab Emirates
Center for Biotechnology (BTC)
Khalifa University of Science and Technology
Abu Dhabi, United Arab Emirates
Center for Membranes and Advanced Water Technology (CMAT)
Khalifa University of Science and Technology
Abu Dhabi, United Arab Emirates
ASPIRE Research Institute for Food Security in the Drylands (ARIFSID)
United Arab Emirates University
Al Ain, United Arab Emirates

Suhaib A. Bandh
Department of Environmental Science
S. P College, Higher Education Department Govt. of Jammu and Kashmir
Srinagar, India

R. Baskaran
Jeppiaar University
Chennai, Tamil Nadu, India

Tesfaye Kassaw Bedru
Wollo University
Dessie and Kombolcha, Ethiopia

Gino La Bella
Universidad Nacional de San Martín, Escuela de Bio y Nanotecnologías
San Martín, Provincia de Buenos Aires, Argentina

Saumya Bharti
Department of Biotechnology, University Institute of Engineering & Technology
Panjab University
Chandigarh, India

Grace Joy Wei Lie Chin
Universiti Malaysia Sabah
Kota Kinabalu, Malaysia

Rosangela Rodrigues Dias
Bioprocess Intensification Group
Federal University of Santa Maria
Santa Maria, Brazil

Mariany Costa Deprá
Bioprocess Intensification Group
Federal University of Santa Maria
Santa Maria, Brazil

Darissa Alves Dutra
Bioprocess Intensification Group
Federal University of Santa Maria, UFSM
Santa Maria, RS, Brazil

Luis Fernando Flores
Club de Biotecnología Agrícola y Ambiental
Villa Regina, Argentina

Workisa Bacha Garuma
Haramaya University
Dire-Dawa, Ethiopia

Juan Andrés Sandoval Herrera
Universidad de América
Bogotá, Colombia, US

Muhammad Uzair Javed
Department of Crop Sciences and Agroforestry, Faculty of Tropical AgriSciences
Czech University of Life Sciences Prague
Prague, Suchdol, Czech Republic

Syafiqah Syazwani Jaffar
Universiti Malaysia Sabah
Kota Kinabalu, Malaysia

Mani Jayakumar
Haramaya University
Dire-Dawa, Ethiopia

Ignatia Justine
Universiti Malaysia Sabah
Kota Kinabalu, Malaysia

Karan Kapoor
Department of Biotechnology, University Institute of Engineering & Technology
Panjab University
Chandigarh, India

Joshophine Pei Ning Kong
Universiti Malaysia Sabah
Kota Kinabalu, Malaysia

Eduardo Jacob-Lopes
Bioprocess Intensification Group
Federal University of Santa Maria
Santa Maria, Brazil

Nyuk Ling Ma
Universiti Malaysia Terengganu
Kuala Terengganu, Terengganu, Malaysia

Richard Luan Silva Machado
Bioprocess Intensification Group
Federal University of Santa Maria
Santa Maria, Brazil

Tehreem Mahmood
Government College University Lahore
Lahore, Pakistan

Afaan A. Malla
Department of Environmental Science, Government Degree College Baramulla
Indira Gandhi National Open University
Srinagar, India

Fayaz A. Malla
Department of Environmental Science, Government Degree College Tral
Higher Education Department Govt. of Jammu and Kashmir
Srinagar, India

Antonela Martelli
CESIMAR-CONICET (Centro para el Estudio de Sistemas Marinos)
Puerto Madryn, Patagonia, Argentina

Mailin Misson
Universiti Malaysia Sabah
Kota Kinabalu, Malaysia

Hamid Mukhtar
Government College University Lahore
Lahore, Pakistan

Hirak Parikh
Department of Earth and Environmental Systems
Indiana State University
Terre Haute, IN, United States

Showkat Rashid
Department of Economics, Government Degree College Tral
Higher Education Department Govt. of Jammu and Kashmir
Srinagar, India

Kenneth Francis Rodrigues
Universiti Malaysia Sabah
Kota Kinabalu, Malaysia

Suryani Saallah
Universiti Malaysia Sabah
Kota Kinabalu, Malaysia

Sudarshan Sahu
Department of Biotechnology, University Institute of Engineering & Technology
Panjab University
Chandigarh, India

Adriane Terezinha Schneider
Bioprocess Intensification Group
Federal University of Santa Maria
Santa Maria, Brazil

T. Selvaganapathy
Tamil Nadu Pollution Control Board
Chennai, Tamil Nadu, India

Pau Loke Show
Department of Chemical and Petroleum Engineering
Khalifa University
Abu Dhabi, United Arab Emirates

Doris Ying Ying Tang
Department of Chemical and Petroleum Engineering
Khalifa University
Abu Dhabi, United Arab Emirates

Archana Tiwari
Amity University Noida
Noida, India

Vun Yee Thien
Xiamen University Malaysia
Sepang, Selangor, Malaysia

Wilson Thau Lym Yong
Universiti Malaysia Sabah
Kota Kinabalu, Malaysia

Leila Queiroz Zepka
Bioprocess Intensification Group
Federal University of Santa Maria
Santa Maria, Brazil

1 Integrating Flue Gas for Agreed Algal Biofuel Production

Luis Fernando Flores, Antonela Martelli, Juan Andrés Sandoval Herrera, and Gino La Bella

1.1 INTRODUCTION

Because of the serious damage they do to the environment, people are starting to worry more and more about flue gas emissions. Some of the pollutants found in flue gases are sulfur dioxide (SO_2), nitrogen oxides (NO_x), carbon dioxide (CO_2), and particulate matter (PM) (Wang et al. 2019; Larki et al. 2023). These gases are usually released when fossil fuels are burned in power plants, industrial facilities, and other sources. Air pollution and its negative effects on human health are both caused by these pollutants. There is strong evidence from studies (Almetwally et al. 2020; Fuller et al. 2022). These pollutants are associated with respiratory and cardiovascular disorders, early mortality, higher hospital admissions, and mortality rates.

One way that flue gas emissions might contribute to acid rain is by releasing sulfur dioxide and nitrogen oxides. Acid rain is the result of these gases reacting with atmospheric water vapor to produce sulfuric and nitric acids. Soil, water, vegetation, and structures are all negatively impacted by this event. Subsequently, acid rain can cause soil acidification, which impacts forest health and agricultural output, and it can disrupt aquatic ecosystems by lowering the pH of water bodies (Grennfelt et al. 2020; Zhang 2023). In contrast, flue gases mostly consist of carbon dioxide, a major greenhouse gas (GHG) that has been found to amplify the effects of climate change and global warming (Mikhaylov et al. 2020). Anthropogenic CO_2 emissions are mainly caused by the combustion of fossil fuels. Global warming, ice cap melting, rising sea levels, and more intense weather events are all consequences of the greenhouse effect, which is worsened by the ever-increasing atmospheric concentration of carbon dioxide. To combat global warming, it is essential to reduce emissions of carbon dioxide (CO_2) (Wang and Song 2020; Gür 2022).

As a result of the increasing number of complaints, authorities have imposed strict rules meant to reduce emissions of flue gas. In order to lessen the impact that industrial and energy production operations have on the environment, cleaner technology should be promoted and emission limits should be implemented (Barla et al. 2024). In order to lessen the severity of these emissions, new technologies have been developed, such as methods to reduce flue gas and to store carbon (Wang and Song 2020; Gür 2022). According to the scenarios put forth, algae have attracted a lot of interest as a potential way to reduce emissions of GHGs (Lage et al. 2018; Singh et al. 2019; Iglina et al. 2022; Frongia et al.2023).

Algae are a promising alternative for carbon capture and utilization due to their numerous desirable characteristics, such as their fast growth rates, high photosynthetic efficiency, and adaptability to diverse environments. This includes marginal lands where other livestock and agricultural activities

DOI: 10.1201/9781003685814-1

are not possible (Gaurav et al. 2024). Algae are superior to land plants in their ability to convert carbon dioxide gas into biomass during photosynthetic processes. According to Siedenburg (2022) and Johnston et al. (2023), this quality makes them perfect for carbon sequestration. Furthermore, unlike plant crops, it can be maintained for significantly longer periods of time due to its cultivation technique and the ability to modify physicochemical parameters, making it independent of environmental seasonality (Jabłońska-Trypuć et al. 2023).

According to Gaurav et al. (2024), algae can be grown in photobioreactors (PBRs) or open ponds to collect carbon dioxide (CO_2) from atmospheric or industrial exhaust gasses and then turned into biomass. Transesterification and other biomass-to-fuel technologies can transform algae biomass into drop-in biofuels because to the algae's high lipid production capacity (Flores et al. 2024). Thus, conventional wastewater treatment technologies produce more GHG emissions; integrating algae farming with these facilities can improve the sustainability of both processes. As a result, according to several studies (Lage et al. 2018; Singh et al. 2019; Bharti et al. 2022; Iglina et al. 2022; Frongia et al. 2023), using biofuels made from algae not only cuts down on GHG emissions, but they could also replace traditional fossil fuels. In addition, the nutrients included in this algal biomass are of excellent quality, including proteins, carbs, and bioactive chemicals. Food, animal feed, pharmaceutical, and cosmeceutical industries are only a few of the many that can benefit from algal biomass (Dolganyuk et al. 2020).

Although there may be some advantages, there are still a number of obstacles that must be overcome before algae can be fully utilized to reduce GHG emissions. Among these obstacles are the following: lowering production costs; improving biomass harvesting and processing procedures; scaling up operations; and optimizing carbon uptake and suitable cultivation systems for flue gas harnessing. New developments in metabolic modeling, bioprocess optimization, and genetic engineering hold great promise for conquering these obstacles. Within the context of a circular economy striving for a more sustainable future, this chapter analyzes the pros, cons, and profitability of utilizing flue gases as feedstock for algae biorefineries, taking into account the aforementioned challenges and the proposed alternative solution by microalgae biorefineries.

1.2 FLUE GAS EMISSIONS

Flue gas emissions are a significant source of air pollution and GHGs, with their composition varying depending on the industry. Understanding the specific pollutants associated with different industries is crucial to developing targeted mitigation strategies (Thomas et al. 2016).

Among the main industries that emit flue gases, the power generation industry, particularly those relying on fossil fuels such as coal, oil, and natural gas, is a major source of flue gas emissions. The primary components of flue gas from power plants include carbon dioxide (CO_2), sulfur dioxide (SO_2), nitrogen oxides (NO_x), PM, and trace amounts of heavy metals and volatile organic compounds (VOCs) (Cheng et al. 2019; Wang et al. 2019; Weltgesundheitsorganisation and Organization 2021). The burning of fossil fuels in power plants, industrial facilities, and vehicles releases vast amounts of CO_2, contributing to the enhanced greenhouse effect. This effect leads to global warming, characterized by rising global temperatures, melting polar ice caps, and rising sea levels, threatening coastal areas and island communities. Elevated temperatures disrupt weather patterns, leading to more frequent and severe weather events such as hurricanes, droughts, and heatwaves. These changes in rainfall patterns and growing seasons impact agriculture and consequently affect food security (Grennfelt et al. 2020; Zhang 2023).

Cement production is another industry characterized by substantial flue gas emissions. The primary emissions include CO_2, resulting from both the combustion of fossil fuels and the calcination process of limestone ($CaCO_3$). NO_x and SO_2 are also significant due to the high-temperature kilns used in production (Sambataro et al. 2024).

Steel manufacturing, particularly through blast furnace operations, produces flue gas emissions rich in CO_2, CO (carbon monoxide), SO_2, NO_x, and PM. The reduction of iron ore using coke generates substantial amounts of CO_2 and CO (Zhang et al. 2021).

Waste incineration plants emit a complex mixture of pollutants due to the combustion of various waste materials. All these primary components of flue gas emissions include CO_2, NO_x, SO_2, CO, PM, dioxins, furans, and heavy metals such as mercury and cadmium, all of which contribute to environmental degradation (Kong et al. 2024).

NO_x and VOCs react in the atmosphere to form ground-level ozone, a key component of smog, which damages crops, forests, and other vegetation. SO_2 and NO_x contribute to acid rain, which acidifies soils and water bodies, harming aquatic life and reducing biodiversity. Acid rain also impacts soil and vegetation in the forest ecosystems, causing disruption through habitat changes and partitioning, leading to species migration, adaptation, or extinction. Furthermore, acid rain negatively affects soil and agricultural productivity and quality. In addition, heavy metals and persistent organic pollutants in flue gases can contaminate soil and water, leading to bioaccumulation in the food chain and posing risks to wildlife and human health (Grennfelt et al. 2020; Mikhaylov et al. 2020; Fuller et al. 2022; Zhang 2023).

In response to the harmful effects of flue gas emissions, governments and regulatory bodies have implemented stringent regulations to limit these emissions, mitigate environmental and health impacts, and promote sustainable industrial practices (Fuller et al. 2022).

1.3 MICROALGAE PRODUCTION FROM FLUE GAS

As photosynthetic organisms, microalgae have a number of biological processes that allow them to convert CO_2 into biomass. As a result, microalgae are becoming more and more acknowledged for their ability to remove CO_2 from exhaust gas. Their metabolic flexibility, unique methods for concentrating carbon, and photosynthetic capacities are utilized by these systems (Zeng et al. 2021). One potential solution to reduce CO_2 emissions from power plants' flue gases is to grow microalgae. Flue gas contains both growth-promoting and biomass-inhibiting chemicals; nonetheless, the high concentrations of CO_2 it contains make it ideal for microalgal growth (Banskota et al. 2024). In a circular economy paradigm, it may be feasible to generate biomass for biofuels and other uses while simultaneously reducing emissions of GHGs (Daneshvar et al. 2022; Siedenburg 2022).

The conversion of carbon dioxide gas from flue gas into organic compounds is facilitated by microalgae through photosynthesis, the principal biological method by which these tiny organisms absorb and use the gas. Al Ketife and AlMomani (2024) state that this process consists of two primary steps: the Calvin cycle and the light-dependent reactions. Chlorophyll and other pigments absorb light energy in the thylakoid membranes of the chloroplasts, where the light-dependent processes take place. Hydrogen and oxygen are produced, as well as Adenosine Triphosphate (ATP) and Nicotinamide Adenine Dinucleotide Phosphate (NADPH), by dividing water molecules using this energy. In the chloroplast stroma, a reaction known as the Calvin cycle occurs, which does not require light. Glucose is converted from carbon monoxide into organic molecules using the ATP and NADPH that are generated in the light-dependent processes (Prasad et al. 2021; Zeng et al. 2021; Al Ketife and AlMomani 2024).

Even in low-CO_2 conditions, microalgae are able to capture and fix CO_2 through their carbon-concentrating processes and other biological systems. In this particular procedure, microalgal bicarbonate transporters are able to actively move bicarbonate (HCO_3^-) across the cell membranes. In order to maintain a consistent supply of carbon monoxide for the Calvin cycle, the enzyme carbonic anhydrase converts bicarbonate to CO_2 within the cell. Just like CO_2 pumps are complexes of specialized proteins that actively transport CO_2 into chloroplasts, Rubisco is an enzyme that catalyzes the first important stage of carbon fixation by increasing the concentration of CO_2 around it. Microalgae rely on the carbonic anhydrase, which catalyzes the reversible conversion of CO_2 to bicarbonate. Enzyme conversion of bicarbonate to CO_2 both within and outside the cell helps maintain an appropriate supply of CO_2, which means carbon fixation for photosynthesis can be more efficiently carried out (Prasad et al. 2021; Zhang and Liu 2021; Zeng et al. 2021).

Biofuel feedstocks such as lipids, carbs, and proteins can be produced by microalgal photosynthesis, which is 10–50 times more efficient than other photosynthetic organisms (Jiao et al. 2024). In the presence of nitrogen limitation and other nutrient stress circumstances, microalgae can shift their carbon flux toward producing biodiesel feedstock in the form of fatty acids and triglycerides (Stavridou et al. 2024). Further, the surplus of carbon that has been fixed can be collected and preserved as starch or other polysaccharides. These can then be utilized to create fuels such as biogas, bioethanol, and biobutanol (Jabłońska-Trypuć et al. 2023). The next part explores into the development of producing these biofuels produced from microalgae.

1.4 MICROALGAE-DERIVED BIOFUELS

Biodiesel, biohydrogen, biogas, bioethanol, and biobutanol are only a few of the biofuels that may be made from microalgal biomass, which is itself produced from flue gas (Jiao et al. 2024). Thermochemical and anaerobic digestion procedures can convert microalgal biomass into biohydrogen and biogas, for example. Microalgal biomass rich in carbohydrates can be fermented to yield bioethanol and biobutanol, and biodiesel can be made via lipid extraction and transesterification (Jiao et al. 2024). As a result, microalgae are being acknowledged as a potential alternative energy source with great promise.

Microalgae are a promising alternative for biofuels due to their high biomass production and lipid content, as previously stated (Alazaiza et al. 2023). Also, they can handle a wide range of pH, sunshine, and nutrient supply fluctuations with ease (Jiao et al. 2024). Another benefit of microalgae for bioenergy applications is their short life cycle, which means there is a steady supply of biomass available all year for biofuel production (Alazaiza et al. 2023). Some of the most common biofuels made from algae biomass are the following.

1.4.1 Bioethanol

Even now, bioethanol is among the most widely used biofuels around the world. Improvements in manufacturing efficiency and reduction of costs are, nonetheless, required by economic considerations. An environmentally friendly and secure alternative is third-generation bioethanol made from microalgae, which is produced from algal biomass, which is a practical raw material for producing ethanol. In order to make bioethanol, the fermentation of algal polysaccharides, especially starch and cellulose, is the most prevalent process (Jabłońska-Trypuć et al. 2023; Chandrasekhar et al. 2023).

There are three main ways to make bioethanol from microalgal biomass, and they all have their own unique qualities. Using microalgae biomass, which is rich in carbohydrates and contains carbon fixed as starch, is the conventional method. Ethanol is produced from this biomass by means of pretreatment, enzymatic hydrolysis, and yeast fermentation. Dark fermentation is another method that makes use of dark metabolic pathways; this process redirects sunlight to produce acids, hydrogen, and alcohols like ethanol. The third approach, photofermentation, makes use of genetic engineering to enhance bioethanol production in microalgae, namely cyanobacteria. This method involves using light energy to produce bioethanol from carbon dioxide and water simultaneously using genetically engineered strains (Lage et al. 2018; Abomohra and Elshobary 2019; Jabłońska-Trypuć et al. 2023).

Blending the produced bioethanol with traditional fossil fuels, such as gasoline, can improve combustion, increase oxygen content, decrease volatility, and increase octane rating (Osman et al. 2023).

1.4.2 Biobutanol

Due to its molecular structure, the four-carbon alcohol biobutanol has a higher heating value and fewer ignition difficulties than bioethanol, resulting in a larger energy content. According to Lose et al. (2018), biobutanol has the ability to be transformed into biodiesel when combined with diesel, and it can also be utilized as a biofuel when mixed with gasoline. The biofuel's increased energy density and gasoline-like properties make it stand out (Bauenova et al. 2024).

Bacterial fermentation of carbohydrate-rich biomass can be employed to generate biobutanol from algal biomass (Lage et al. 2018). Low conversion efficiency occurs, however, since only a small number of species, including *Clostridium* spp., are capable of converting microalgae into biobutanol. Faster hydrolysis and increased yields are achieved through the correct pretreatment of biomass, which converts complex polymer structures such as starch and cellulose into simple fermentable sugars (Abomohra and Elshobary 2019). Alternatively, it has been observed that butanol productivity is low when intact algal cells are used. Pretreatment of biomass, fermentation, and recovery are the three main phases of the ABE (acetone-butanol-ethanol) fermentation process. Minimizing inhibitor release that affects later hydrolysis and fermentation processes is another benefit of selecting the most appropriate pretreatment procedure. Biofuel production, however, necessitates taking into account the price of the chosen saccharification pretreatment method (Abomohra and Elshobary 2019).

Biobutanol is superior than ethanol and methanol for cold-weather engine start-ups because of its much lower heat of vaporization. Storage and transportation are also made easier by its decreased volatility and lower vapor pressure, which reduces hazards including vapor lock, cavitation, and pipe breakage. Abomohra and Elshobary (2019) and Bauenova et al. (2024) both agree that biobutanol is the best biofuel kind.

1.4.3 Biohydrogen

According to Hassan et al. (2023), biohydrogen is perfect for fuel cells since it is a clean fuel that produces only water as a byproduct of combustion. The production of biohydrogen from microalgae has garnered a lot of attention since it is a clean and sustainable fuel source that can be integrated with wastewater treatment, which helps with carbon neutrality and environmental sustainability (Sayago 2024). Microalgae have the potential to increase production efficiency thanks to their ability to produce hydrogen through several biological mechanisms. Under certain stress conditions, like sulfur deficiency, low oxygen levels, and intense light exposure, some types of algae can use photobiological processes to create hydrogen gas (Danial et al. 2023; Jiao et al. 2024).

Biophotolysis and fermentation are the two most important pathways in the multi-stage process by which microalgae produce biohydrogen. Enzymes such as hydrogenase and nitrogenase help microalgae produce gaseous hydrogen and oxygen during biophotolysis (Jiao et al. 2024). There are a number of biological methods that have been investigated for the purpose of producing biohydrogen; these methods include photofermentation, dark fermentation, and electro-bio-hydrogenation (Alazaiza et al. 2023).

1.4.4 Biogas

Algal biomass has the ability to produce biogas, fertilizers, and biofuels, making it an attractive alternative to traditional biogas production methods. About 65% of biogas produced by algae is methane and 35% is carbon dioxide. Produced in trace quantities (less than 1%), other gases like hydrogen, nitrogen, ammonia, hydrogen sulfide, and nitrogen oxides can be utilized as fuel for gas turbines and engines (Paul et al. 2023; Jabłońska-Trypuć et al. 2023). Furthermore, biogas has the potential to be utilized for both heating and power generation (Abusweireh et al. 2023).

The remaining microalgal biomass can be converted into biogas by means of anaerobic digestion following lipid extraction. Due to their lower or nonexistent lignin content, algae are typically not pretreated before digestion, making them more biodegradable than conventional plants (Jabłońska-Trypuć et al. 2023; Kong et al. 2024). Several important steps, including hydrolysis, acidogenesis, acetogenesis, and methanogenesis, are involved in the anaerobic digestion process that builds biogas. Algae such as *Macrocystis pyrifera*, *Gracilaria*, *Sargassum*, *Ascophyllum*, *Laminaria*, *Ulva*, *Chaetomorpha*, and *Cladophora* are commonly utilized in biogas generation due to their high methane concentrations and gas yields. Anaerobic digestion is well-suited to other species, like *Dunaliella salina* and *Chlamydomonas reinhardtii*, which have cell walls that are rich in proteins that can be rapidly biodegraded. Hydrolysis is more of a challenge for species with more robust cell walls, such as *Scenedesmus* and *Chlorella kessleri*. Biogas generation from algae with a higher resistance may necessitate further pretreatment to dissolve or break down their cell walls. A possible cost-effective way to increase methane outputs during anaerobic digestion is to co-ferment algal biomass with sewage sludge or other waste biomass. Many advantages can be gained by growing algae in wastewater, such as treatment of the wastewater, removal of nutrients, and generation of biomass for energy (Paul et al. 2023; Jabłońska-Trypuć et al. 2023; Abusweireh et al. 2023). On the other hand, according to Jabłońska-Trypuć et al. (2023), the raw material supply's seasonality and transportation costs can impact the feasibility of biogas production from algae. To overcome these challenges, biogas plants should be placed near fresh algae sources, and more efficient biomass collection technologies should be implemented.

1.4.5 Biodiesel

The capacity of microalgae to synthesize large quantities of triacylglycerol (TAG) and other lipids makes them attractive substrates for biodiesel generation. The process of transesterification allows these lipids to transform into biodiesel precursors known as fatty acid methyl esters (FAME). The biodiesel's quality is greatly affected by the composition and quality of FAME, which includes fatty acids such as palmitic acid (C16:0), linoleic acid (C18:2), and γ-linolenic acid (C18:3). The biodegradability, high energy production, and environmental benefits of fuels produced from microalgae are major pluses. In addition, compression-ignition engines can make use of them with or without modifications, either on their own or in combination with petroleum diesel (Alazaiza et al. 2023; Jabłońska-Trypuć et al. 2023). Furthermore, it is possible to transform complete algal biomass into bio-oil, a material similar to crude oil, by means of procedures such as pyrolysis and hydrothermal liquefaction (Liiv et al. 2024). Diesel, jet fuel, gasoline, and other refined fuels are all possible from this biofuel (Denson et al. 2024).

Biodiesel can be most effectively produced from microalgae like *Dunaliella primolecta*, *Botryococcus braunii*, *Crypthecodinium cohnii*, *Nannochloropsis* sp., and *Chlorella* sp. because of their high lipid and hydrocarbon content. These microalgae typically contain 20%–50% oil, although this can go as high as 80%. As an example, *B. braunii* can have an oil content of up to 80% and *Chlorella* up to 50%. Notable among these microorganisms is *B. braunii*, which allows for an easy and inexpensive biodiesel conversion by directly secreting long-chain lipids into the growth media. One limitation of *B. braunii* for large-scale biodiesel production is its modest growth rate (Alazaiza et al. 2023).

1.5 INTEGRATING FLUE GAS FOR ALGAL BIOFUEL PRODUCTION

Microalgae have a lot of potential as a source of renewable biofuels because of their fast growth rates, high photosynthetic efficiency, and adaptability to different types of water—including wastewater—which helps reduce carbon dioxide emissions. Most commonly utilized cultivation strategies include open ponds and closed PBRs (Al-Dailami et al. 2022). Sunlight and atmospheric carbon dioxide are the fuels for algal blooms in open ponds, which are vast, shallow bodies of water. Although they are inexpensive, they are less productive due to environmental limitations including light and temperature swings and the dangers of contamination. While more expensive to construct and run, enclosed systems like PBRs allow for more precise control over growth conditions and greater production. Despite the greater initial and ongoing expenses, they are more suitable for large-scale biofuel production because of their enhanced biomass productivity and reduced contamination hazards (Al-Dailami et al. 2022). The growth can also take place under photoautotrophic, mixotrophic, or heterotrophic circumstances. Mixotrophic growth, which maximizes biomass output by combining light-dependent autotrophic and organic carbon-dependent processes, is the most employed method (Yun et al. 2021).

At large scales, the design and operation of PBRs are critical for the success of integrating algal biofuel production with flue gas industries. Tubular PBRs are often preferred for their ability to handle large volumes of gas, allowing for efficient CO_2 absorption and biomass production. However, issues such as oxygen accumulation and maintaining uniform light distribution must be addressed through effective design modifications, such as implementing segmented flow systems or using light guides. Flat panel PBRs can also be employed for their scalability and ease of maintenance, although they require careful management of light and temperature. Key parameters in large-scale PBRs include maintaining optimal light intensity, temperature, pH, and CO_2 concentration to maximize biomass productivity. Additionally, integrating advanced monitoring and control systems is essential for ensuring consistent environmental conditions and minimizing energy consumption (Al-Dailami et al. 2022; Penloglou et al. 2024). Moreover, the impact of light on indoor growth can be improved by employing affordable LED lights at suitable wavelengths. Thus, the selection of material for reactor design is equally important, as it affects light transmission within the reactor, aiding the photosynthesis process, CO_2 intake, and carbon fixation (Priyadharsini et al. 2022).

Recent advancements in microalgae cultivation and bioreactor design have significantly improved the efficiency of CO_2 capture from flue gases. PBRs, which provide a controlled environment for algae growth, have been optimized for better light distribution, gas exchange, and temperature regulation (Zhang and Liu 2021). PBRs are categorized into several types based on their design and operational mechanisms, including flat panel, tubular, bubble column, and airlift reactors. Each type offers specific advantages depending on the desired application, with tubular PBRs being popular for their high surface area-to-volume ratio, which maximizes light absorption and thus microalgae growth. Flat panel reactors are known for their ease of scaling up, while bubble column and airlift reactors are efficient in terms of mass transfer and energy consumption (Abdur Razzak et al. 2023). The controlled physicochemical parameters of culture such as temperature, pH, salinity, light, and essential nutrients can modulate the algae growth and consequently maximize the CO_2 fixation by microalgae. Moreover, microalgae cells can adhere to surfaces, forming dense layers that increase the surface area for CO_2 absorption forming an interaction matrix of cells such as biofilms. This ability can be utilized in the PBR culture to maximize the CO_2 fixation or can be replicated using other manipulated immobilization techniques such as alginate beads or other materials that can enhance gas exchange and CO_2 capture efficiency. There are optimal ranges for each microalgae, being specie-specific process. Particularly, the light utilization efficiency of microalgae cells can be improved by controlling the appropriate light intensity and uniform light exposure. Consequently, to improve the microalgae performance in resource utilization, it is important to enhance the capacity of carbon fixation while fully photosynthesizing (Zeng et al. 2021; Daneshvar et al. 2022). Moreover, continuous cultivation methods in PBRs have shown superior performance in terms of biomass yield and energy efficiency compared to batch cultures. However, challenges remain in scaling up these systems for industrial applications, particularly in balancing cost and energy inputs with biomass output (Rehman et al. 2022).

To leverage the CO_2 emissions from industrial flue gas as a carbon source for microalgae cultivation, a closed-loop system that recycles water, nutrients, and inorganic carbon might enhance the sustainability and efficiency of this integrated process. Therefore, the integration of algal biofuel production with flue gas industries starts with the capture and utilization of CO_2-rich flue gas. The flue gas can be directly fed into PBRs where microalgae absorb CO_2 during photosynthesis, converting it into biomass that can be processed into biofuels. Water recycling is crucial in this system, as it minimizes freshwater usage and reduces operational costs. Additionally, nutrient recycling, particularly nitrogen and phosphorus, is vital for maintaining algal growth without depleting external resources. To maintain a continuous supply, these nutrients can be sourced from industrial effluents or recovered from algal biomass residues after biofuel-feedstock recovery. Wastewater from industrial processes or the algal biomass extraction process can be treated and reused in the cultivation process, enhancing the process's circularity. As a result, this closed-loop system may ensure a continuous supply of essential resources while minimizing waste.

Optimal bioprocessing parameters are essential for maximizing the efficiency and productivity of algal biofuel production systems when integrated with flue gas industries. Engineering approaches focus on enhancing mass and energy transfer within PBRs, improving light distribution, and optimizing the mixing of gas and liquid phases (Guo and Song 2024). Computational fluid dynamics (CFD) models are often employed to simulate and optimize PBR performance, helping to identify the most efficient reactor designs and operational parameters. Additionally, integrating automation and real-time monitoring systems enables precise control over cultivation conditions, leading to improved consistency and scalability of the process (Kubelka et al. 2024).

The first step after microalgae production using flue gas as a carbon source consists of the separation of biomass from the liquid medium through different methods such as flocculation (adding chemicals to aggregate algal cells for easier collection), centrifugation (using centrifugal forces to separate algae from water, a method suitable for large-scale operations but energy intensive), and filtration (passing the algal culture through filters to retain biomass, suitable for small-scale or specific algal strains). For subsequent further processing, great advantages include the simple cellular structure of some species of microalgae, which necessitate only mild pretreatment (Jiao et al. 2024).

Solvents like hexane allow the lipid extraction from algal biomass to be refined into biodiesel. More environmentally friendly methods use CO_2 in a supercritical state to extract lipids without harmful solvents (De Bhowmick et al. 2024). After lipid extraction, algal lipids can be converted into biodiesel by reacting them with an alcohol (usually methanol) in the presence of a homogeneous or heterogeneous catalyst. Thus, the efficient separation of the catalyst and biodiesel is required, and such reaction conditions have to be optimized to achieve a maximum yield (Farouk et al. 2024).

Biological methods are also important to convert algae biomass into biofuels such as biogas, bioethanol, and biohydrogen. By fermentation, algal carbohydrates are converted into bioethanol using microorganisms like yeast. Its principal challenge is the pretreatment of algal biomass to break down complex carbohydrates and improve fermentation efficiency (Khammee et al. 2021). Alternatively, in the absence of oxygen, algae biomass can be decomposed by microbial action to produce biogas composed of methane, principally, and hydrogen, in a minor percentage (Paul et al. 2023; Abusweireh et al. 2023). In addition, as mentioned, certain algae can produce hydrogen under specific conditions (e.g., anaerobic and sulfur-deprived environments). One of the principal limitations of this process is that the production rate is generally low, and scaling up remains a significant challenge; however, logistics is a crucial topic, especially because of the low density of hydrogen (Bora et al. 2023).

Thermochemical conversion encompasses three primary types of methods for energetic valorization of biomass, pyrolysis, gasification, and hydrothermal liquefaction (Fatih et al. 2023; Liiv et al. 2024). Among them, the principal difference is related to time and temperature conditions. In pyrolysis, algal biomass is heated until 700–800°C in the absence of oxygen to produce bio-oil, which can be further refined into transportation fuels. On the contrary, hydrothermal liquefaction consists of submerging algae in water at high temperatures, but lower than 350°C, and medium to high pressures to produce bio-oil without prior drying, making it more energy-efficient (Nogueira et al. 2023).

1.6 BIOLOGICAL APPROACHES FOR ENHANCING ALGAL BIOFUEL PRODUCTION FROM FLUE GASES

As industrial off-gas commonly contain NO_x and SO_2, along with metals that may inhibit microalgae metabolism and alter productivity, many potential solutions to overcome this inhibition have been explored. Most of these solutions are based on bioprospecting for the most suitable strains, genetic modification, chemical additions, and various bioreactor designs, as well as optimized operating procedures (Comley et al. 2024). Biological approaches play an essential role in enhancing the capabilities of microalgae for flue gas utilization from coal-fired power plants. Flue gas, with its high CO_2 concentrations (10%–20% v/v) and toxic compounds such as NO and SO_2, presents a stark contrast to atmospheric conditions. Therefore, effective CO_2 fixation relies on strains that can withstand harsh environments, high carbon concentrations, and achieve rapid growth. Developing varieties that can thrive under these conditions, tolerate high carbon levels, and exhibit high enzymatic activity for rapid growth is essential. Strategies such as genetic engineering, random mutagenesis, and adaptive evolution are pivotal in creating these organisms, aiming to meet industrial demands and enhance CO_2 sequestration (Cheng et al. 2019; Shokravi et al. 2021). To improve microalgae's utilization of flue gas, modifying metabolic pathways is key. By modifying the metabolic pathways of microalgae, researchers have increased their growth rates, lipid content, and tolerance to high concentrations of CO_2 and other pollutants. In the context of the circular economy, these genetically modified strains can more effectively capture CO_2 and produce higher yields of biofuels and other valuable bioproducts (Zhang and Liu 2021; Hu et al. 2023; Stavridou et al. 2024).

Integrating advanced genetic tools with high-throughput screening methods is vital for optimizing microalgae. Techniques such as intrinsic chlorophyll fluorescence (Best et al. 2016) and magnetic fields (Sung et al. 2017) enable precise, label-free cell separation based on density, aiding in the development of organisms that thrive in challenging flue gas environments while maintaining rapid growth. Advances in gene-editing technologies, including CRISPR-Cas9 and TALENs, allow for precise metabolic pathway modifications, enhancing biomass productivity and product yields while reducing costs (Cheng et al. 2019). In addition, synthetic biology offers the possibility to design metabolic processes and opens up innovative possibilities for flue-gas cultured microalgae biofuel industrialization. To boost both CO_2 sequestration efficiency and biomass production, the enhancement of photosynthesis is essential, typically achieved by increasing the efficiency of CO_2-fixing enzymes, expanding the range of photosynthetically active radiation to improving light utilization, minimizing energy loss, and replacing existing carbon fixation pathways with more efficient alternatives, while decreases the release of captured CO_2. These modifications can be implemented through transformation and gene editing. The introduction of new gene constructs into microalgae has been a particularly challenging endeavor over the past decade. Recent studies suggest that digesting the microalgae cell wall, a major barrier to transformation could be a promising strategy. Additionally, the use of preassembled Cas9 protein-gRNA ribonucleoproteins, which eliminate the need for vector constructs, has been proposed as an effective approach for gene editing (Barati et al. 2021).

Microalgae metabolic pathways modification involves targeting the biosynthesis of TAGs and fatty acids, both of them crucial for lipid accumulation. TAG synthesis follows the Kennedy pathway, with fatty acid synthase (FAS) playing a significant role. Genetic engineering efforts focus on overexpressing genes related to TAG and FAS production or inhibiting competing pathways through gene modifications (Shokravi et al. 2021). Despite advancements, challenges remain in optimizing microalgal metabolic pathways for enhanced biofuel production. Traditional metabolic engineering methods have had mixed success due to the complexity of lipid metabolism. Transcriptional engineering (TE) emerges as a promising alternative by using transcription factors (TFs) to regulate multiple metabolic nodes simultaneously. For example, the GAME9 TF regulates cholesterol biosynthesis enzymes, while the bHLH TF affects lipid accumulation in *Nannochloropsis salina*. In *C. reinhardtii*, the PSR1 TF was upregulated under nitrogen starvation (Ngan et al. 2015), increasing lipid content. Similarly, knocking out the Zn(II)2Cys6 TF in *Nannochloropsis gaditana* resulted in significantly higher lipid accumulation (Xue et al. 2021).

1.7 BENEFITS AND LIMITATIONS OF HARNESSING FLUE GAS FOR ALGAL BIOFUEL PRODUCTION

Microalgae cultivation using flue gas from power plants offers several environmental benefits. In the context of climate change resilience, the conversion of CO_2 into biofuel products through microalgal metabolism offers the dual benefits of reducing air pollution and providing an economically viable method for cutting GHG emissions (Alazaiza et al. 2023). By capturing CO_2 through photosynthesis, microalgae contribute to a lower or close to neutral carbon footprint. In this regard, third-generation biofuels—or microalgae-based biofuels—have garnered substantial attention due to their potential to replace fossil fuels without competing with food or feed crops (Bauenova et al. 2024). Also, domestic production of algal biofuels can reduce dependence on imported fossil fuels, enhancing energy security (Mahmood et al. 2023). Moreover, in addition to mitigating carbon dioxide emissions, microalgae can address water pollution through simultaneous nutrient utilization and wastewater treatment (Jabłońska-Trypuć et al. 2023; Bauenova et al. 2024). Water and nutrient recycling further enhance the sustainability of the process by reducing the need for external inputs and minimizing waste generation. Economically, this integration can lower production costs by utilizing waste streams from industrial processes as inputs for algal cultivation (Penloglou et al. 2024; Clack et al. 2024).

Algae can produce significantly more biomass per unit area than terrestrial crops, making them an attractive option for biofuel production. Certain species can be harvested multiple times a year, providing a continuous feedstock supply. Some authors consider that the principal benefit is the non-competition of microalgae cultivation with food crops because algae can be grown on non-arable land and do not require freshwater, reducing competition with food crops and conserving freshwater resources (Ezhumalai et al. 2024). Besides, as mentioned, microalgae can be converted into various types of biofuels, offering flexibility depending on market demands and technological advancements. Finally, microalgae can grow in nutrient-rich wastewater, helping to treat wastewater while producing biofuels, leading to integrated biorefinery concepts (Al Ketife and AlMomani 2024). Successful algal biofuel strategies must focus on maximizing coproduct value and efficient biomass utilization (Wiatrowski et al. 2022). Therefore, flue gas-cultured microalgae offers a renewable source of proteins, carbohydrates, fatty acids, vitamins, and pigments, among other bioactive compounds (Olabi et al. 2022). As a result, the biomass produced can be harvested and used as biofuels, animal feed, or other bioproducts, effectively storing carbon and reducing net emissions (Dolganyuk et al. 2020; Daneshvar et al. 2022). These characteristics make microalgae promising candidates not only for closing nutrient and carbon cycles in numerous processes but also for waste valorization across various industries. Therefore, microalgae have the potential to transition from a fossil fuel-based economy to one dependent on renewable energy (Klepacz-Smolka et al. 2023).

Although microalgae-based biofuels are renewable and environmentally friendly, current production methods may result in higher GHG emissions due to low microalgal yields and high energy demands. Therefore, several factors hinder the economic viability of microalgal biodiesel (Olabi et al. 2022). These include enhancing the productivity of algal cultures, developing cost-effective methods for harvesting and dehydrating algal cells, and reducing the energy and costs associated with flue gas-based cultivation processes such as mixing, pumping, or gas compression. Another challenge is the high cost of essential nutrients like nitrate, potassium, and freshwater for microalgae growth. This expense significantly impacts production, making microalgal biodiesel more costly than fossil fuels. Research indicates that nutrient supply can contribute 20%–30% of the total biodiesel production cost, affecting both production and distribution (Sundaram et al. 2023). Using wastewater as a nutrient source for microalgae risks introducing bacteria, pathogens, and pollutants. Moreover, scaling up algal biofuel production is also challenged by economic factors, notably the trade-off between biomass production and lipid accumulation. Algae species with high biomass often have low lipid content, while those with high lipid content typically grow more slowly. Typically, microalgae contain 10%–30% oil by dry weight, but species like *Nannochloris* sp. and *Schizochytrium* sp. can achieve lipid levels up to 56% and 80%, respectively, though their slower growth rates limit commercial viability. Conversely, fast-growing species such as *Scenedesmus* and *Chlorella* tend to have lower lipid content (Shokravi et al. 2020). Additionally, harvesting algae is complicated by their small cell size (less than 10 μm in diameter) and density similar to that of water, necessitating energy-intensive processes like microfiltration or disc-stack centrifugation for large-scale production (Jabłońska-Trypuć et al. 2023). Removing water from diluted cell suspensions is complex, with biomass harvesting accounting for 20%–30% of the total downstream processing cost, presenting a significant hurdle to commercialization. Energy consumption varies among cultivation systems. Open ponds are the most energy-efficient, while tubular PBRs consume more energy but often achieve higher biomass concentrations. Compound parabolic concentrators (CPCs) can enhance light delivery and reduce lighting costs. For instance, cultivating *Chlorella vulgaris* in glass tubes with CPCs resulted in a 351% increase in irradiance and a 462% rise in light intensity compared to plain tubes, boosting growth during the exponential phase (Olabi et al. 2022).

Biofuel production processes, including transesterification, harvesting, and drying, further contribute to emissions. Transesterification is particularly energy-intensive and environmentally impactful, while dewatering has fewer environmental concerns (Olabi et al. 2022). Another point to consider is that large-scale algal cultivation could lead to biodiversity loss, water usage issues, and the release of nutrients into the environment, potentially causing eutrophication. The environmental impact of microalgae cultivation also affects biodiversity. Moreover, cultivating microalgae in raceway ponds generally uses less energy compared to PBRs. However, microalgae release pollutants like methane, nitrogen compounds, biogenic halogenated substances, sulfur compounds, isoprene, and volatile organic carbon (Olabi et al. 2022). Large-scale cultivation can harm ecosystems by causing nutrient overloads in aquatic systems, leading to reduced biodiversity. The decomposition of microalgal biomass depletes oxygen in water bodies, endangering aquatic organisms. Over-exploitation of microalgal species in coastal ecosystems exacerbates these issues, reducing water quality, and oxygen levels, and increasing water opacity, all of which threaten biodiversity. Thus, large-scale microalgae cultivation can lead to significant biodiversity loss due to habitat degradation and increased nutrient loads (Olabi et al. 2022).

1.8 FUTURE PERSPECTIVES

The potential to expand microalgae production using flue gas represents a prominent strategy for converting renewable gaseous waste into energy, with microalgae playing a crucial role as an intermediary (Hoang et al. 2023). This method adheres to the principles of the circular economy, transforming waste into valuable resources, and promotes eco-friendly and economical solutions for waste management (Al Ketife and AlMomani 2024). Efforts are underway to improve the economic feasibility of this technology. Integrating microalgae cultivation with existing wastewater treatment facilities or utilizing waste heat from power plants can reduce operational costs and enhance sustainability. In the future, previous challenges should be addressed by developing algal strains that balance lipid content with growth rates for economic viability when flue gas is used as feedstock for biomass and biofuel production. Advances in genetic engineering are promising, nonetheless, challenges such as an incomplete understanding of microalgal metabolism, and complex genome manipulation should attract attention. Concerns include toxic algal blooms, ecological impacts, gene transfer, health and environmental risks, management difficulties, and ethical issues. GM algae can be cultivated in open ponds or enclosed PBRs, but managing risks of engineered algae escaping into natural environments is crucial. Genetic biocontainment strategies include programming cells to produce toxic proteins that kill them if they escape or disabling native genes to reduce fitness outside controlled conditions (Sebesta et al. 2022). Addressing these issues is crucial for advancing genetically modified microalgae in sustainable energy production and reducing reliance on traditional resources (Cheng et al. 2019). Continued advancements in genetic modification and bioprocessing will be key to overcoming current challenges and unlocking the full potential of flue gas-sourced algal biofuels. Increased collaboration between academia, industry, and government agencies is essential for advancing this field. Policy support and financial incentives for research and development can accelerate the adoption of microalgae-based CO_2 capture and utilization systems (Llamas et al. 2021). Additionally, developing cost-effective harvesting and processing techniques for microalgal biomass is crucial for making this technology commercially viable (Kumar et al. 2023).

Continued research and development in emission control technologies and sustainable energy practices are essential for addressing the harmful effects of flue gas emissions. While reducing energy use on a global scale is unrealistic, improving energy efficiency is a highly challenging task necessary to meet climate change mitigation goals. Advancements in renewable energy sources, such as wind, solar, and bioenergy, offer promising alternatives to fossil fuels. International cooperation and policy frameworks play a crucial role in tackling the transboundary nature of air pollution and climate change, ensuring a coordinated global response to these pressing challenges (Asghar et al. 2021). In this way, the biorefineries of microalgae culture could be a strategic and sustainable resource to mitigate some flue gas emissions (Wang and Song 2020; Llamas et al. 2021).

1.9 CONCLUSION

Traditional methods of reducing flue gas emissions often focus on capturing and storing CO_2 and other pollutants, but these methods can be costly and have limitations in scalability and sustainability. The integration of microalgae cultivation with flue gas treatment presents a promising and innovative approach to addressing these challenges. Microalgae, through their natural photosynthetic processes, can efficiently capture and utilize CO_2 and other pollutants from flue gases, converting them into valuable biomass. This biomass can be further processed into biofuels, animal feed, and various bioproducts, providing an additional economic incentive for the adoption of this technology. The use of microalgae for CO_2 sequestration aligns with broader goals of sustainability and circular economy by utilizing waste products (flue gases) to generate valuable resources such as biofuels. This approach not only mitigates the environmental impact of flue gas emissions but also contributes to the development of renewable energy sources and reduces reliance on fossil fuels.

In conclusion, the combination of flue gas treatment and microalgae cultivation represents a synergistic strategy to address the dual challenges of industrial emissions and renewable energy production. By capturing CO_2 and other pollutants from flue gases and converting them into useful biomass, this system offers a sustainable, scalable, and economically viable solution to reduce industrial carbon footprints and combat climate change. Continued research and development in this field, supported by policy and financial incentives, are crucial to realizing the full potential of this innovative approach and achieving long-term environmental and economic benefits.

DECLARATION

The authors have not used any generative AI tools during the writing of this manuscript.

REFERENCES

Abdur Razzak S, Bahar K, Islam KMO, Haniffa AK, Faruque MO, Hossain SMZ, Hossain MM (2023) Microalgae cultivation in photobioreactors: Sustainable solutions for a greener future. *Green Chemical Engineering*. https://doi.org/10.1016/j.gce.2023.10.004

Abomohra AE-F, Elshobary M (2019) Biodiesel, bioethanol, and biobutanol production from microalgae. In: Alam MdA, Wang Z (eds) *Microalgae Biotechnology for Development of Biofuel and Wastewater Treatment*. Springer Singapore, pp 293–321.

Abusweireh RS, Rajamohan N, Sonne C, Vasseghian Y (2023) Algae biogas production focusing on operating conditions and conversion mechanisms – A review. *Heliyon* 9(7):e17757. https://doi.org/10.1016/j.heliyon.2023.e17757

Al Ketife AMD, AlMomani F (2024) Optimizing biofuel production from algae using four-element framework: Insights for maximum economic returns. *Energy Reports* 12:1254–1268. https://doi.org/10.1016/j.egyr.2024.06.066

Alazaiza MYD, Albahnasawi A, Al Maskari T, Abujazar MSS, Bashir MJK, Nassani DE, Abu Amr SS (2023) Biofuel production using cultivated algae: Technologies, economics, and its environmental impacts. *Energies* 16(3):1316. https://doi.org/10.3390/en16031316

Al-Dailami A, Iwamoto K, Ahmad I, Goto M (2022) Potential of photobioreactors (PBRs) in cultivation of microalgae. *Journal of Advanced Research in Applied Sciences and Engineering Technology* 27:32–44. https://doi.org/10.37934/araset.27.1.3244

Almetwally AA, Bin-Jumah M, Allam AA (2020) Ambient air pollution and its influence on human health and welfare: An overview. *Environmental Science and Pollution Research* 27(20):24815–24830. https://doi.org/10.1007/s11356-020-09042-2

Asghar U, Rafiq S, Anwar A, Iqbal T, Ahmed A, Jamil F, Khurram MS, Akbar MM, Farooq A, Shah NS, Park Y-K (2021) Review on the progress in emission control technologies for the abatement of CO_2, SO_x and NO_x from fuel combustion. *Journal of Environmental Chemical Engineering* 9(5):106064. https://doi.org/10.1016/j.jece.2021.106064

Banskota AH, Stefanova R, Hui JPM, Bermarija T, Stemmler K, McGinn PJ, O'Leary SJB (2024) Comprehensive analysis of biomass from *Chlorella sorokiniana* cultivated with industrial flue gas as the carbon source. *Molecules* 29(14):3368. https://doi.org/10.3390/molecules29143368

Barati B, Zeng K, Baeyens J, Wang S, Addy M, Gan S-Y, El-Fatah Abomohra A (2021) Recent progress in genetically modified microalgae for enhanced carbon dioxide sequestration. *Biomass and Bioenergy* 145:105927. https://doi.org/10.1016/j.biombioe.2020.105927

Barla RJ, Raghuvanshi S, Gupta S (2024) A comprehensive review of flue gas bio-mitigation: Chemolithotrophic interactions with flue gas in bio-reactors as a sustainable possibility for technological advancements. *Environmental Science and Pollution Research* 31(23):33165–33189. https://doi.org/10.1007/s11356-024-33407-6

Bauenova MO, Sadvakasova AK, Kossalbayev BD, Yilmaz G, Huang Z, Wang J, Balouch H, Zaletova DE, Lyaguta MA, Alharby HF, Allakhverdiev SI (2024) Optimising microalgae-derived butanol yield. *International Journal of Hydrogen Energy* 49:593–601. https://doi.org/10.1016/j.ijhydene.2023.11.065

Best RJ, Lyczakowski JJ, Abalde-Cela S, Yu Z, Abell C, Smith AG (2016) Label-free analysis and sorting of microalgae and cyanobacteria in microdroplets by intrinsic chlorophyll fluorescence for the identification of fast growing strains. *Analytical Chemistry* 88(21):10445–10451. https://doi.org/10.1021/acs.analchem.6b02364

Bharti RK, Singh A, Dhar DW, Kaushik A (2022) Chapter 6 – Biological carbon dioxide sequestration by microalgae for biofuel and biomaterials production. In: Thakur IS, Pandey A, Ngo HH, Soccol CR, Larroche C (eds) *Biomass, Biofuels, Biochemicals*. Elsevier, pp 137–153.

Bora J, Ghosh S, Das S, Malik S (2023) Chapter 13 – Photobiological production of hydrogen from biomass. In: Srivastava R, Chattopadhyay J, Santos DMF (eds) *Solar-Driven Green Hydrogen Generation and Storage*. Elsevier, pp 259–285.

Chandrasekhar T, Varaprasad D, Gnaneswari P, Swapna B, Riazunnisa K, Anu Prasanna V, Korivi M, Wee Y-J, Lebaka VR (2023) Algae: The reservoir of bioethanol. *Fermentation* 9(8):712. https://doi.org/10.3390/fermentation9080712

Cheng J, Zhu Y, Zhang Z, Yang W (2019) Modification and improvement of microalgae strains for strengthening CO_2 fixation from coal-fired flue gas in power plants. *Bioresource Technology* 291:121850. https://doi.org/10.1016/j.biortech.2019.121850

Clack K, Rajagopal D, Hoek EMV (2024) Life cycle and techno-economic assessment of bioresource production from wastewater. *npj Clean Water* 7(1):22. https://doi.org/10.1038/s41545-024-00314-9

Comley JG, Scott JA, Laamanen CA (2024) Utilizing CO_2 in industrial off-gas for microalgae cultivation: considerations and solutions. *Critical Reviews in Biotechnology*. 44(5):910–923.

Daneshvar E, Wicker RJ, Show P-L, Bhatnagar A (2022) Biologically-mediated carbon capture and utilization by microalgae towards sustainable CO_2 biofixation and biomass valorization – A review. *Chemical Engineering Journal* 427:130884. https://doi.org/10.1016/j.cej.2021.130884

Danial AW, Abdel-Basset R, Abdel-Kader HAA (2023) Tuning photosynthetic oxygen for hydrogen evolution in synergistically integrated, sulfur deprived consortia of *Coccomyxa chodatii* and *Rhodobium gokarnense* at dim and high light. *Photosynthesis Research* 155(2):203–218. https://doi.org/10.1007/s11120-022-00961-4

De Bhowmick G, Plouviez M, Reis MG, Guieysse B, Everett DW, Agnew MP, Maclean P, Thum C (2024) Evaluation of extraction techniques for recovery of microalgal lipids under different growth conditions. *ACS Omega* 9(26):27976–27986. https://doi.org/10.1021/acsomega.4c00221

Denson MD, Manrique R, Olarte M, Garcia-Perez M (2024) Co-hydrotreatment of bio-oil and waste cooking oil to produce transportation fuels. *Energy Fuels* 38(8):6982–6991. https://doi.org/10.1021/acs.energyfuels.3c05176

Dolganyuk V, Belova D, Babich O, Prosekov A, Ivanova S, Katserov D, Patyukov N, Sukhikh S (2020) Microalgae: A promising source of valuable bioproducts. *Biomolecules* 10(8):1153. https://doi.org/10.3390/biom10081153

Ezhumalai G, Arun M, Manavalan A, Rajkumar R, Heese K (2024) A holistic approach to circular bioeconomy Through the sustainable utilization of microalgal biomass for biofuel and other value-added products. *Microbial Ecology* 87(1):61. https://doi.org/10.1007/s00248-024-02376-1

Farouk SM, Tayeb AM, Abdel-Hamid SMS, Osman RM (2024) Recent advances in transesterification for sustainable biodiesel production, challenges, and prospects: A comprehensive review. *Environmental Science and Pollution Research*. https://doi.org/10.1007/s11356-024-32027-4

Fatih G, Anburajan P, Umenweke G, Musa U, Williams O, Mortezaei Y, Suk-Oh H, Lester E, Ogbaga C, Gunes B, Okolie J (2023) Progress in lignocellulosic biomass valorization for biofuels and value-added chemical production in the EU: A focus on thermochemical conversion processes. *Biofuels, Bioproducts and Biorefining* 18. https://doi.org/10.1002/bbb.2544

Flores LF, Osorio-Gonzalez CS, Saini R, Brar SK (2024) Renewable residues as feedstock for drop-in biofuel production. In: Brar SK, Osorio Gonzalez CS, Soccol CR, Saini R (eds) *The Microbiology of the Drop-in Biofuel Production*. Springer Nature Switzerland, pp 41–74.

Frongia F, Cramarossa MR, Arru L, Forti L (2023) Integrated wastewater treatment and CO_2 capture by microalgae-based system. *Journal of Biotech Research,* 14:227–233. https://iris.unimore.it/retrieve/5c960d4f-c5b8-443f-8f7b-e7382019c5d6/Frongia%20et%20al.pdf

Fuller R, Landrigan PJ, Balakrishnan K, Bathan G, Bose-O'Reilly S, Brauer M, Caravanos J, Chiles T, Cohen A, Corra L, Cropper M, Ferraro G, Hanna J, Hanrahan D, Hu H, Hunter D, Janata G, Kupka R, Lanphear B, Lichtveld M, Martin K, Mustapha A, Sanchez-Triana E, Sandilya K, Schaefli L, Shaw J, Seddon J, Suk W, Téllez-Rojo MM, Yan C (2022) Pollution and health: A progress update. *The Lancet Planetary Health* 6(6):e535–e547. https://doi.org/10.1016/S2542-5196(22)00090-0

Gaurav K, Neeti K, Singh R (2024) Microalgae-based biodiesel production and its challenges and future opportunities: A review. *Green Technologies and Sustainability* 2(1):100060. https://doi.org/10.1016/j.grets.2023.100060

Grennfelt P, Engleryd A, Forsius M, Hov Ø, Rodhe H, Cowling E (2020) Acid rain and air pollution: 50 years of progress in environmental science and policy. *Ambio* 49(4):849–864. https://doi.org/10.1007/s13280-019-01244-4

Guo W, Song Y (2024) Chapter 3 – Improvement of mass transfer in microalgal photobioreactors. In: Lafarga T, Acién G (eds) *Sustainable Industrial Processes Based on Microalgae*. Elsevier, pp 41–57.

Gür TM (2022) Carbon dioxide emissions, capture, storage and utilization: review of materials, processes and technologies. *Progress in Energy and Combustion Science* 89:100965. https://doi.org/10.1016/j.pecs.2021.100965

Hassan Q, Algburi S, Zuhair A, Salman H, Jaszczur M (2023) Green hydrogen: A pathway to a sustainable energy future. *International Journal of Hydrogen Energy* 50. https://doi.org/10.1016/j.ijhydene.2023.08.321

Hoang AT, Sirohi R, Pandey A, Nižetić S, Lam SS, Chen W-H, Luque R, Thomas S, Arıcı M, Pham VV (2023) Biofuel production from microalgae: Challenges and chances. *Phytochemistry Reviews* 22(4):1089–1126. https://doi.org/10.1007/s11101-022-09819-y

Hu J, Wang D, Chen H, Wang Q (2023) Advances in genetic engineering in improving photosynthesis and microalgal productivity. *International Journal of Molecular Sciences* 24(3):1898. https://doi.org/10.3390/ijms24031898

Iglina T, Iglin P, Pashchenko D (2022) Industrial CO_2 capture by algae: A review and recent advances. *Sustainability* 14(7):3801. https://doi.org/10.3390/su14073801

Jabłońska-Trypuć A, Wołejko E, Ernazarovna MD, Głowacka A, Sokołowska G, Wydro U (2023) Using algae for biofuel production: A review. *Energies* 16(4):1758. https://doi.org/10.3390/en16041758

Jiao H, Tsigkou K, Elsamahy T, Pispas K, Sun J, Manthos G, Schagerl M, Sventzouri E, Al-Tohamy R, Kornaros M, Ali SS (2024) Recent advances in sustainable hydrogen production from microalgae: Mechanisms, challenges, and future perspectives. *Ecotoxicology and Environmental Safety* 270:115908. https://doi.org/10.1016/j.ecoenv.2023.115908

Johnston KG, Abomohra A, French CE, Zaky AS (2023) Recent advances in seaweed biorefineries and assessment of their potential for carbon capture and storage. *Sustainability* 15(17):13193. https://doi.org/10.3390/su151713193

Khammee P, Ramaraj R, Whangchai N, Bhuyar P, Unpaprom Y (2021) The immobilization of yeast for fermentation of macroalgae *Rhizoclonium* sp. for efficient conversion into bioethanol. *Biomass Conversion and Biorefinery* 11(3):827–835. https://doi.org/10.1007/s13399-020-00786-y

Klepacz-Smolka A, Shah MR, Jiang Y, Zhong Y, Chen P, Pietrzyk D, Szelag R, Ledakowicz S, Daroch M (2023) Microalgae are not an umbrella solution for power industry waste abatement but could play a role in their valorization. *Critical Reviews in Biotechnology*: 1–29. https://doi.org/10.1080/07388551.2023.2284644

Kong W, Kong J, Feng S, Yang T, Xu L, Shen B, Bi Y, Lyu H (2024) Cultivation of microalgae–bacteria consortium by waste gas–waste water to achieve CO_2 fixation, wastewater purification and bioproducts production. *Biotechnol Biofuels* 17(1):26. https://doi.org/10.1186/s13068-023-02409-w

Kubelka BG, Abreu PC, Pinto WT (2024) Balancing energy efficiency and biomass yield through Computational Fluid Dynamics of a commercial scale bubble-column photobioreactor. https://doi.org/10.21203/rs.3.rs-4444605/v1

Kumar N, Banerjee C, Negi S, Shukla P (2023) Microalgae harvesting techniques: Updates and recent technological interventions. *Critical Reviews in Biotechnology. 43*(3), 342–368. https://doi.org/10.1080/07388551.2022.2031089

Lage S, Gojkovic Z, Funk C, Gentili F (2018) Algal biomass from wastewater and flue gases as a source of bioenergy. *Energies* 11(3):664. https://doi.org/10.3390/en11030664

Larki I, Zahedi A, Asadi M, Forootan MM, Farajollahi M, Ahmadi R, Ahmadi A (2023) Mitigation approaches and techniques for combustion power plants flue gas emissions: A comprehensive review. *Science of the Total Environment* 903:166108. https://doi.org/10.1016/j.scitotenv.2023.166108

Liiv J, Mäeorg U, Vaino N, Rikmann E (2024) Low-temperature and Low-pressure HydroThermal Liquefaction (L-HTL) of biomass using ultrasonic cavitation to achieve a local supercritical state in water. *Science and Technology for Energy Transition* 79:3. https://doi.org/10.2516/stet/2023043

Llamas B, Suárez-Rodríguez MC, González-López CV, Mora P, Acién FG (2021) Techno-economic analysis of microalgae related processes for CO_2 bio-fixation. *Algal Research* 57:102339. https://doi.org/10.1016/j.algal.2021.102339

Mahmood T, Hussain N, Shahbaz A, Mulla SI, Iqbal HMN, Bilal M (2023) Sustainable production of biofuels from the algae-derived biomass. *Bioprocess and Biosystems Engineering* 46(8):1077–1097. https://doi.org/10.1007/s00449-022-02796-8

Mikhaylov A, Moiseev N, Aleshin K, Burkhardt T (2020) Global climate change and greenhouse effect. *Entrepreneurship and Sustainability Issues* 7:2897–2913. https://doi.org/10.9770/jesi.2020.7.4(21)

Ngan CY, Wong C-H, Choi C, Yoshinaga Y, Louie K, Jia J, Chen C, Bowen B, Cheng H, Leonelli L, Kuo R, Baran R, García-Cerdán JG, Pratap A, Wang M, Lim J, Tice H, Daum C, Xu J, Northen T, Visel A, Bristow J, Niyogi KK, Wei C-L (2015) Lineage-specific chromatin signatures reveal a regulator of lipid metabolism in microalgae. *Nature Plants* 1(8):15107. https://doi.org/10.1038/nplants.2015.107

Nogueira RC, Neto FS, Junior PGDS, Valério RBR, Serpa JDF, Lima AMDS, De Souza MCM, De Lima RKC, Lopes AAS, Guimarães AP, Melo RLF, Rios MADS, Dos Santos JCS (2023) Research trends and perspectives on hydrothermal gasification in producing biofuels. *Energy Nexus* 10:100199. https://doi.org/10.1016/j.nexus.2023.100199

Olabi AG, Alami AH, Alasad S, Aljaghoub H, Sayed ET, Shehata N, Rezk H, Abdelkareem MA (2022) Emerging technologies for enhancing microalgae biofuel production: Recent progress, barriers, and limitations. *Fermentation* 8(11):649. https://doi.org/10.3390/fermentation8110649

Osman S, Sapunaru OV, Sterpu AE, Chis TV, Koncsag CI (2023) Impact of adding bioethanol and dimethyl carbonate on gasoline properties. *Energies* 16(4):1940. https://doi.org/10.3390/en16041940

Paul R, Silkina A, Melville L, Suhartini S, Sulu M (2023) Optimisation of ultrasound pretreatment of microalgal biomass for effective biogas production through anaerobic digestion process. *Energies* 16(1):553. https://doi.org/10.3390/en16010553

Penloglou G, Pavlou A, Kiparissides C (2024) Recent advancements in photo-bioreactors for microalgae cultivation: A brief overview. *Processes* 12(6). https://doi.org/10.3390/pr12061104

Prasad R, Gupta SK, Shabnam N, Oliveira CYB, Nema AK, Ansari FA, Bux F (2021) Role of microalgae in global CO_2 sequestration: Physiological mechanism, recent development, challenges, and future prospective. *Sustainability* 13(23):13061. https://doi.org/10.3390/su132313061

Priyadharsini P, Nirmala N, Dawn SS, Baskaran A, SundarRajan P, Gopinath KP, Arun J (2022) Genetic improvement of microalgae for enhanced carbon dioxide sequestration and enriched biomass productivity: Review on CO_2 bio-fixation pathways modifications. *Algal Research* 66:102810. https://doi.org/10.1016/j.algal.2022.102810

Rehman M, Kesharvani S, Dwivedi G, Gidwani Suneja K (2022) Impact of cultivation conditions on microalgae biomass productivity and lipid content. *Materials Today: Proceedings* 56:282–290. https://doi.org/10.1016/j.matpr.2022.01.152

Sambataro L, Bre F, Ukrainczyk N, Koenders EAB (2024) Environmental benchmarks for the European cement industry. *Sustainable Production and Consumption* 45:429–449. https://doi.org/10.1016/j.spc.2024.01.020

Sayago UFC (2024) The design of a sustainable industrial wastewater treatment system and the generation of biohydrogen from *E. crassipes*. *Polymers* 16(7):893. https://doi.org/10.3390/polym16070893

Sebesta J, Xiong W, Guarnieri MT, Yu J (2022) Biocontainment of genetically engineered algae. *Frontiers in Plant Science* 13:839446. https://doi.org/10.3389/fpls.2022.839446

Shokravi H, Shokravi Z, Heidarrezaei M, Ong HC, Rahimian Koloor SS, Petrů M, Lau WJ, Ismail AF (2021) Fourth generation biofuel from genetically modified algal biomass: Challenges and future directions. *Chemosphere* 285:131535. https://doi.org/10.1016/j.chemosphere.2021.131535

Shokravi Z, Shokravi H, Chyuan OH, Lau WJ, Koloor SSR, Petrů M, Ismail AF (2020) Improving 'Lipid Productivity' in microalgae by bilateral enhancement of biomass and lipid contents: A review. *Sustainability* 12(21):9083. https://doi.org/10.3390/su12219083

Siedenburg J (2022) Could microalgae offer promising options for climate action via their agri-food applications? *Frontiers in Sustainable Food Systems* 6. https://doi.org/10.3389/fsufs.2022.976946

Singh HM, Kothari R, Gupta R, Tyagi VV (2019) Bio-fixation of flue gas from thermal power plants with algal biomass: Overview and research perspectives. *Journal of Environmental Management* 245:519–539. https://doi.org/10.1016/j.jenvman.2019.01.043

Stavridou E, Karapetsi L, Nteve GM, Tsintzou G, Chatzikonstantinou M, Tsaousi M, Martinez A, Flores P, Merino M, Dobrovic L, Mullor JL, Martens S, Cerasino L, Salmaso N, Osathanunkul M, Labrou NE, Madesis P (2024) Landscape of microalgae omics and metabolic engineering research for strain improvement: An overview. *Aquaculture* 587:740803. https://doi.org/10.1016/j.aquaculture.2024.740803

Sundaram T, Rajendran S, Gnanasekaran L, Rachmadona N, Jiang J-J, Khoo KS, Show PL (2023) Bioengineering strategies of microalgae biomass for biofuel production: recent advancement and insight. *Bioengineered* 14(1):2252228. https://doi.org/10.1080/21655979.2023.2252228

Sung YJ, Kim JYH, Choi HI, Kwak HS, Sim SJ (2017) Magnetophoretic sorting of microdroplets with different microalgal cell densities for rapid isolation of fast growing strains. *Scientific Reports* 7(1):10390. https://doi.org/10.1038/s41598-017-10764-6

Thomas DM, Mechery J, Paulose SV (2016) Carbon dioxide capture strategies from flue gas using microalgae: A review. *Environmental Science and Pollution Research* 23(17):16926–16940. https://doi.org/10.1007/s11356-016-7158-3

Varela Villarreal J, Burgués C, Rösch C (2020) Acceptability of genetically engineered algae biofuels in Europe: Opinions of experts and stakeholders. *Biotechnol Biofuels* 13(1):92. https://doi.org/10.1186/s13068-020-01730-y

Wang H, Yuan B, Hao R, Zhao Y, Wang X (2019) A critical review on the method of simultaneous removal of multi-air-pollutant in flue gas. *Chemical Engineering Journal* 378:122155. https://doi.org/10.1016/j.cej.2019.122155

Wang X, Song C (2020) Carbon capture from flue gas and the atmosphere: A perspective. *Frontiers in Energy Research* 8. https://doi.org/10.3389/fenrg.2020.560849

Weltgesundheitsorganisation, Organization WH (2021) WHO global air quality guidelines: particulate matter (PM2.5 and PM10), ozone, nitrogen dioxide, sulfur dioxide and carbon monoxide. World Health Organization.

Wiatrowski M, Klein BC, Davis RW, Quiroz-Arita C, Tan ECD, Hunt RW, Davis RE (2022) Techno-economic assessment for the production of algal fuels and value-added products: opportunities for high-protein microalgae conversion. *Biotechnol Biofuels* 15(1):8. https://doi.org/10.1186/s13068-021-02098-3

Xue J, Balamurugan S, Li T, Cai J-X, Chen T-T, Wang X, Yang W-D, Li H-Y (2021) Biotechnological approaches to enhance biofuel producing potential of microalgae. *Fuel* 302:121169. https://doi.org/10.1016/j.fuel.2021.121169

Yun H-S, Kim Y-S, Yoon H-S (2021) Effect of different cultivation modes (photoautotrophic, mixotrophic, and heterotrophic) on the growth of *Chlorella* sp. and Biocompositions. *Frontiers in Bioengineering and Biotechnology* 9:774143. https://doi.org/10.3389/fbioe.2021.774143

Zeng J, Wang Z, Chen G (2021) Biological characteristics of energy conversion in carbon fixation by microalgae. *Renewable and Sustainable Energy Reviews* 152:111661. https://doi.org/10.1016/j.rser.2021.111661

Zhang L (2023) Effects of acid rain on forest organisms and countermeasures. *Highlights in Science, Engineering and Technology* 69:292–298. https://doi.org/10.54097/hset.v69i.12041

Zhang S, Liu Z (2021) Advances in the biological fixation of carbon dioxide by microalgae. *Journal of Chemical Technology & Biotechnology* 96(6):1475–1495. https://doi.org/10.1002/jctb.6714

Zhang X, Jiao K, Zhang J, Guo Z (2021) A review on low carbon emissions projects of steel industry in the World. *Journal of Cleaner Production* 306:127259. https://doi.org/10.1016/j.jclepro.2021.127259

2 Role of Microalgae in Sustainable Development Goals and Climate Change Resilience

Ashfaq Ahmad, Doris Ying Ying Tang, Pau Loke Show, and Syed Salman Ashraf

2.1 INTRODUCTION

Microalgae are single-celled photosynthetic organisms that are crucial to various ecosystems and have significant industrial uses. These microorganisms are found in different aquatic environments, from freshwater to marine, and serve as the foundation of the food web, supporting numerous aquatic species. They effectively convert sunlight, carbon dioxide, and nutrients into biomass, making them primary producers in marine ecosystems. Notably, microalgae are significant contributors to global oxygen production, with estimates suggesting that they generate approximately half of Earth's atmospheric oxygen. Oxygen production results from photosynthetic activity, which also helps to sequester carbon dioxide, aiding in climate change mitigation. Additionally, the ability to thrive in diverse and extreme environments further emphasizes ecological importance (Olabi *et al.*, 2023; Kholssi *et al.*, 2021; Lau *et al.*, 2022).

Microalgae have garnered interest for their potential use in various industries and for their ecological significance. Due to their high lipid content and rapid growth rate, microalgae are being studied as a sustainable source of biofuels, including biodiesel and bioethanol. Microalgae contain proteins, lipids, carbohydrates, vitamins, and antioxidants and are utilized in the production of nutraceuticals, pharmaceuticals, and cosmetics. Microalgae have been used in bioremediation to remove pollutants from wastewater, demonstrating their versatility and environmental benefits (Olabi *et al.*, 2023) . Microalgae are known to provide numerous benefits, but their large-scale commercialization faces challenges, such as elevated production expenses and the need for enhanced cultivation methods. Continuous research and technological developments are essential to overcome these challenges and tap into the full potential of microalgae across a range of sectors.

The Sustainable Development Goals (SDGs) established by the United Nations General Assembly in 2015 provide a comprehensive and integrated framework for achieving a sustainable and equitable future. These 17 global goals address a wide range of interrelated challenges such as poverty, inequality, climate change, environmental degradation, and social justice (Pathak *et al.*, 2024). Climate change and its effects are of utmost importance, as highlighted by SDG 13. The relationship between climate change and sustainable development is crucial because climate change threatens the achievement of several SDGs. Building resilience to climate change involves developing the capacity of systems, communities, and societies to withstand, adjust, and recover from the unfavorable consequences of climate change (Mishra *et al.*, 2024). Climate-related disruptions can significantly impede progress toward achieving SDGs. Therefore, it is essential to enhance resilience for

 DOI: 10.1201/9781003685814-2

sustainable development. To achieve this, it is crucial to integrate climate change resilience into SDGs, which requires adopting holistic strategies that encompass both mitigation and adaptation measures. Mitigation efforts primarily involve reducing greenhouse gas emissions, whereas adaptation strategies aim to effectively enhance the capacity of communities to manage climate-related impacts (Mensah, 2019).

Research has emphasized the significance of connections between climate change initiatives and other SDGs. Implementing strategies such as promoting renewable energy sources, enhancing energy efficiency, and constructing sustainable urban infrastructure can address climate change, while advancing objectives involving health, economic growth, and sustainable cities (Pathak *et al.*, 2024). However, realizing these synergies involves several obstacles, including data limitations, capacity constraints, and the necessity of integrated policy frameworks. In summary, SDGs and climate change resilience are deeply intertwined, with climate action being a crucial component of sustainable development. Enhancing resilience and integrating climate action with other developmental objectives can help to create a more sustainable and fair society.

Microalgae have gained recognition as a potential solution for tackling multiple SDGs and improving climate change resilience. These tiny photosynthesizing organisms can transform carbon dioxide (CO_2) into biomass through photosynthesis, making them useful tools for decreasing greenhouse gas emissions and reducing the impacts of climate change (Olabi *et al.*, 2023). The ability of microalgae to flourish in diverse environments, including freshwater and marine ecosystems, highlights their potential for sustainable development. A significant benefit of microalgae is their ability to generate renewable energies. Biofuel production can be achieved through the cultivation of organisms, such as microalgae, which produce biodiesel and bioethanol—cleaner alternatives to traditional fossil fuels. This process aligns with SDG 7, which aims to ensure access to affordable, reliable, sustainable, modern energy sources. Furthermore, the integration of microalgae cultivation with wastewater treatment processes contributes to Clean Water and Sanitation (SDG 6) by removing pollutants and improving water quality (Olabi *et al.*, 2023).

Microalgae have a significant influence on sustainable agriculture and food security. They can be used as biofertilizers, such as chemical fertilizers, to improve soil fertility and crop yields without causing negative environmental consequences; promoting sustainable agricultural practices and enhancing food production supports SDG 2 (Zero Hunger) (Gao & Beardall, 2022). Microalgae are an excellent source of essential nutrients and can help address nutritional deficiencies in animal and human diets, promoting better health outcomes under SDG 3 (Good Health and Well-being) (Olabi *et al.*, 2023). Microalgae play a critical role in carbon sequestration and in reducing atmospheric carbon dioxide levels in terms of adaptation to climate change. Their rapid growth rates and high photosynthetic efficiency render them effective in capturing and storing carbon, positioning them as valuable tools in combating climate change in accordance with SDG 13. Furthermore, the integration of microalgae into circular economy models can improve resource efficiency and reduce waste, thereby contributing to Responsible Consumption and Production (SDG 12). In summary, microalgae provide numerous benefits that align with various SDGs and enhance resilience to climate change. Their applications in renewable energy, wastewater treatment, sustainable agriculture, and carbon sequestration demonstrate potential as sustainable solutions to global challenges.

2.2 MICROALGAE AND SUSTAINABLE DEVELOPMENT GOALS

Microalgae play a significant role in achieving various SDGs, including Zero Hunger, Good Health and Well-being, Clean Water and Sanitation, Affordable and Clean Energy, Industry, Innovation, Infrastructure, Life Below Water, and Life on Land.

2.2.1 Zero Hunger: Role of Microalgae in Food and Nutrition Security

Microalgae are small organisms that thrive in both freshwater and seawater. They are well-known for their high nutritional value, which includes a wealth of proteins, vitamins, essential fatty acids, and other beneficial compounds. This makes them a promising solution for tackling food and nutrition security challenges, particularly in regions in which agriculture is difficult. Microalgae are loaded with essential nutrients that contain up to 70% protein by dry weight, which is significantly higher than that found in traditional plant and animal sources. Furthermore, they are rich in vital amino acids, vitamins (such as vitamin B12, which is often absent in plant-based diets), and omega-3 fatty acids, which are crucial for human health (Caporgno & Mathys, 2018). The nutritional profile of microalgae is further enhanced by their high content of antioxidants and other bioactive compounds, which makes them a superfood. Malnutrition and food insecurity are pressing global issues in regions with low agricultural productivity, such as developing countries. Microalgae can be cultivated in various environments, including arid and saline conditions, in which conventional crops struggle to grow, making them ideal for improving food security in resource-limited settings. They can be integrated into various food products such as supplements, snacks, and staple foods to enhance their nutritional value. For instance, *Spirulina* and *Chlorella*, two commonly cultivated microalgae, are already used in dietary supplements and health foods because of their high nutrient contents. Incorporating microalgae into daily diets can help combat malnutrition by providing essential nutrients, which are often lacking in many populations.

Microalgal cultivation is an environmentally sustainable practice that requires fewer resources than traditional agriculture does. It can be grown on land that is not arable and in saltwater, making it a practical and eco-friendly choice for producing food and bioactive compounds (Sporchia *et al.*, 2024). This helps alleviate the strain on scarce freshwater resources and arable land, which is becoming more limited owing to climate change and population increase. Furthermore, microalgae cultivation can be combined with wastewater treatment techniques, which enable the elimination of contaminants while simultaneously generating biomass that can serve as a source of food or feed (Sporchia *et al.*, 2024). The use of microalgae in food production presents a dual advantage: first, it enhances the sustainability of food systems and contributes to the circular economy. Microalgae have the potential to improve food and nutrition security. By utilizing advancements in biotechnology and cultivation methods, it is possible to scale up microalgal production to meet the global demand. Ongoing research aims to increase the efficiency and cost-effectiveness of microalgal cultivation, making it a practical solution for large-scale food production without modifying numbers, references, or citations (Caporgno & Mathys, 2018). Microalgae have great potential to address the issue of achieving Zero Hunger. Their nutrient-dense profile, ability to thrive in diverse environments, and sustainable farming methods make them valuable tools for combating malnutrition and food insecurity, particularly in regions with scarce agricultural resources.

2.2.2 Good Health and Well-Being: Health Benefits of Microalgae

Microalgae are a diverse group of photosynthetic microorganisms that have gained significant attention due to potential health benefits. This group comprises various bioactive compounds, including proteins, essential fatty acids, vitamins, and antioxidants, which contribute to nutritional and therapeutic properties. For instance, microalgae such as *Spirulina* and *Chlorella* are rich in protein and are frequently referred to as "superfoods" because of their high nutrient content. A notable health benefit of microalgae is their capacity to maintain cardiovascular health. Research has demonstrated that bioactive compounds in microalgae, including omega-3 fatty acids, may aid in reducing cholesterol levels and lowering the risk of heart disease (Ampofo & Abbey, 2022). Furthermore, microalgae possess antioxidant properties that enable them to combat oxidative stress, which has been linked to several chronic diseases, including cardiovascular disorders (Ampofo &

Abbey, 2022). Microalgae have anti-inflammatory and immune-boosting properties. Compounds such as phycocyanin in *Spirulina* have been shown to decrease inflammation and improve immune function. Because of these properties, microalgae are considered a useful dietary supplement for addressing inflammatory issues and supporting general immune health (Basheer *et al.*, 2020). Microalgae have garnered attention owing to their potential for cancer prevention and therapeutic applications. Certain microalgae strains generate bioactive substances with anticancer activities, including carotenoids and polysaccharides, which have been demonstrated to impede cancer cell proliferation and trigger apoptosis (Su *et al.*, 2023). This study highlights the potential of microalgae as a natural source of anticancer agents. In summary, microalgae have extensive and varied health advantages, including cardiovascular health, anti-inflammatory effects, immune system support, and potential anticancer properties. Further research and development in this field are crucial to fully realize the potential of microalgae to promote good health and well-being.

2.2.3 Clean Water and Sanitation: Microalgae in Wastewater Treatment

Microalgae have been identified as a viable solution for treating wastewater due to their environmental and financial advantages. These tiny photosynthesizing organisms can efficiently remove pollutants and nutrients from wastewater, thereby enhancing water quality. The ability of microalgae to absorb nutrients such as nitrogen and phosphorus renders them particularly effective in preventing eutrophication, which occurs when excessive algal growth in water bodies leads to a decrease in dissolved oxygen levels (Liu & Hong, 2021). The application of microalgae in wastewater treatment provides significant benefits in reducing biochemical oxygen demand (BOD) and chemical oxygen demand (COD), both of which are critical measures of water pollution [16]. Microalgae can also eliminate heavy metals and other harmful substances from wastewater, making them a valuable resource for environmental restoration. Certain species, such as *Chlorella* and *Scenedesmus*, have been shown to effectively adsorb heavy metals, including cadmium, lead, and mercury (Ahmad, Buang, & Bhat, 2016; Ahmad, Banat, & Taher, 2022; Ahmad *et al.*, 2022).

In addition to their capacity for pollutant removal, microalgae-based wastewater treatment systems provide the benefit of generating valuable biomass that can be utilized for various purposes, such as the production of biofuels, animal feed, and fertilizers (Srimongkol *et al.*, 2022). Integrating microalgal cultivation with wastewater treatment facilities offers the dual advantage of enhancing the sustainability and cost-effectiveness of the process. The cultivation of microalgae during wastewater treatment not only improves the quality of the effluent but also produces biomass. Moreover, the utilization of microalgae in wastewater treatment facilities contributes to carbon sequestration as organisms absorb carbon dioxide during photosynthesis. Therefore, the combination of wastewater treatment with microalgal cultivation represents a sustainable and cost-effective solution (Kumar *et al.*, 2022). Furthermore, the incorporation of microalgae into wastewater treatment aligns with the principles of the circular economy, which emphasize reducing waste and optimizing resource utilization. Transforming wastewater into a resource for microalgae cultivation addresses water pollution and promotes the generation of renewable energy and other valuable products (Elangovan *et al.*, 2023). In summary, microalgae provide a viable and affordable solution for treating wastewater that is both environmentally friendly and effective in improving water quality. Continued research and development are necessary to achieve optimal efficiency and scalability in microalgae-based wastewater treatment systems.

2.2.4 Affordable and Clean Energy: Microalgae as a Source of Biofuel

Microalgae have gained recognition as an attractive option for biofuel production due to their rapid growth rate, high lipid content, and capacity to thrive in various environments. Unlike conventional fossil fuels, biofuels derived from microalgae are renewable and have a substantially reduced environmental impact. By utilizing their ability to convert sunlight, carbon dioxide, and water into biomass through photosynthesis, microalgae cultivation for biofuel production offers an efficient and sustainable energy source (Wang *et al.*, 2024). Recent advances in biotechnology have significantly enhanced the feasibility of large-scale production of microalgal biofuels. Genetic modifications and omics technologies have led to an increase in lipid accumulation in microalgae, which is essential for biodiesel production (Olabi *et al.*, 2023). The development of photobioreactors (PBRs) has provided a controlled environment for optimizing microalgal growth and lipid yield, thereby addressing some challenges associated with outdoor cultivation (Wang *et al.*, 2024). Microalgal biofuels have garnered attention in both economic and environmental assessments, owing to their potential to support national energy strategies. The Department of Energy underscored the significant biomass production capacity of microalgae, which could be utilized for sustainable aviation fuels and other eco-friendly products. Furthermore, integrating microalgae cultivation with renewable electricity sources, such as wind or solar power, can decrease greenhouse gas emissions, making microalgal biofuels a promising solution for achieving clean energy objectives (El-Moustaqim *et al.*, 2025; Archana & Thomas, 2018). Microalgae have the potential to be a dependable and environment-friendly source of biofuels, which can help in the shift toward inexpensive and renewable energy. Further research and technological progress are necessary to overcome these obstacles and maximize the use of microalgae as a biofuel source.

2.2.5 Industry, Innovation, and Infrastructure: Industrial Applications of Microalgae

Microalgae have gained considerable attention for various industrial applications due to their exceptional characteristics, including rapid growth, high biomass production, and the capability to produce a diverse range of valuable bioactive compounds. As unicellular photosynthetic organisms, microalgae are utilized in multiple industries, including food, nutraceuticals, pharmaceuticals, animal feed, cosmetics, and biofertilizers. The nutrient-rich composition of microalgae, characterized by high levels of proteins, lipids, carbohydrates, vitamins, and antioxidants, makes them an appealing resource for the development of functional foods and nutraceuticals (Choudhary *et al.*, 2023) . Microalgae have been investigated in the pharmaceutical sector for their ability to generate bioactive compounds with therapeutic potential. Several microalgae species produce compounds possessing anti-inflammatory, antioxidant, and antimicrobial activities, which can be used to develop novel medications and health supplements (Basheer *et al.*, 2020; Tan & Lee, 2016). Furthermore, the cosmetic industry profits from microalgae-derived ingredients such as pigments and fatty acids, which are utilized in skincare products for moisturizing and anti-aging effects (Choudhary *et al.*, 2023).

The potential of microalgae in animal feed is promising because they can enhance the nutritional value of feed and improve livestock health and growth rates. Furthermore, microalgae-based biofertilizers are becoming increasingly popular owing to their ability to enhance soil fertility and crop yields while reducing the need for chemical fertilizers (Choudhary *et al.*, 2023). Advances in biotechnological methods have facilitated the creation of microalgae-based bioplastics that are sustainable substitutes for conventional plastics and help minimize plastic pollution (Rani & Roy, 2021). Despite the encouraging potential of microalgal products, significant obstacles remain in their widespread commercialization. To fully harness the power of microalgae in different sectors, it is crucial to address challenges, such as high production expenses, scalability, and regulatory limitations. Overcoming these hurdles requires ongoing research and innovation to devise cost-effective and sustainable methods for producing microalgae-based products (Rani & Roy, 2021; Ahmad *et al.*, 2023; Shanthakumar *et al.*, 2023).

2.2.6 Life Below Water: Impact of Microalgae on Marine Ecosystems

Microalgae are vital components of marine ecosystems, functioning as primary producers at the base of the aquatic food web. Through photosynthesis, these organisms convert sunlight, carbon dioxide, and nutrients into organic matter, which supports a wide variety of marine organisms ranging from zooplankton to larger fish and marine mammals. This essential primary production maintains the health and stability of marine ecosystems by forming the foundation of the food web and supporting marine biodiversity, without which the ecosystem would collapse (Lau *et al.*, 2022). Microalgae play a crucial role in regulating the global carbon cycle by absorbing substantial amounts of carbon dioxide from the atmosphere, thereby helping to reduce the impact of climate change. These microorganisms transfer fixed carbon through the food web and can be sequestered in deep-ocean sediments, acting as long-term carbon sinks. Additionally, they contribute significantly to the food web (Kholssi *et al.*, 2021). This process is vital for maintaining carbon balance in marine ecosystems and broader environments. Although microalgae positively affect marine ecosystems, they are not always beneficial. In some instances, harmful algal blooms (HABs) caused by the overgrowth of specific microalgae species can negatively impact marine life and water quality. These blooms can generate toxins that are detrimental to fish, shellfish, and other marine organisms, resulting in large-scale die-offs and disruptions in food webs. Furthermore, the decay of algal blooms can lead to decreased oxygen levels in water, creating hypoxic conditions that pose a threat to marine life (Trainer & Hardy, 2015). This process is vital for maintaining carbon balance in marine ecosystems and broader environments. Human activity, including the release of nutrients from farming and wastewater, has increased the frequency and intensity of HABs. Measures must be taken to control and reduce the amount of nutrients entering marine ecosystems to limit the adverse consequences of HABs and safeguard marine environments (Trainer & Hardy, 2015). The implementation of sustainable practices and policies to minimize nutrient pollution is crucial for preserving the ecological balance and well-being of marine ecosystems. In summary, microalgae play a vital role in maintaining healthy marine ecosystems by supporting biodiversity and participating in the carbon cycle. Despite their numerous ecological advantages, the emergence of HABs emphasizes the need for careful nutrient input management and ongoing research on the intricate dynamics of microalgae in marine environments.

2.2.7 Life on Land: Role of Microalgae in Soil Fertility and Carbon Sequestration

Microalgae have shown considerable promise in fostering soil fertility and participating in carbon sequestration, making them valuable assets for sustainable agriculture. These single-celled organisms can improve soil quality by providing essential nutrients and fortifying soil structure. When applied as biofertilizers, microalgae can increase the accessibility of nitrogen, phosphorus, and other vital nutrients, thereby stimulating plant growth and enhancing crop yields. Furthermore, microalgae can bolster soil organic matter content, which is vital for preserving soil health and fertility (Maity *et al.*, 2021). Utilization of microalgae in agriculture offers a crucial advantage in their capacity to capture atmospheric carbon dioxide through photosynthesis and transform it into organic carbon. This action not only helps diminish the quantity of greenhouse gases but also augments the organic carbon pool in the soil (Rani & Roy, 2021). Microalgae can fix and store carbon in soil for extended periods, thereby contributing to long-term carbon sequestration and mitigating the effects of climate change. In addition, these microorganisms produce exopolysaccharides that enhance soil structure by improving soil aggregation and water retention (Singh *et al.*, 2020).

Research has demonstrated that microalgae can also support the growth of beneficial soil microorganisms, further enhancing soil fertility and health (Singh *et al.*, 2020). Microorganisms are vital for nutrient cycling, organic matter decomposition, and disease suppression, which result in healthier and more productive soil ecosystems. Farmers can create sustainable and environment-friendly systems by incorporating microalgae-based biofertilizers into their agricultural practices. Despite these benefits, the widespread use of microalgae in agriculture faces several obstacles such as high production costs and the need for optimized cultivation techniques. To overcome these challenges, continuous research and technological advancements are necessary to develop cost-effective and scalable solutions using microalgae to enhance soil fertility and sequester carbon (Maity *et al.*, 2021; Rani & Roy, 2021).

2.3 MICROALGAE AND CLIMATE CHANGE RESILIENCE

Climate change significantly affects the trophic structure and dynamics of aquatic ecosystems by altering the phenology, physiology, and life-history traits of various organisms and disrupting their balance and health (Strecker *et al.*, 2004). Climate change also threatens the agricultural sector and food security significantly. Erratic rainfall and extreme weather conditions increasing in frequency and severity can damage crops, reduce yields, and affect the food supply chain. This can lead to food shortages, higher prices, and increased vulnerability of agricultural communities (Siedenburg *et al.*, 2021). Resilience is a system that manages hazardous social, economic, and environmental disruptions to maintain core functions, identity, and structure while preserving the ability to adapt, learn, and transform (IPCC, 2014). Climate-resilient pathways are development routes that integrate adaptation and mitigation efforts to achieve SDGs (IPCC, 2014).

Microalgae, which are photosynthetic unicellular microorganisms, have shown significant potential for climate action in terms of mitigating and adapting to climate change. Key adaptation benefits include rapid growth under controlled conditions, water reuse, high nutritional value for humans and animals, and enhanced crop resilience (Siedenburg *et al.*, 2021). Microalgae have an impressive ability to aid in climate change mitigation by capturing atmospheric CO_2 from the atmosphere, which offers significant potential for sustainable applications to synthesize value-added products such as biofuels, foods, beverages, and pharmaceuticals (Smetana *et al.*, 2017). Song, Xie (2021)revealed the carbon sequestration capacity of microalgae, which enabled the creation of bio-factories that could efficiently capture and convert CO_2 into bioenergy and other valuable bioproducts. Microalgae can capture 273.75 g of CO_2/year from 1 L of wastewater (Nwoba *et al.*, 2022)and can serve as a natural carbon sink. In addition, microalgae can tackle other environmental challenges such as bioremediation and biomonitoring agents for water pollution. This approach can contribute to a circular economy (Ahmad *et al.*, 2022).

Climate change is expected to augment the severity and probability of its negative impacts on food security. The risks posed by climate change to food security involve disruption of food systems and interruptions in food distribution. These issues are related to global warming, droughts, floods, and extreme variations in precipitation (Schmidhuber and Tubiello, 2007). Currently, agri-food systems are struggling to fulfill global demand and are responsible for 33% of greenhouse gas emissions. Conventional agriculture is constrained by increasing greenhouse gas emissions and limited arable land and freshwater availability (IPCC, 2021).. Therefore, microalgae are a viable alternative to produce food, feed, and feedstock because they can grow on non-arable lands, for example, by cultivating them in brackish or seawater rather than on land (Gao *et al.*, 2021). In the agriculture sector, microalgae offer numerous benefits, including nitrogen and carbon dioxide fixation, nutrient cycling, and plant growth promotion (Kumar *et al.*, 2020). Similarly, Siedenburg (Siedenburg *et al.*, 2021) reported that microalgae are promising candidates for combating climate change, such as carbon sequestration, biofertilizers, animal feed, and nutrient recycling. Siedenburg (Siedenburg *et al.*, 2021) summarized the key characteristics of microalgae and their connections to promote climate-resilient agriculture and food systems. The features are as follows: (1) tolerance at high temperatures is one of the attributes of climate change, and crops or microalgae with high-temperature tolerance can cope better. (2) Regulated water supply due to climate change impacts, such as water scarcity and unpredictable rainfall, but evaporation of the culture medium can be minimized, allowing for water reuse. (3) Microalgae are unaffected by degraded land because land degradation increases vulnerability to climate change and reduces soil fertility. (4) Efficient land use as climate change threatens agriculture and microalgae can be cultivated in small spaces rapidly, which can help offset potential shortfalls. (5) Minimized pest and disease risks: Climate change increases the threat of pests and diseases to crops and livestock; however, microalgal cultivation sidesteps many of these challenges.

However, they are still vulnerable to aquatic pathogens (Siedenburg *et al.*, 2021). Collectively, these traits demonstrated the resilience of microalgae to climate change.

Incorporating microalgae into animal feed can reinforce the resilience of livestock farms. First, it ensures a steady supply of protein supplements, as microalgal biomass production is likely to be climate-resilient. Second, this feed can positively affect animals by improving their ability to cope with the effects of climate change. For instance, it has the potential to enhance ruminant digestive function, particularly protein synthesis in the rumen, thereby improving animal performance, given the available feed resources (Ramin *et al.*, 2020). Research has also indicated that the milk supply and digestive health of goats are enhanced by *Chlorella* sp. (Becker, 2013). Microalgae, such as *Chlorella* and *Spirulina*, have a comparable amino acid composition to soybeans, which are a major source of feed protein (Yoo *et al.*, 2012)and contribute to a significant carbon footprint owing to deforestation and transportation (Lamminen *et al.*, 2019; Makkar *et al.*, 2014). Lamminen, Lamminen *et al.* (2019)showed that in terms of the nutrition of dairy cows, microalgae are probably a protein source similar to soybean meal, with *Spirulina* showing the highest yields. *Spirulina*-based diets enhance milk fat percentage, potentially through enhanced body lipid mobilization or rumen fermentation, which is more acetate-intensive. The inclusion of *Nannochloropsis gaditana* in the diet increased milk Eicosapentaenoic Acid (EPA) concentration without causing milk fat depression. Third, microalgae-based supplements can potentially mitigate ruminant enteric methane emissions, which are associated with impaired digestive efficiency and reduced feed energy. The potential to improve ruminant digestion and reduce methane emissions is particularly significant in tropical regions (Behera *et al.*, 2021; Kumari *et al.*, 2022). Climate change is posing an escalating threat to the world. Research indicates that microalgae, when used for application as food, feed, or crops, show promise for potential climate change mitigation benefits, such as absorbing CO_2 and converting it into biomass.

2.3.1 Microalgae for Wastewater Treatment

The United Nations World Water Development Report indicates that the world produces more than 2212 km^3 of wastewater annually, accounting for 56% of global water withdrawals (UNESCO, 2023). Untreated wastewater can pose health problems and affect aquatic environments. Additionally, conventional wastewater treatment methods have been a challenge owing to their high chemical and energy demands as well as greenhouse gas emissions. In China, the recent enforcement of stricter effluent standards (Class 1-grade A wastewater treatment) has complicated further issues. Delayed aeration processes and the need for extra operating units have significantly increased both operating costs and chemical usage (Xie *et al.*, 2021; Zhang *et al.*, 2020). Researchers have sought new and improved wastewater treatment methods. Microalgae-based wastewater treatment has gained considerable attention owing to its numerous benefits. The use of microalgae to treat wastewater is energy-efficient, as it requires low energy input. Microalgae are highly adaptable and thrive in various environmental conditions, making them suitable for diverse wastewater treatment applications. Microalgae can recover and convert nutrients from wastewater into bioactive chemicals. This coordinated strategy optimizes wastewater treatment, allows resource recovery, and creates valuable products (Sutherland *et al.*, 2020). Microalgae-based wastewater treatment also enables carbon capture, utilization, and storage through photosynthetic CO_2 fixation by microalgae (Chen *et al.*, 2020). This chapter explores innovative techniques to improve the efficacy of wastewater treatment using microalgae. These strategies include pretreatment, optimization of cultivation parameters, genetic engineering, and random mutagenesis (Figure 2.1).

2.3.2 Pretreatment Using Advanced Oxidation Processes

The color, turbidity, suspended solids, and elemental composition of wastewater make it unsuitable for direct use in the treatment of microalgae. Therefore, proactive pretreatment approaches are

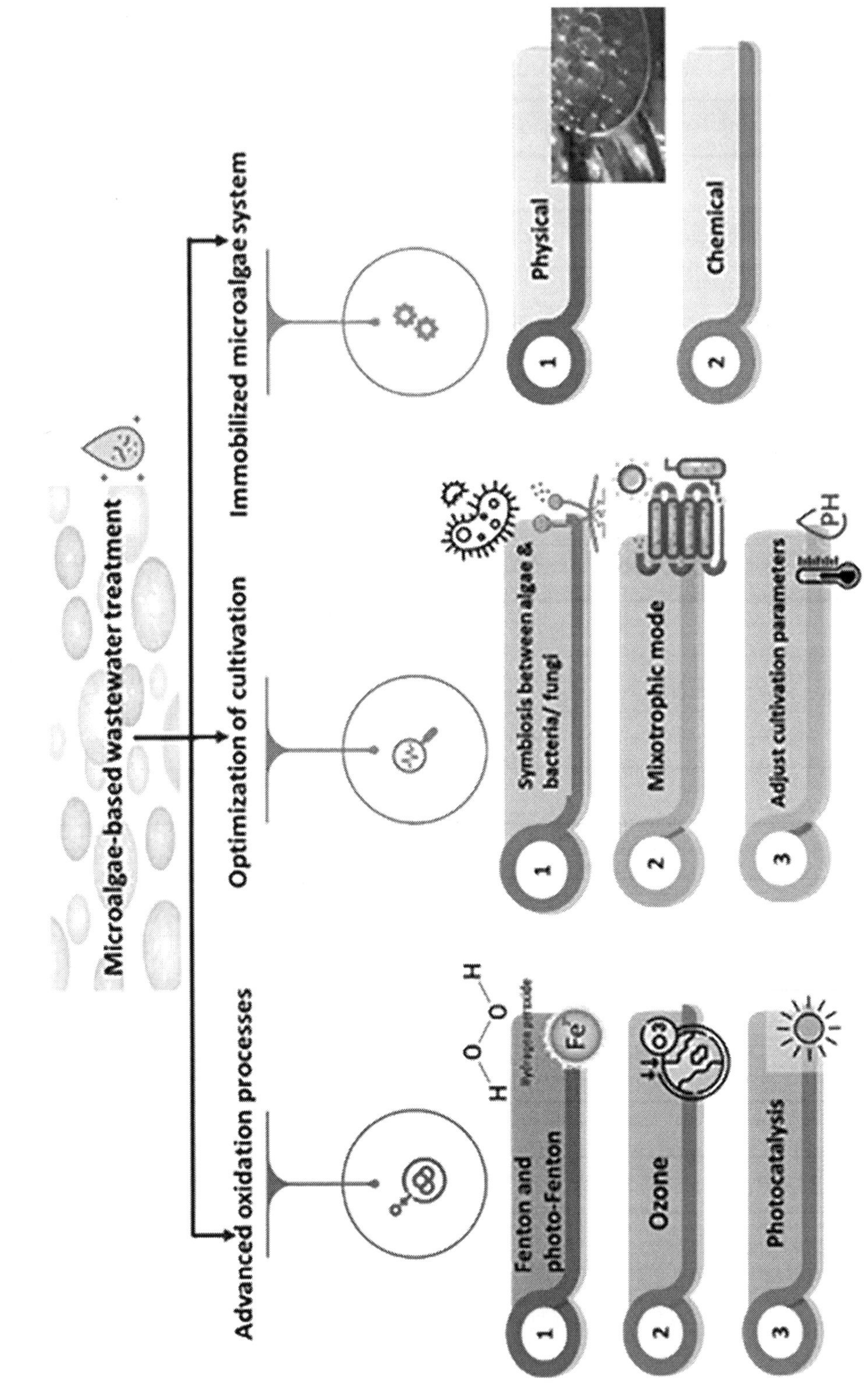

FIGURE 2.1 Microalgae-based wastewater treatment.

necessary to improve the adaptability of microalgae to wastewater treatment (Chen *et al.*, 2020). Filtration is the most commonly used pretreatment method and is suitable for large-scale applications. However, the filtration method is expensive, particularly for large-scale filtration, maintenance, and filter replacement. Therefore, advanced oxidation processes (AOPs), an aqueous phase oxidation technique that relies on the core principles of highly reactive species, have been introduced for wastewater treatment. AOPs utilize potent oxidants such as ozone, hydrogen peroxide, or ultraviolet radiation to decompose complex organic and inorganic contaminants in wastewater (Miklos *et al.*, 2018). Common AOP methods include (1) Fenton (H_2O_2 and Fe^{2+} ions) and photo-Fenton, (2) ozone (O_3), (3) photocatalysis, (4) electrooxidation, and (5) electrocoagulation (Chen *et al.*, 2020; Li *et al.*, 2021). Fenton's reaction is mostly used to treat wastewater because of the advantages of a short reaction time, reasonable price of Fenton's reagent, and no energy input as hydroxyl radicals are generated chemically (Chowdhury *et al.*, 2019). Wang and Zeng (2021)synthesized a novel iron–carbon composite to treat high-concentration antibiotic wastewater. Through the incorporation of Fenton oxidation to increase Fe^{2+} levels, the treatment efficacy increased, with a remarkable reduction in the COD. This synergistic application resulted in a removal efficiency of 66.9% and reduced the cost of the solid waste treatment. Photocatalysts convert photonic energy into chemical energy, such as sunlight, UV light, and visible light Kim and Kan, 2016; Fujishima *et al.*, 2008; Hisatomi and Domen, 2017).

Safari and Seyed Ahmadian (2022) synthesized Gd-doped $MTiO_3$ perovskites (M = Co, Cu, and Ni) to degrade methylene blue under a xenon lamp (visible light source). The results showed that adding 2% Gd ions to the crystalline lattice of the perovskites enhanced the photocatalytic activity with a degradation efficiency of 88.64% for methylene blue within 120 min by 2% Gd-doped $NiTiO_3$. The next sophisticated treatment method that works well at eliminating environmental contaminants from wastewater is ozone (O_3) treatment. Because oxygen is a raw treatment material, no additional chemicals are required. It can also perform advanced processes, such as decolorization and deodorization, which are difficult to conduct using traditional wastewater treatment methods (Zhang *et al.*, 2016; Oulego *et al.*, 2016; Azuma and Usui, 2020). Azuma and Usui (2020)revealed that direct ozone treatment is effective for disinfecting hospital wastewater, particularly by eliminating antimicrobials and antimicrobial-resistant bacteria. Combining O_3 with UV or $O_3/UV/H_2O_2$ demonstrated higher efficiencies in treating hospital wastewater.

Nevertheless, there are drawbacks to using AOP to treat wastewater, such as side reactions forming various by-products, which may be more toxic or hazardous than the original pollutants. Some of these by-products react with reactive oxygen species to form carbon dioxide (Miklos *et al.*, 2018; Bijlsma *et al.*, 2012; Marti *et al.*, 2011). Therefore, it is essential to perform toxicity assessments of the by-products produced by AOPs and their impact on human health prior to the introduction of this technology in the real world, as these by-products may be toxic, mutagenic, or exhibit estrogenic activity (Miklos *et al.*, 2018).Furthermore, most interactions between the parent compounds, transformed products, and environmental conditions are unknown. The lack of experimental data makes it challenging to develop models that can predict the fate of these pollutants when treated with AOPs (Marti *et al.*, 2011).

2.3.3 Optimization of Cultivation Parameters

A more economical approach would be to modify the culture mode by promoting symbiosis between algae, bacteria, and fungi or by switching to a mixotrophic mode. The mixotrophic mode had superior biomass productivity and nutrient removal efficiency among the three main types of microalgae cultures (autotrophic, heterotrophic, and mixotrophic). Thus, it can assimilate both organic and inorganic carbon, opening new pathways for microalgal growth and metabolism (Rawat *et al.*, 2011). Table 2.1 illustrates the optimization of microalgae cultivation to improve wastewater treatment efficiency.

TABLE 2.1
Optimization of Microalgae Cultivation in Treating Wastewater

Microalgae	Cultivation Mode	Wastewater Type	Degradation Efficiency	References
***Chlorella sorokiniana* SU-1**	Integrated anaerobic digestion with algae cultivation	High-strength dairy wastewater	Removal of 86.8% of chemical oxygen demand (COD), 94.6% of total phosphorus, and 80.7% of total nitrogen.	Kusmayadi *et al.*, 2020
***C. sorokiniana* SXAU-04**	Combined Fenton oxidation and algal cultivation system	Chicken farm flushing wastewater	Single Fenton oxidation pretreatment removed 70.5%, 96.7%, 86.1%, and 96.2% of total nitrogen, ammonia nitrogen, phosphorus, and COD, respectively.	Zhang *et al.*, 2022
Chlorella vulgaris	Bacterial-algal coupling system which combines acidogenic fermentation with microalgae cultivation	Mariculture wastewater	54.0% ammonium, 78.6% of total phosphorus, 82.8% of COD, and 86.0% of volatile fatty acids (VFAs) were removed.	Zhang *et al.*, 2023
C. vulgaris	Bacterial-algal coupling system which combines acidogenic fermentation with microalgae cultivation: Impact of temperature	High-strength mariculture wastewater	The efficiency of removing COD and ammonium can reach 99.0% and 68.8%, respectively. The optimal temperature for cultivation is 25°C.	Zhang *et al.*, 2021
Chlorella pyrenoidosa	Co-cultivation and flocculation of microalgae with filamentous fungi (*Aspergillus niger* and *Aspergillus oryzae*)	Simulated potato starch wastewater (PSW)	At a low initial ratio of algae to fungi (2000:1), the removal efficiencies for COD, TN, and TP were 81.75%, 79.70%, and 83.12%, respectively. When the initial algal ratio was increased to 4000:1, the removal efficiencies improved to 88.44%, 82.59%, and 96.07%, respectively. The color of PSW was effectively decolorized.	Zhao *et al.*, 2021
C. pyrenoidosa	Heterotrophic and mixotrophic cultivation mode	Furfural wastewater	The removal rate of total nitrogen was greater in the mixotrophic mode than in the heterotrophic mode. However, COD and TP removal rates are lower in mixotrophic conditions than in heterotrophic conditions.	Zhao *et al.*, 2022

2.3.4 Immobilized Microalgal System

Immobilization of microalgae involves the use of physical and chemical methods to prevent free-living microalgal cells from moving independently at their original site. This approach can maintain specific biological activities and enable reusability in aqueous systems. Immobilized microalgae have a high population density, making wastewater treatment simple and effective. This reduces carbon emissions and improves treatment effectiveness, without the need for energy-intensive recycling. Moreover, similar to suspended microalgae, immobilized microalgal cultures have a higher potential for nutrient recovery and are more economically valuable in the chemical and agricultural industries (Boonsong *et al.*, 2016; Mallick, 2002; Wang *et al.*, 2018). There are several microalgal immobilization techniques, such as biofilm formation, alginate beads, capsules, and cell entrapment in hydrogels. *Desmodesmus subspicatus* was immobilized in alginate beads, demonstrating good stability in sugarcane vinasse. The growth of immobilized *D. subspicatus* in alginate beads was enhanced, resulting in significant removal of potassium, carbon, and nitrogen from vinasse compared to free microalgae cultivation (de Jesus & Gaspar Bastos, 2022).

Alhumairi and Hamouda (2020)showed that immobilized microalgae *Chlorella* sp. were more effective than suspended *Chlorella* sp. in reducing organic compounds in contaminated areas in the Dhiba Port Kingdom of Saudi Arabia. Another study by Mujtaba and Rizwan (2018)showed that the removal of nutrients by immobilized *Chlorella* sp. was 100% for nitrogen and 99.8% for phosphorus over two days. Immobilized microalgae to treat wastewater are currently underdeveloped, including efficiency enhancement strategies and optimization of cultivation parameters. Current immobilized microalgal technologies primarily focus on cell attachment using affordable biological matrix materials suitable only for mild hydrodynamic environments and weak cell-substrate adhesion. Extended cultivation times frequently result in leakage of connected or entrapped immobilized microalgal cells, which could harm the ecosystem. Immobilized microalgal systems have been explored for a restricted range of contaminants.

2.3.5 Microalgae in Carbon Capture and Sequestration

Climate-resilient development involves the implementation of strategies for greenhouse gas mitigation and adaptation to promote sustainable development(IPCC, 2018). Carbon capture and utilization technologies have attracted significant attention (Dassey *et al.*, 2014) for mitigating global warming and greenhouse gas emissions. One of them is biological approaches, which are cost-effective and have low energy consumption. Microalgae are promising candidates for carbon capture because they can fix CO_2 and utilize nutrients through photosynthesis. Microalgae are significantly more efficient than terrestrial plants in capturing carbon by up to 50 times because of their superior photosynthetic capabilities and high carbon capture rates (Kumar *et al.*, 2010). Additionally, microalgae do not compete for agricultural land and grow rapidly. For most algae, CO_2 fixation accounts for approximately 50% of the biomass by weight, resulting in a CO_2-to- biomass weight ratio of approximately 2:1, with a typical value of 1.88 (Wang *et al.*, 2008).Microalgal cultivation is an excellent method for large-scale carbon capture. However, the existing microalgal cultivation systems have not yet reached the production levels necessary for effective carbon capture and storage. Microalgae are generally cultivated in two systems: open ponds and closed PBR systems. PBRs have been shown to enhance microalgal biomass growth and carbon capture rates in both systems (Zhou *et al.*, 2017). Microalgae can metabolize inorganic carbon through a photoautotrophic mechanism involving multiple enzymes including carbonic anhydrase and RuBisCo (ribulose-1,5-bisphosphate carboxylase/oxygenase). These enzymes facilitate the conversion of CO_2 to carbohydrates and support glucose synthesis via the Calvin cycle, which is essential for CO_2 capture by algae. First, RuBisCo catalyzes the reaction between CO_2 and ribulose-1,5-bisphosphate to form 3-phosphoglycerate. This compound is then converted into 1,3-diphosphoglycerate by incorporating inorganic phosphate from ATP and is further reduced to 3-phosphoglyceraldehyde by the enzyme phosphoglyceraldehyde dehydrogenase. Finally, 3-phosphoglyceraldehyde is converted into ribulose-5-phosphate along with by-products, such as sugars, fatty acids, and amino acids (Kaplan *et al.*, 1980; Bassham, 1977).

Various factors can influence the cultivation of microalgae in PBRs, such as light intensity, temperature, carbon source, pH, and mass transfer, which can affect algal growth rate and CO_2 fixation efficiency. Lim, Ilankoon (2021)enhanced the microalgal growth and carbon capture in flat-panel PBRs by generating microbubbles. This can be achieved by assessing the effects of various gas sparger designs on the mixing performance. Enhancing the sparger design can improve gas distribution, create more favorable growth conditions, and boost carbon capture efficiency. The results revealed that customized sintered metal spargers featuring a novel double-sparger orientation demonstrated superior mixing performance and excellent microbubble generation compared to both perforated and commercial spargers. However, the cost of producing customized sintered metal spargers is relatively high. Moreover, Song and Xie (2022)*et al*demonstrated that the carbon sequestration capability of *Scenedesmus obliquus*, *Phaeodactylum tricornutum*, and *Chlamydomonas reinhardtii* increased by 17.18%, 38.89%, and 35.71%, respectively, when aggregation-induced emission (AIE) material was added. This shows the promise of AIE light-converting films in cultivating microalgae, which efficiently absorb large amounts of carbon dioxide to create carbon-neutral energy and valuable active compounds through photosynthesis (Song & Xie, 2022). Ratomski and Hawrot-Paw (2020)also explored the use of sodium bicarbonate as a cost-effective CO_2 source for cultivating *Chlorella vulgaris* in 100 L PBRs. The result revealed that maximum productivity was 7.0 ± 1.0 mg/L/day with lipid content increased to 26 ± 4%, carbon content in biomass to 1322 ± 0.062 g/dw, and CO_2 fixation rate to 0.925 ± 0.073 g/L/day. Wang and Li (2021developed a kinetic growth model for nutrient-supplemented microalgae. A new optimal nutrient supplementation strategy using BG-11 was developed, which enhanced the carbon sequestration rate by 56.2%. As part of the microalgae cultivation strategy, an initial 0.2 BC nutrient concentration (1 BC being the normal BG-11 culture medium concentration) was followed by additional supplementation at 100 h, and then every 72 h at a concentration of 0.3 BC.

Flue gases are a valuable carbon source for algal photosynthesis (Ramanna *et al.*, 2014). There are two methods of flue gas-based algal cultivation: direct and indirect. Direct capture involves bubbling flue gas into open algal ponds (PBRs) while adding inorganic carbon to water. However, CO_2 has limited solubility in water, reaching only 0.5833 mg/L at 25°C when exposed to the atmosphere. This results in significant CO_2 loss during direct capture (Ratomski & Hawrot-Paw, 2020; Sharma & Sarkar, 2020; Banerjee & Guria, 2021; Nayak & Dhanarajan, 2022).Indirect carbon capture transforms CO_2 into bicarbonate via a reversible reaction with carbonate (Pan *et al.*, 2011). The bicarbonate produced can be used to promote algal growth and has high solubility in water (93.2 g/L) at room temperature and atmospheric pressure (Carvalho *et al.*, 2011; Lu *et al.*, 2015; Ota *et al.*, 2016). A process integrating flue gas CO_2 sequestration and wastewater treatment to enhance *Scenedesmus* sp. biomass production was developed using an artificial neural network combined with a genetic algorithm optimization tool. The optimal parameters were identified as a light intensity of 124 μmol/m^2/s, a photoperiod of 17:7 (h), a temperature of 27.5°C, and an initial pH of 9.5. These conditions improved algal biomass productivity by approximately 57%, with a CO_2 sequestration rate of 578.1 mg/L/d (Nayak & Dhanarajan, 2022). Sharma and Sarkar (2020). Sharma, Sarkar [96] evaluated the influence of flue gas on the cultivation of *C. vulgaris* using wastewater at the cathode of a microbial fuel cell (MFC) and assessed its carbon capture efficiency. The results revealed that sodium carbonate-enriched wastewater absorbs CO_2, forming flue gas-derived bicarbonates (FGDBs), which are introduced into the PBRs connected to the MFC for absorption by *C. vulgaris*. Biomass productivity increased to 0.677 kg/m^3/day compared with that of the cathode without FGDB. The algae efficiently utilized 76.84 ± 1.23% of the bicarbonates. Additionally, over 50% of the sodium carbonate can be recycled for subsequent CO_2 capture cycles. Banerjee and Guria (2021) cultivated *Nannochloropsis* sp. using flue gas and synthetic fertilizers. Variations in the carbon content (32%) and C/N ratio (100%) indicated metabolic shifts under stress. Cultivation using response surface methodology improved the yields of biomass, lipids, and carbon bio-fixation by 43%, 31%, and 79%, respectively, compared to the standard f/2 medium. Light intensity and carbon availability significantly influenced carbon sequestration. Additionally, using a nitrogen-limiting fertilizer medium boosted volumetric yield, which in turn enhanced lipid production, biomass growth, and carbon fixation (Banerjee & Guria, 2021). García-Cubero and Moreno-Fernández (2021)modeled the growth and CO_2 fixation rate of *Scenedesmus vacuolatus* by examining the effects of various culture parameters. Under optimal conditions, *S. vacuolatus* achieved a biomass productivity of 0.63 g/L/day and a CO_2 assimilation rate of 1.15 g/L/day. The potential of *S. vacuolatus* for carbon capture was further evaluated using synthetic flue gas as a CO_2 source. For non-diluted flue gas (12% CO_2), *S. vacuolatus* demonstrated a CO_2 fixation rate of 1.17 g/L/day, comparable to the 1.15 g/L/day observed with 12 times diluted flue gas or pure CO_2. These findings highlight the capacity of *S. vacuolatus* to convert flue gas CO_2 into valuable biomass, thereby supporting its application in

CO_2 capture systems. In summary, microalgae cultivation can serve as an effective tool for carbon capture due to the high CO_2 fixation capability of microalgae. This represents a significant step toward mitigating climate change, aligning with the Paris Agreement and the United Nations SDG 13 (Climate Action).

2.3.6 Biofuel Production from Microalgae

Microalgae represent a promising source of biodiesel due to their high lipid content and rapid growth rates. Unlike conventional crops, they can be cultivated on non-arable land and in wastewater, thereby reducing competition for food production and conserving freshwater resources. Various cultivation systems, including open ponds and closed PBRs, have been developed to optimize growth conditions and maximize biomass yield. Factors such as light intensity, temperature, nutrient availability, and CO_2 concentration play crucial roles in influencing the growth and lipid accumulation of microalgae. Environmental stressors, including nutrient limitation and salinity, can enhance lipid production, thereby improving the efficiency of the process. Furthermore, genetic engineering and metabolic manipulation are being explored to increase lipid content in microalgae (Kumar *et al.*, 2010; Xie *et al.*, 2020).Converting microalgae lipids into biodiesel involves several steps, including harvesting, lipid extraction, and transesterification. Commonly used harvesting methods to collect microalgal biomass include centrifugation, flocculation, and filtration. Lipid extraction techniques encompass solvents, supercritical fluids, and mechanical methods. The extracted lipids are then converted into biodiesel through transesterification, where they react with alcohols (usually methanol) in the presence of a catalyst to produce biodiesel and glycerol. Despite this potential, several challenges remain in making microalgal biodiesel commercially viable, including high production costs, energy-intensive processes, and scalability. Future research is focused on improving cultivation techniques, developing cost-effective harvesting and extraction methods, and enhancing lipid productivity through genetic and metabolic engineering (Kumar *et al.*, 2010; Xie *et al.*, 2020).

Biodiesel production can benefit significantly from the use of microalgae as a raw material, owing to its substantial lipid content and rapid growth. Several microalgae species have been recognized as appropriate for biodiesel manufacturing, including *C. vulgaris*, *Nannochloropsis oceanica*, *Dunaliella salina*, *Botryococcus braunii*, *Desmodesmus* sp., *Neochloris oleoabundans*, *S. obliquus*, and *Tetraselmis suecica* (Pruvost *et al.*, 2011; Taleb *et al.*, 2021).These strains are favorable for their ability to accumulate significant amounts of lipids under specific cultivation conditions. For instance, *C. vulgaris* is known for its high lipid productivity, achieving up to 58% lipid content under nitrogen-starved conditions [22] (Figure 2.2).

Similarly, *N. oceanica* produces high-quality biodiesel with favorable fatty acid methyl ester (FAME) profiles (Dong *et al.*, 2013).The productivity and lipid content of microalgae strains are affected by several factors such as nutrient availability, light intensity, and environmental stress factors. As an illustration, *D. salina* can accumulate lipids by up to 60% when exposed to high-salinity conditions (Liang *et al.*, 2019).Researchers are currently investigating genetic and metabolic engineering methods for boosting lipid production in microalgal strains. The application of omics technologies and genetic modifications has led to substantial increases in microalgal lipid yields, enhancing their potential for industrial-scale biodiesel manufacturing. However, despite these improvements, obstacles such as elevated production expenses and difficulties in scaling up operations persist, highlighting the need for ongoing research to enhance cultivation and processing methodologies (Xiong *et al.*, 2021).

Bioethanol production has been found to be a promising source of microalgae owing to its substantial carbohydrate content and swift growth patterns. Researchers have extensively investigated several microalgal species for their bioethanol production capabilities, including *C. vulgaris*, *S. obliquus*, *C. reinhardtii*, and *Dunaliella tertiolecta* (Singh *et al.*, 2011; Ugoala *et al.*, 2012).These strains are favored for their ability to accumulate significant amounts of carbohydrates, which can be

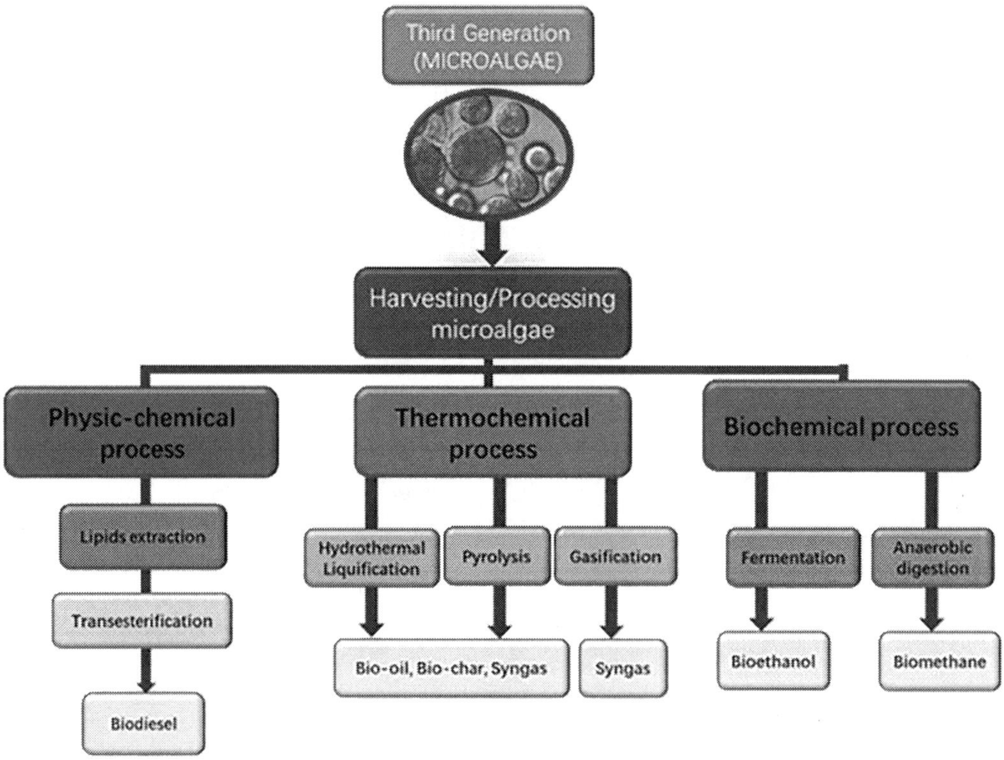

FIGURE 2.2 Microalgal-based biofuel production.

converted into fermentable sugars. For instance, *C. vulgaris* has been reported to achieve a carbohydrate content up to 55% of its dry weight under optimal conditions (Van Den Hende *et al.*, 2014).The selection of appropriate microalgal strains is crucial for optimizing bioethanol production because different strains exhibit varying levels of carbohydrate productivity and fermentation efficiency (Singh *et al.*, 2011; Ugoala *et al.*, 2012). The productivity of microalgal strains for bioethanol production is influenced by several factors including cultivation conditions, nutrient availability, and environmental stressors. Strains such as *S. obliquus* and *C. reinhardtii* have high carbohydrate content and biomass yield when cultured in PBRs, with carbohydrate productivity reaching up to 0.4 g/L/day (Choi *et al.*, 2017). Additionally, the pretreatment and hydrolysis processes are critical for the conversion of carbohydrates into fermentable sugars. Techniques such as acid hydrolysis, enzymatic hydrolysis, and thermal pretreatment are commonly used to break down the cell walls and release sugars (Singh *et al.*, 2011; Ugoala *et al.*, 2012). Despite this potential, challenges such as high production costs and scalability issues remain, necessitating further research to optimize the cultivation and processing techniques (Ugoala *et al.*, 2012; Choi *et al.*, 2017).

Bio-hydrogen (H_2) is a renewable and clean energy source owing to its sustainability, high energy potential, and conversion efficiency. With approximately 120 MJ of energy/kg, hydrogen surpasses most conventional hydrocarbon-based fuels (Mandal *et al.*, 2019; Chen *et al.*, 2019).Biohydrogen is generated through biological methods that utilize various microorganisms such as bacteria, cyanobacteria, and microalgae. The production of biohydrogen from microalgae is considered a promising method for carbon neutrality and sustainable bioenergy production (Asada *et al.*, 2019).Hence, various studies have focused on utilizing microalgae, such as *C. reinhardtii*, *Spirulina platensis*, *C. vulgaris*, *S. obliquus*, and *Anabaena* sp., for biohydrogen production because microalgae have a rapid growth rate, high tolerance, adaptability to diverse environmental stresses, and the ability to grow on non-arable land. In addition, the high lipid content of microalgal biomass makes it a suitable candidate for biofuel production. Table 2.1 illustrates the hydrogen production from microalgae via different pathways.

Hydrogen production in microalgae can be accomplished either directly or indirectly via biophotolysis and fermentation (dark fermentation and photofermentation) (Zhou *et al.*, 2018). Direct biophotolysis: In microalgae, solar energy is converted into chemical energy using water as the substrate and the acellular chloroplast–ferredoxin–hydrogenase system. However, the reaction time for this transient process was short. Light energy captured by PSI and PSII reduces water by transferring electrons to ferredoxin, thereby initiating oxygenic photosynthesis to produce hydrogen (Chen *et al.*, 2020; Khetkorn *et al.*, 2020).The benefits of this process include its environmental accessibility and high availability in nature owing to the use of sunlight as an energy source. However, its practical application remains challenging because of the oxygen sensitivity of hydrogenase reactions and the complexity of the pathways involved (Rahimnejad *et al.*, 2020).

Indirect biophotolysis: CO_2 is initially converted into carbohydrates, which are then used in separate steps to produce H_2 (Chen *et al.*, 2020), demonstrating the sustainable production of H_2. This process separates the H_2 and O_2 evolution reactions into different stages linked by CO_2 fixation and evolution (Jian *et al.*, 2019). However, the drawbacks of this reaction are as follows: (1) microalgae growth can be influenced by various cultivation conditions, (2) transient H_2 production, and (3) sulfide deprivation (Aini *et al.*, 2020).

Photofermentation: Owing to the low efficiency of direct H_2 generation from microalgae, microalgae are used as feedstock for microbial fermentation to produce biohydrogen (Kang *et al.*, 2021). Photofermentation utilizes light energy coupled with photo catabolism and oxidative carbon metabolism to generate energy without pyruvate degradation or limitations associated with respiratory ATP generation (Yao *et al.*, 2017).The disadvantage of this approach is that the reaction requires light energy; thus, the efficiency is affected by the amount of light present. Despite the presence of hydrogenases, nitrogenase is the primary enzyme that helps photosynthetic microorganisms to produce hydrogen (Tiwari *et al.*, 2018).

Dark fermentation: In microalgae, hydrogen is produced anaerobically through dark conversion of organic substrates without solar energy (Yao *et al.*, 2017).Organic substances undergo various stages such as hydrolysis, acidogenesis, and acetogenesis (Yao *et al.*, 2017). The advantages of this method are its high energy efficiency, environmental sustainability, and favorable carbon balance (Lee *et al.*, 2019).To overcome the kinetic and thermodynamic limitations of anaerobic fermentation reactions, boosting the biological activity and electron transfer of microbial fermentation is essential (Lee *et al.*, 2019).To boost hydrogen production, substrates must be pretreated prior to anaerobic fermentation (Zhou *et al.*, 2021). Utilizing microalgal biomass for dark fermentation is a viable and promising method for producing biohydrogen.

Biomethane (CH_4), a carbon-neutral renewable fuel, is a crucial strategy for addressing global climate change, given the pressing need to reduce greenhouse gas emissions. CH_4 is the end product of anaerobic digestion, by which methanogenic bacteria anaerobically digest organic-rich waste streams (Sivaramakrishnan & Gangadharan, 2014).The four main processes involved in anaerobic digestion are hydrolysis, acetogenesis, acidogenesis, and methanogenesis (Kumar *et al.*, 2015). The utilization of biomass and organic waste to generate CH_4 mitigates the environmental impact and promotes a circular economy (González-Fernández *et al.*, 2021).Microalgae has recently emerged and been identified as a promising platform for the production of CH_4. However, in the sustainable production of CH_4, harvesting and rigid cell walls, which require additional pretreatment steps, are major technical barriers to large-scale production.

Thus, research on a novel, cost-effective cell wall disruption and harvesting methods to enhance microalgae capability for green CH_4 production is ongoing. Kannah Ravi *et al.* (2021)investigated a dual pretreatment process to increase biomethane production. In this process, microalgae biomass is pretreated by thermochemical disintegration (TC) at different temperatures and times,

TABLE 2.2
Biohydrogen Production from Different Strains of Microalgae

Algae Species	Process Involved	Cultivation Condition	Biohydrogen	References
Chlorella vulgaris ESP-6	Dark fermentation	Cultivation condition: Dairy manure wastewater (25%, 50%, 75%, and 100%) in 1 L photobioreactor, employing two-stage bacteria-microalgae cultivation strategy (Bacteria: indigenous bacteria in the wastewater). Pretreatment method: Thermal-acid pretreatment. Biohydrogen production: Hydrogen-producing bacterial strain *C. butyricum* CGS5 involved. The algae biomass is used as the carbon source.	Total H_2 production: 353 mL; H_2 yield: 1.43 mol/mol RS (179 mL/g VS) and maximum H_2 productivity: 391.1 mL/L/day (RS: reducing sugars, VS: volatile solids).	Vijayaraghavan & Hemanathan, 2022
Chlorella kessleri VKPM AI–11 ARW	Dark fermentation	Cultivation condition: Algae is cultivated in nutrient medium mixed with wastewater. Biohydrogen production: A mixture of hydrogen-producing bacteria (*Aquifex pyrophilus*, *Calderobacterium hydrogenophilum*, and *Acidithiobacillus ferrooxidans*) is used. Anaerobic conditions are created in dark condition at 38°C for six days.	Maximum amount: 44.24 mL/L suspension.	Vijayaraghavan & Hemanathan, 2022
C. vulgaris	Dark fermentation	Cultivation condition: Immobilized microalgae-alginate beads are applied with different sodium alginate concentration. Algae is cultivated in simulated municipal wastewater. Biohydrogen production: Using wastewater and immobilized microalgae-alginate beads. Nitrogen gas is supplied for creating an anaerobic environment in the absence of light.	The synthesis of biohydrogen is enhanced by two to three times in immobilized microalgae. Immobilized microalgae-alginate beads (8% sodium alginate) have the highest biohydrogen production, followed by 6%, 4% and 2%. Maximum biohydrogen production with 8% sodium alginate was 19.54 mL at the 8th day followed by 19.01 mL for 6% and 13.92 mL for 4% at the 7th day. The efficiency of immobilized microalgae-alginate beads decreased with each successive cycle of dark fermentative biohydrogen production.	Vijayaraghavan & Hemanathan, 2022

Chlamydomonas reinhardtii	Dark fermentation	Pretreatment: Autoclaving and ultrasonication to boost the biohydrogen production. Carbohydrate source for fermentation: Acidogenic culture from cow dung and *C. reinhardtii*. A hydrogen generation investigation using batch fermentation method was carried out.	Dark fermentation: 0.73 L/L of biohydrogen (biomass concentration of 18 g/L). Batch fermentation with pretreated algal biomass: 2.08 L/L of hydrogen with an energy recovery rate of around 8%. Biophotolysis: 0.1 L/L of biohydrogen.	Goud *et al.*, 2021
Anabaena sp. (UTEX 1448)	Biophotolysis	First step (Microalgae cultivation and formation of heterocysts): The culture was kept in an aerobic environment in BG-11 with glucose (organic carbon source) with limited nitrogen to induce heterocyst formation, thereby enhancing hydrogen production through nitrogenase activity. Second step (Hydrogen production): The culture was moved to a photobioreactor, where the concentrations of ammonium and glucose were varied, and argon gas was added.	*Exponential phase* Glucose concentration of 1.05 g/L a) Without ammonium: 13.15 mmol H_2 mg Chl/a with 0.317 h^{-1} b) With ammonium: 9.40 mmol H_2 mg Chl/a with 0.246 h^{-1} Glucose concentration of 2.1 g/L a) Without ammonium: 13.36 mmol H_2 mg Chl/a with 0.269 h^{-1} b) With ammonium: 11.19 mmol H_2 mg Chl/a with 0.225 h^{-1} *Stationary phase* Glucose concentration of 1.05 g/L a) Without ammonium: 3.24 mmol H_2 mg Chl/a with 0.237 h^{-1} b) With ammonium: 1.39 mmol H_2 mg Chl/a with 0.379 h^{-1} Glucose concentration of 2.1 g/L a) Without ammonium: 2.90 mmol H_2 mg Chl/a with 0.290 h^{-1} b) With ammonium: 1.25 mmol H_2 mg Chl/a with 0.272 h^{-1} Hydrogen production rate: h^{-1} Maximum production of hydrogen by chlorophyll a: mmol H_2 mg Chl/a.	Kanwal *et al.*, 2021
C. reinhardtii FACHB-265, *Chlorella protothecoides* FACHB-3, *Chlorella pyrenoidosa* FACHB-5, *and Scenedesmus obliquus* FACHB-416	Photolysis	Co-cultivation of algae and bacteria to produce hydrogen.	Direct water photolysis is the primary contributor to hydrogen production in algae. *C. reinhardtii* Pure algal culture: 6 mL/L of H_2 Combined with the bacterial flora: Hydrogen accumulation began on the 3rd day and reached a maximum rate of 21.5 mL/L/d by the 7th day. Over this period, a total of 154 mL/L of hydrogen was produced, which is approximately 26 times more than the pure algal culture alone. Maximum H_2 accumulation (mL/L H_2) by co-cultures of: a) *Chlorella* sp. WFY: 40.1 b) *Scenedesmus* sp. WFY: 11.1 c) *C. protothecoides*: 39.8 d) *C. pyrenoidosa*: 29.2 e) *S. obliquus*: 3.6.	Hossain *et al.*, 2020

followed by bacterial disintegration (BD). This combined method resulted in a higher disintegration rate (33%) than BD alone (24%), TC alone (8.5%), and control (7%). This dual process also resulted in the highest CH_4 production among all the methods, with a maximum biomethane yield of 232.4 mL/g COD. The prospects for commercialization of this dual-disintegration method are intriguing. Vu, Vu *et al.* (2021)also explored the potential of expired isopropanol (IPA) COVID-19 disinfectant (64% w/w) for the pretreatment of *Scenedesmus* sp. microalgae and different harvesting methods (centrifugation and polymer flocculation) to boost CH_4 production. The findings revealed that the harvesting methods had a minimal impact on CH_4 yield. CH_4 production from centrifuged and polymer-harvested biomass was low, at 66 and 74 L/kg volatile solids, respectively. The drying process was found to increase the CH_4 yield compared to wet biomass. IPA pretreatment significantly increased CH_4 production by 212%. Kusmayadi *et al.* (2021)also cultivated *Chlorella sorokiniana* SU-1 in dairy wastewater integrated with combined anaerobic digestion that can improve the methane content and production rate to 65.2% and 0.16 L CH_4/L/d, respectively. Hasanath and Chatterjee (2021) cultivated *S. obliquus* and *Chlorella pyrenoidosa* in BG-11 media and high-strength beverage wastewater to assess their biochemical methane potential (BMP). For *Scenedesmus* sp., the BMP increased from 185.53 mL CH_4/g VSadded (BG-11) to 336.35 mL CH_4/ g VSadded (beverage wastewater). For *Chlorella*, VSadded BMP increased to 179.64 mL CH_4/g VSadded (BG-11) to 245.51 mL CH_4/g VSadded (BG-11). Surprisingly, methane production for both microalgae species improved after lipid extraction which were with 393.66 and 421.86 mL CH_4/g VSadded for *Scenedesmus* and *Chlorella*, respectively, demonstrating the possibility of prolonged energy recovery following lipid extraction. Overall, the optimization of cell wall disruption and harvesting methods, along with the anaerobic digestion process to boost the production of CH_4 from microalgae, is still in the refinement stage.

2.4 CHALLENGES AND OPPORTUNITIES

Microalgae have significant prospects for achieving SDGs and enhancing climate change resilience through the mitigation of climate change (sequestration of CO_2), sustainable agricultural practices (as biofertilizers to enhance soil health and crop productivity), food security (able to survive in non-arable lands, and the nutrients of microalgae-based food products are comparable with those of commercial food products), and wastewater treatment (removal of contaminants, pollutants, and heavy metals). Despite extensive research on microalgae over the past few decades, microalgal biotechnology still encounters several technical challenges, including complex and expensive upstream (selection of the most suitable microalgal strain and cultivation stage) and downstream (harvesting, pretreatment, and biomolecule extraction) processes. The rigid cell walls standardize cost- and time-efficient downstream processing to disrupt the cell walls and recover intracellular biomolecules. Microalgal biomass is considered one of the most energy-intensive steps, contributing up to 30% of the total production cost (Tavanandi *et al.*, 2021).Most microalgae studies and research are limited to laboratory scales because scaling up to industrial or commercial levels is difficult due to the high cost and the variation or complexity in the environment and seasonal conditions in the real world. Finally, despite the favorable climatic conditions for microalgae cultivation in many low-income countries such as Africa, South America, and South Asia, research on microalgae cultivation in these regions remains scarce (Wijffels & Barbosa, 2010).Through collaboration or government grants, some microalgae-based technologies have already been successfully applied in the real world, and microalgae-based products are already available in the market. For instance, full-scale demonstrations of wastewater treatment in high-rate algal ponds have proven successful. They are ready for implementation in areas lacking a complete and advanced wastewater treatment infrastructure, although cost-effective biomass harvesting remains challenging (Christenson & Sims, 2011).. Hence, emerging research and innovations in microalgae utilization are necessary to harness the full potential of microalgae.

I. Bioprospecting for novel microalgae strains with desirable properties for various applications, such as biofuels, pharmaceuticals, nutraceuticals, bioplastics, food, and beverages. At the laboratory stage, the experiment should comply with real-world conditions, focusing on engineering aspects, costs, and cultivation conditions.
II. Innovations in PBRs and open pond design, as well as the integration of microalgae cultivation with other activities, such as wastewater treatment and carbon sequestration, enable multiple benefits of contaminant removal, carbon capture, and, at the same time, the production of biomass-rich biomolecules. This integrated approach is characterized by low technological requirements and minimal capital and operational costs (Tavanandi *et al.*, 2021).
III. Biorefinery approaches allow for the extraction of multiple biomolecules from microalgal biomass to produce biofuels, pharmaceuticals, foods, beverages, and other high-value by-products, making the process more economically viable.
IV. Advanced genetic and metabolic engineering research integrated with a multi-omics approach to produce and engineer microalgal cells with enhanced traits, such as stress resistance, and to boost the production of specific compounds to enhance productivity and resilience.
V. Machine learning and artificial intelligence, particularly in the optimization of microalgae cultivation, to monitor growth and predict yield.
VI. Planning and implementation of legislation, policy, and economic frameworks related to commercializing microalgae-based products and meeting SDG goals.

2.5 CONCLUSION AND FUTURE PERSPECTIVES

In this chapter, we carefully examined microalgae and their substantial impact on sustainable development and resilience to climate change. This study highlights the pivotal role of microalgae in achieving the United Nations' SDGs and enhancing the circular economy. Microalgae are instrumental in aiding all SDGs either directly or indirectly. They serve as sustainable energy sources that do not compete with food crops or necessitate freshwater, making them indispensable for reducing poverty and hunger (SDGs 1 and 2). Furthermore, their rapid growth rate and minimal cultivation costs render them a potential food source. Microalgae produce carotenoids and other natural products, thereby improving their health and well-being (SDG 3). Additionally, they are commercially utilized to generate renewable energy sources such as biodiesel, biohydrogen, bioethanol, and biogas, which are vital for clean energy (SDG 7). Companies leverage microalgae for economic growth, industry, and innovation (SDG 8, 9, and 12). Microalgae can be cultivated using atmospheric CO_2 and wastewater, which makes them effective for carbon capture and wastewater treatment. They play a crucial role in fostering green communities by reducing pollution and resource depletion (SDG 6, 11, 13, 14, and 15). Moreover, microalgae indirectly support quality education, gender equality, and other SDGs (SDG 4, 5, 10, and 16). Their uncomplicated cultivation requirements and capacity to produce bioactive compounds, biofuels, and biofertilizers underscore their importance to the circular economy.

ACKNOWLEDGMENT

The authors extend their gratitude to Khalifa University, Abu Dhabi, United Arab Emirates, for the support received through ASPIRE, the Technology Program Management pillar of Abu Dhabi's Advanced Technology Research Council (ATRC), under the ASPIRE Research Institute program (VRI-FS grant 8434000459).

DECLARATION

The authors have not used any generative AI tools during the writing of this manuscript.

REFERENCES

Ahiahonu, E. K., *et al.* (2023). Bioprospecting wild South African microalgae as a potential third-generation biofuel feedstock, biological carbon-capture agent and for nutraceutical applications. *Biomass Conversion and Biorefinery, 13*(8), 6897–6912.

Ahmad, A., Banat, F., & AlBlooshi, H. (2022). Algal Biotechnology: Integrated algal engineering for bioenergy, bioremediation, and biomedical applications. Elsevier.

Ahmad, A., Banat, F., & Taher, H. (2022). Algal engineering for bioremediation, bioenergy production, and biomedical applications. In A. Ahmad, F. Banat, & H. Taher (Eds.), *Algal Biotechnology* (pp. 3–32). Elsevier.

Ahmad, A., Banat, F., Taher, H., AlBlooshi, H. (2022). Algae biotechnology for industrial wastewater treatment, bioenergy production, and high-value bioproducts. *Science of the Total Environment, 806*, 150585.

Ahmad, A., Buang, A., & Bhat, A. H. (2016). Renewable and sustainable bioenergy production from microalgal co-cultivation with palm oil mill effluent (POME): A review. *Renewable and Sustainable Energy Reviews, 65*, 214–234.

Ahmad, A., Hassan, S. W., & Banat, F. (2022). An overview of microalgae biomass as a sustainable aquaculture feed ingredient: Food security and circular economy. *Bioengineered, 13*(4), 9521–9547.

Alami, A. H., *et al.* (2021). Investigating algae for CO2 capture and accumulation and simultaneous production of biomass for biodiesel production. *Science of the Total Environment, 759*, 143529.

Alhumairi, A., Hamouda, R., & Saddiq, A. (2022). Comparative study between immobilized and suspended Chlorella sp in treatment of pollutant sites in Dhiba port Kingdom of Saudi Arabia. *Heliyon, 8*(9), e10766.

Amorim, M. L., *et al.* (2021). Microalgae proteins: Production, separation, isolation, quantification, and application in food and feed. *Critical Reviews in Food Science and Nutrition, 61*(12), 1976–2002.

Ampofo, J., & Abbey, L. (2022). Microalgae: Bioactive composition, health benefits, safety and prospects as potential high-value ingredients for the functional food industry. *Foods, 11*(12), 1744.

Archana, T., & Thomas, K. (2018). Biofuels from microalgae. In N.-R. Madhugiri & R. S. Jaya (Eds.), *Advances in Biofuels and Bioenergy*. IntechOpen: Rijeka. 239–249. http://dx.doi.org/10.5772/intechopen.73012

Azuma, T., Usui, M., & Hayashi, T. (2024). Inactivation of antibiotic-resistant bacteria in hospital wastewater by ozone-based advanced water treatment processes. *Science of the Total Environment, 906*, 167432.

Ban, S., *et al.* (2018). Algal-bacterial cooperation improves algal photolysis-mediated hydrogen production. *Bioresource Technology, 251*, 350–357.

Banerjee, A., *et al.* (2019). Carbon bio-fixation, effect of physicochemical factors and carbon supply strategies by Nannochloropsis sp. using flue gas and fertilizer. *Biomass and Bioenergy, 125*, 95–104.

Basak, N., & Das, D. (2007). The prospect of purple non-sulfur (PNS) photosynthetic bacteria for hydrogen production: The present state of the art. *World Journal of Microbiology and Biotechnology, 23*(1), 31–42.

Basheer, S., Ali, Z., Alam, M. A., Xu, J.-L., & Wang, Z. (2020). Microalgae in human health and medicine. In M. A. Alam, J.-L. Xu, & Z. Wang (Eds.), *Microalgae Biotechnology for Food, Health and High Value Products* (pp. 149–174). Springer Singapore.

Beltran, A. B., *et al.* (2018). Evaluation of Ankistrodesmus falcatus for Bicarbonate-Based Integrated Carbon Capture System (BICCAPS). *MATEC Web of Conferences, 156,* 1–5. www.matec-conferences.org/articles/matecconf/pdf/2018/15/matecconf_rsce2018_03016.pdf.

Benemann, J. R. (2000). Hydrogen production by microalgae. *Journal of Applied Phycology, 12*(3), 291–300.

Benemann, J. R., *et al.* (1973). Hydrogen evolution by a chloroplast-ferredoxin-hydrogenase system. *Proceedings of the National Academy of Sciences, 70*(8), 2317–2320.

Bhatia, S. K., *et al.* (2021). Wastewater based microalgal biorefinery for bioenergy production: Progress and challenges. *Science of the Total Environment, 751*, 141599.

Bhola, V., *et al.* (2014). Overview of the potential of microalgae for CO2 sequestration. *International Journal of Environmental Science and Technology, 11*(7), 2103–2118.

Caporgno, M. P., & Mathys, A. (2018). Trends in microalgae incorporation into innovative food products with potential health benefits. *Frontiers in Nutrition, 5*, Article 58.

Caroppo, C., & Pagliara, P. (2022). Microalgae: A promising future. *Microorganisms, 10*, Article 1488. https://doi.org/10.3390/microorganisms10081488

Casanova, L. M., *et al.* (2023). Development of microalgae biodiesel: Current status and perspectives. *Microorganisms, 11*(1), 34.

Cheah, W. Y., *et al.* (2015). Biosequestration of atmospheric CO2 and flue gas-containing CO2 by microalgae. *Bioresource Technology, 184*, 190–201.

Chen, W.-H., *et al.* (2021). Current status of biohydrogen production from lignocellulosic biomass, technical challenges and commercial potential through pyrolysis process. *Energy, 226*, 120433.

Cheng, J., *et al.* (2020). Improving hydrogen and methane co-generation in cascading dark fermentation and anaerobic digestion: The effect of magnetite nanoparticles on microbial electron transfer and syntrophism. *Chemical Engineering Journal, 397*, 125394.

Cheng, P., *et al.* (2022). Heterotrophic and mixotrophic cultivation of microalgae to simultaneously achieve furfural wastewater treatment and lipid production. *Bioresource Technology, 349*, 126888.

Chislock, M. F., Doster, E., Zitomer, R. A., & Wilson, A. E. (2013). Eutrophication: Causes, consequences, and controls in aquatic ecosystems. *Nature Education Knowledge, 4*(4), 10.

Comninellis, C., *et al.* (2008). Advanced oxidation processes for water treatment: Advances and trends for R&D. *Journal of Chemical Technology & Biotechnology, 83*(6), 769–776.

Cui, H., *et al.* (2021). Advanced treatment of chicken farm flushing wastewater by integrating Fenton oxidation and algal cultivation process for algal growth and nutrients removal. *Journal of Environmental Management, 298*, 113543.

de Jesus, G. C., Gaspar Bastos, R., & Altenhofen da Silva, M. (2019). Production and characterization of alginate beads for growth of immobilized Desmodesmus subspicatus and its potential to remove potassium, carbon and nitrogen from sugarcane vinasse. *Biocatalysis and Agricultural Biotechnology, 22*, 101438.

de-Bashan, L. E., & Bashan, Y. (2010). Immobilized microalgae for removing pollutants: Review of practical aspects. *Bioresource Technology, 101*(6), 1611–1627.

Elangovan B, Detchanamurthy S, Senthil Kumar P, Rajarathinam R, Deepa VS. (2023). Biotreatment of Industrial Wastewater using Microalgae: A Tool for a Sustainable Bioeconomy. *Molecular Biotechnology*. doi: 10.1007/s12033-023-00971-0. Epub ahead of print. PMID: 37999921.

El-Moustaqim, K., Sbai, S. E., Mabrouki, J., & Hmouni, D. (2025). Critical review of green microalgal biomass development through CO_2 sequestration. *Euro-Mediterranean Journal for Environmental Integration, 10*(1), 1–13. https://doi.org/10.1007/s41207-024-00548-6

Fabris M, Abbriano RM, Pernice M, Sutherland DL, Commault AS, Hall CC, Labeeuw L, McCauley JI, Kuzhiuparambil U, Ray P, Kahlke T, Ralph PJ. (2020). Emerging Technologies in Algal Biotechnology: Toward the Establishment of a Sustainable, Algae-Based Bioeconomy. *Frontiers in Plant Science;11*:279. doi: 10.3389/fpls.2020.00279. PMID: 32256509; PMCID: PMC7090149

Fatta-Kassinos, D., Vasquez, M. I., & Kümmerer, K. (2011). Transformation products of pharmaceuticals in surface waters and wastewater formed during photolysis and advanced oxidation processes – Degradation, elucidation of byproducts and assessment of their biological potency. *Chemosphere, 85*(5), 693–709.

Gao, K., & Beardall, J. (2022). Using macroalgae to address UN Sustainable Development Goals through CO2 remediation and improvement of the aquaculture environment. *Applied Phycology, 3*(1), 360–367.

García-Cubero, R., Moreno-Fernández, J., & García-González, M. (2017). Modelling growth and CO2 fixation by Scenedesmus vacuolatus in continuous culture. *Algal Research, 24*, 333–339.

Gaurav, K., Neeti, K., & Singh, R. (2024). Microalgae-based biodiesel production and its challenges and future opportunities: A review. *Green Technologies and Sustainability, 2*(1), 100060.

Gerrity, D., & Snyder, S. (2011). Review of ozone for water reuse applications: Toxicity, regulations, and trace organic contaminant oxidation. *Ozone: Science & Engineering, 33*(4), 253–266.

Hallenbeck, P. C., & Benemann, J. R. (2002). Biological hydrogen production; fundamentals and limiting processes. *International Journal of Hydrogen Energy, 27*(11), 1185–1193.

Halmemies-Beauchet-Filleau, A., *et al.* (2018). Review: Alternative and novel feeds for ruminants: Nutritive value, product quality and environmental aspects. *Animal, 12*(s2), s295–s309.

Hanssen, S. V., *et al.* (2020). The climate change mitigation potential of bioenergy with carbon capture and storage. *Nature Climate Change, 10*(11), 1023–1029.

Harte, C. R., Jr., Baker, E. M., & Purcell, H. H. (1933). Absorption of carbon dioxide in sodium carbonate-bicarbonate solutions. *Industrial & Engineering Chemistry, 25*(5), 528–531.

Hasanath Md, A., & Chatterjee, P. (2025). Biochemical methane potential of microalgae cultivated in high-strength beverage wastewater: Implications for energy enhancement. *Journal of Hazardous, Toxic, and Radioactive Waste, 29*(1), 04024026.

Ike, I. A., *et al.* (2019). Oxidation byproducts from the degradation of dissolved organic matter by advanced oxidation processes – A critical review. *Water Research, 164*, 114929.

IPCC. (2022). Climate change: Impacts, adaptation, and vulnerability. Contribution of Working Group II to the Sixth Assessment Report of the Intergovernmental Panel on Climate Change (D. C. R. H.-O. Pörtner, M. Tignor, E. S. Poloczanska, K. Mintenbeck, A. Alegría, M. Craig, S. Langsdorf, S. Löschke, V. Möller, A. Okem, & B. Rama, Eds.). Cambridge University Press: Cambridge, UK and New York. p. 3056.

Jiang, L., Wang, Y., & Feng, C. (2012). Application of photocatalytic technology in environmental safety. *Procedia Engineering, 45*, 993–997.

Kahoush, M., *et al.* (2018). Bio-Fenton and bio-electro-Fenton as sustainable methods for degrading organic pollutants in wastewater. *Process Biochemistry, 64*, 237–247.

Kannah Ravi, Y., *et al.* (2024). Dual disintegration of microalgae biomass for cost-effective biomethane production: Energy and cost assessment. *Bioresource Technology, 399*, 130630.

Kholif, A. E., *et al.* (2017). Dietary Chlorella vulgaris microalgae improves feed utilization, milk production and concentrations of conjugated linoleic acids in the milk of Damascus goats. *The Journal of Agricultural Science, 155*(3), 508–518.

Kholssi, R., *et al.* (2022). Green agriculture: A review of the application of micro- and macroalgae and their impact on crop production on soil quality. *Journal of Soil Science and Plant Nutrition, 22*(4), 4627–4641.

Kholssi, R., Ramos, P. V., Marks, E. A. N., Montero, O., & Rad, C. (2021). Biotechnological uses of microalgae: A review on the state of the art and challenges for the circular economy. *Biocatalysis and Agricultural Biotechnology, 36*, 102114.

Kim, E. J., *et al.* (2020). Co-production of biodiesel and bioethanol using psychrophilic microalga Chlamydomonas sp. KNM0029C isolated from Arctic sea ice. *Biotechnology for Biofuels, 13*(1), 20.

Knapp, J. R., *et al.* (2014). Invited review: Enteric methane in dairy cattle production: Quantifying the opportunities and impact of reducing emissions. *Journal of Dairy Science, 97*(6), 3231–3261.

Kongkeitkajorn, M. B. (2023). Bioethanol production from microalgae: Potentials and challenges. In E. Betiku & M. M. Ishola (Eds.), *Bioethanol: A Green Energy Substitute for Fossil Fuels* (pp. 161–192). Springer International Publishing: Cham.

Kumar, N., *et al.* (2022). Microalgae in wastewater treatment and biofuel production: Recent advances, challenges, and future prospects. In V. Kumar & I. S. Thakur (Eds.), *Omics Insights in Environmental Bioremediation* (pp. 237–271). Springer Nature Singapore.

Kusmayadi, A., *et al.* (2022). Integrating anaerobic digestion and microalgae cultivation for dairy wastewater treatment and potential biochemicals production from the harvested microalgal biomass. *Chemosphere, 291*, 133057.

Kusmayadi, A., *et al.* (2023). Integration of microalgae cultivation and anaerobic co-digestion with dairy wastewater to enhance bioenergy and biochemicals production. *Bioresource Technology, 376*, 128858.

Lakatos, G. E., *et al.* (2019). Bioethanol production from microalgae polysaccharides. *Folia Microbiologica, 64*(5), 627–644.

Lamminen, M., *et al.* (2019). Different microalgae species as a substitutive protein feed for soya bean meal in grass silage based dairy cow diets. *Animal Feed Science and Technology, 247*, 112–126.

Lau, Z. L., Abdullah, I. A. P., Tan, X. Y., Lim, Y. L., Ong, H. C., & Show, P. L. (2022). A review on the diverse interactions between microalgae and nanomaterials: Growth variation, photosynthetic performance and toxicity. *Bioresource Technology, 351*, 127048.

Leong, Y. K., *et al.* (2024). Integrated biohydrogen production and dairy manure wastewater treatment via a microalgae platform. *International Journal of Hydrogen Energy, 52*, 404–417.

Li, K., *et al.* (2019). Microalgae-based wastewater treatment for nutrients recovery: A review. *Bioresource Technology, 291*, 121934.

Li, S., *et al.* (2022). Biohydrogen production from microalgae for environmental sustainability. *Chemosphere, 291*, 132717.

Lim, Y. A., *et al.* (2023). Improving microalgae growth and carbon capture through micro-size bubbles generation in flat-panel photobioreactors: Impacts of different gas sparger designs on mixing performance. *Renewable and Sustainable Energy Reviews, 171*, 113001.

Limongi, A. R., *et al.* (2021). Biohydrogen from microalgae: Production and applications. *Applied Sciences, 11*(4), 1616. https://doi.org/10.3390/app11041616

Liu, X.-Y., & Hong, Y. (2021). Microalgae-based wastewater treatment and recovery with biomass and value-added products: A brief review. *Current Pollution Reports, 7*(2), 227–245.

Lorentz, J. F., *et al.* (2024). Microalgae biomass as a conditioner and regulator of soil quality and fertility. *Environmental Monitoring and Assessment, 196*(2), 198.

Luo, Z., *et al.* (2024). Editorial: The response of microalgae and plankton to climate change and human activities. *Frontiers in Marine Science, 11*, Article 1366749. https://doi.org/10.3389/fmars.2024.1366749

Mallick, N. (2002). Biotechnological potential of immobilized algae for wastewater N, P and metal removal: A review. *Biometals, 15*(4), 377–390.

Mandotra, S. K., *et al.* (2023). Current prospects and future developments in algal bio-hydrogen production: A review. *Biomass Conversion and Biorefinery, 13*(10), 8575–8592.

Mensah, J. (2019). Sustainable development: Meaning, history, principles, pillars, and implications for human action: Literature review. *Cogent Social Sciences, 5*(1), 1653531.

Mishra, M., Tiwari, R., Sharma, S., & Pandey, G. (2024). A bibliometric analysis of Sustainable Development Goals (SDGs): A review of progress, challenges, and opportunities. *Environment, Development and Sustainability, 26*(5), 11101–11143.

Mishra, S., & Sundaram, B. (2023). A review of the photocatalysis process used for wastewater treatment. *Materials Today: Proceedings, [volume needed]*, [page range needed].

Mona, S., *et al.* (2020). Green technology for sustainable biohydrogen production (waste to energy): A review. *Science of the Total Environment, 728*, 138481.

Mondal, K., & Sharma, A. (2014). Photocatalytic oxidation of pollutant dyes in wastewater by TiO2 and ZnO nano-materials – a mini-review. *Nanoscience & Technology for Mankind*, 36–72.

Mondal, M., *et al.* (2017). Production of biodiesel from microalgae through biological carbon capture: A review. *3 Biotech, 7*(2), 99.

Moreno-Garrido, I. (2008). Microalgae immobilization: Current techniques and uses. *Bioresource Technology, 99*(10), 3949–3965.

Mujtaba, G., *et al.* (2018). Removal of nutrients and COD through co-culturing activated sludge and immobilized Chlorella vulgaris. *Chemical Engineering Journal, 343*, 155–162.

Nagarajan, D., *et al.* (2021). Biohydrogen production from microalgae—Major bottlenecks and future research perspectives. *Biotechnology Journal, 16*(5), 2000124.

Nascimento, I. A., *et al.* (2013). Screening microalgae strains for biodiesel production: Lipid productivity and estimation of fuel quality based on fatty acids profiles as selective criteria. *BioEnergy Research, 6*(1), 1–13.

Nayak, M., *et al.* (2018). Artificial intelligence driven process optimization for cleaner production of biomass with co-valorization of wastewater and flue gas in an algal biorefinery. *Journal of Cleaner Production, 201*, 1092–1100.

Neves, R. A. F., & Rodrigues, E. T. (2020). Harmful algal blooms: Effect on coastal marine ecosystems. In W. Leal Filho *et al.* (Eds.), *Life Below Water* (pp. 1–31). Springer International Publishing: Cham.

O'Neill, B. C., *et al.* (2022). Key risks across sectors and regions. (Chapter 16). In H.-O. Pörtner, D. C. Roberts, M. Tignor, E. S. Poloczanska, K. Mintenbeck, A. Alegría, M. Craig, S. Langsdorf, S. Löschke, V. Möller, A. Okem, & B. Rama (Eds.), Climate Change 2022: Impacts, Adaptation and Vulnerability. Contribution of Working Group II to the Sixth Assessment *Report of the Intergovernmental Panel on Climate Change* (pp. 2411–2538). Cambridge University Press. https://doi.org/10.1017/9781009325844.025.16

Olabi, A. G., Shehata, N., Sayed, E. T., Rodriguez, C., Anyanwu, R. C., Russell, C., & Abdelkareem, M. A. (2023). Role of microalgae in achieving sustainable development goals and circular economy. *Science of the Total Environment, 854*, 158689.

Olabi, A. G., Shehata, N., Sayed, E. T., Rodriguez, C., Anyanwu, R. C., Russell, C., & Abdelkareem, M. A. (2023). Role of microalgae in achieving sustainable development goals and circular economy. *Science of the Total Environment, 854*, 158689.

Oliveira, C. Y. B., *et al.* (2022). An overview on microalgae as renewable resources for meeting Sustainable Development Goals. *Journal of Environmental Management, 320*, 115897.

Pal, D., *et al.* (2024). The generation of biohydrogen from pretreated algal biomass in batch fermentation mode. *International Journal of Hydrogen Energy, 107*, 321–331. https://doi.org/10.1016/j.ijhydene.2024.01.078

Pathak, M., Patel, S., & Some, S. (2024). Climate change mitigation and Sustainable Development Goals: Evidence and research gaps. *PLOS Climate, 3*(3), e0000366.

Peter, A. P., *et al.* (2022). Microalgae's prospects in attaining sustainable economic and environmental development. *Journal of Biotechnology, 357*, 18–27.

Petersen, J., Rredhi, A., Szyttenholm, J., Oldemeyer, S., Kottke, T., & Mittag, M. (2021). The World of Algae Reveals a Broad Variety of Cryptochrome Properties and Functions. *Frontiers in Plant Science, 12*, 766509. https://doi.org/10.3389/fpls.2021.766509

Qu, J., *et al.* (2022). Emerging trends and prospects for municipal wastewater management in China. *ACS ES&T Engineering, 2*(3), 323–336.

Qu, J., Fan, M., Zhang, X., & Zhang, Y. (2019). Municipal wastewater treatment in China: Development history and future perspectives. *Frontiers of Environmental Science & Engineering, 13*(6), 88.

Ramakrishnan, B., *et al.* (2023). Potential of microalgae and cyanobacteria to improve soil health and agricultural productivity: A critical view. *Environmental Science: Advances, 2*(4), 586–611.

Rambhiya, S. J., Magar, C. S., & Deodhar, M. A. (2021). Using seawater-based Na2CO3 medium for scrubbing the CO2 released from Bio-CNG plant for enhanced biomass production of Pseudanabaena limnetica. *SN Applied Sciences, 3*(2), 276.

Rame, R., Purwanto, P., & Sudarno, S. (2023). Sustainable energy harnessing: Microalgae as a potential biofuel source and carbon sequestration solution. *Renewable Energy Focus, 47*, 100498.

Ratomski, P., Hawrot-Paw, M., & Koniuszy, A. (2021). Utilisation of CO2 from sodium bicarbonate to produce Chlorella vulgaris biomass in tubular photobioreactors for biofuel purposes. *Sustainability, 13*, Article 6918.

Rayaroth, M. P., *et al.* (2022). Advanced oxidation processes (AOPs) based wastewater treatment – Unexpected nitration side reactions – A serious environmental issue: A review. *Chemical Engineering Journal, 430*, 133002.

Rekhate, C. V., & Srivastava, J. K. (2020). Recent advances in ozone-based advanced oxidation processes for treatment of wastewater — A review. *Chemical Engineering Journal Advances, 3*, 100031.

Safari, S., Seyed Ahmadian, S. M., & Amani-Ghadim, A. R. (2020). Visible light photocatalytic activity enhancing of MTiO3 perovskites by M cation (M = Co, Cu, and Ni) substitution and gadolinium doping. *Journal of Photochemistry and Photobiology A: Chemistry, 394*, 112461.

Saravanakumar, K., *et al.* (2024). Recent progress in biotechnological approaches for diverse applications of algae: An overview. *International Journal of Environmental Science and Technology, 21*(3), 3453–3474.

Sarkar, O., Katakojwala, R., & Venkata Mohan, S. (2021). Low carbon hydrogen production from a waste-based biorefinery system and environmental sustainability assessment. *Green Chemistry, 23*(1), 561–574.

Sharma, A., Ahmad, J., & Flora, S. J. S. (2018). Application of advanced oxidation processes and toxicity assessment of transformation products. *Environmental Research, 167*, 223–233.

Sharma, A., *et al.* (2023). Carbon capture from petrol-engine flue gas: Reviving algae-based sequestration with integrated microbial fuel cells. *Chemical Engineering Journal, 476*, 146578.

Shi, J., *et al.* (2015). Enzymatic conversion of carbon dioxide. *Chemical Society Reviews, 44*(17), 5981–6000.

Show, K.-Y., *et al.* (2019). State of the art and challenges of biohydrogen from microalgae. *Bioresource Technology, 289*, 121747.

Siedenburg, J. (2022). Could microalgae offer promising options for climate action via their agri-food applications? *Frontiers in Sustainable Food Systems, 6,* Article 976946. https://doi.org/10.3389/fsufs.2022.976946

Song, X., Bo, Y., Feng, Y., Tan, Y., Zhou, C., Yan, X., Ruan, R., Xu, Q., & Cheng, P. (2022). Potential applications for multifunctional microalgae in soil improvement. *Frontiers in Environmental Science, 10*, Article 1035332. https://doi.org/10.3389/fenvs.2022.1035332

Song, Y., *et al.* (2024). Enhancement of biomass, lipid accumulation, and carbon sequestration potential in microalgae via cultivation with aggregation-induced emission light-conversion films. *Chemical Engineering Journal, 483*, 149148.

Sporchia, F., *et al.* (2024). Zero hunger: Future challenges and the way forward towards the achievement of Sustainable Development Goal 2. *Sustainable Earth Reviews, 7*(1), 10.

Srimongkol, P., Sangtanoo, P., Songserm, P., Watsuntorn, W., & Karnchanatat, A. (2022). Microalgae-based wastewater treatment for developing economic and environmental sustainability: Current status and future prospects. *Frontiers in Bioengineering and Biotechnology, 10*, Article 904046. https://doi.org/10.3389/fbioe.2022.904046

Su, M., *et al.* (2023). Applications of microalgae in foods, pharma and feeds and their use as fertilizers and biostimulants: Legislation and regulatory aspects for consideration. *Foods, 12*(20), 3878.

Sutherland, D. L., *et al.* (2021). How microalgal biotechnology can assist with the UN Sustainable Development Goals for natural resource management. *Current Research in Environmental Sustainability, 3*, 100050.

Tahir, F., *et al.* (2024). Emerging trends in algae farming on non-arable lands for resource reclamation, recycling, and mitigation of climate change-driven food security challenges. *Reviews in Environmental Science and Bio/Technology, 23*(3), 869–896.

Thakur, P., Majumder, S., & Thakur, A. (2024). Microalgae synthesis for biodiesel production. In D. B. Pal, A. K. Rai, & S. Siddiqui (Eds.), *Sustainable Clean Energy Production Using Waste Biomass: Sustainable Energy Production and Utilization* (pp. 273–298). Springer Nature Singapore.

Tyoker Kukwa, D., & Chetty, M. (2021). Microalgae: The Multifaceted Biomass of the 21st Century. IntechOpen. doi: 10.5772/intechopen.94090

Udayan, A., *et al.* (2021). Emerging industrial applications of microalgae: Challenges and future perspectives. *Systems Microbiology and Biomanufacturing, 1*(4), 411–431.

UN-Water. (2021). *Summary progress update 2021: SDG 6 – water and sanitation for all* (54 pp.). Geneva, Switzerland: UN-Water. www.unwater.org/app/uploads/2021/02/SDG-6-Summary-Progress-Update-2021_Version-2021-03-03.pdf

Vargas, S. R., Zaiat, M., & Calijuri, M. D. C. (2020). Influence of culture age, ammonium and organic carbon in hydrogen production and nutrient removal by Anabaena sp. in nitrogen-limited cultures. *International Journal of Hydrogen Energy, 45*(55), 30222–30231.

Vasistha, S., *et al.* (2023). Exploring the pivotal significance of microalgae-derived sustainable lipid production: A critical review of green bioenergy development. *Energies, 16*, Article 10531. https://doi.org/10.3390/en16010531

Velmozhina, K., *et al.* (2023). Production of biohydrogen from microalgae biomass after wastewater treatment and air purification from CO2. *Processes, 11*, Article 2978. https://doi.org/10.3390/pr11102978

Veza, I., *et al.* (2022). Microalgae and macroalgae for third-generation bioethanol production. In C. R. Soccol *et al.* (Eds.), *Liquid Biofuels: Bioethanol* (pp. 301–331). Springer International Publishing: Cham.

Vu, H. P., *et al.* (2024). Enhanced biomethane production from Scenedesmus sp. using polymer harvesting and expired COVID-19 disinfectant for pretreatment. *Chemosphere, 356*, 141869.

Walsh, B. J., *et al.* (2015). New feed sources key to ambitious climate targets. *Carbon Balance and Management, 10*(1), 26.

Wang, L., *et al.* (2023). Dissimilatory manganese reduction facilitates synergistic cooperation of hydrolysis, acidogenesis, acetogenesis and methanogenesis via promoting microbial interaction during anaerobic digestion of waste activated sludge. *Environmental Research, 218*, 114992.

Wang, M., *et al.* (2024). Microalgae biofuels: Illuminating the path to a sustainable future amidst challenges and opportunities. *Biotechnology for Biofuels and Bioproducts, 17*(1), 10.

Wang, R.-L., & Li, M.-J. (2024). The study of novelty nutrients supplementary scheme and methodology on advanced carbon sequestration of microalgae. *Results in Engineering, 22*, 102178.

Wang, S.-K., *et al.* (2022). One-step co-cultivation and flocculation of microalgae with filamentous fungi to valorize starch wastewater into high-value biomass. *Bioresource Technology, 361*, 127625.

Wang, Y., *et al.* (2024). Advances in microalgae-based livestock wastewater treatment: Mechanisms of pollutants removal, effects of inhibitory components and enhancement strategies. *Chemical Engineering Journal, 483*, 149222.

Wang, Z., *et al.* (2024). Novel combination of iron-carbon composite and Fenton oxidation processes for high-concentration antibiotic wastewater treatment. *Journal of Environmental Management, 354*, 120383.

Woon, J. M., *et al.* (2024). Rederiving kinetics to model biohydrogen production from immobilized microalgae alginate beads at various polymerization degrees of alginate under dark fermentative environment. *Algal Research, 82*, 103684.

Wu, X., *et al.* (2012). Current status and prospects of biodiesel production from microalgae. *Energies, 5*(8), 2667–2682.

Xu, X., *et al.* (2019). Progress, challenges and solutions of research on photosynthetic carbon sequestration efficiency of microalgae. *Renewable and Sustainable Energy Reviews, 110*, 65–82.

Xu, X., Martin, G. J. O., & Kentish, S. E. (2019). Enhanced CO2 bio-utilization with a liquid–liquid membrane contactor in a bench-scale microalgae raceway pond. *Journal of CO2 Utilization, 34*, 207–214.

Xu, X.-J., *et al.* (2022). Enhanced methane production in anaerobic digestion: A critical review on regulation based on electron transfer. *Bioresource Technology, 364*, 128003.

Yadav, A., *et al.* (2016). Pilot project at Hazira, India, for capture of carbon dioxide and its biofixation using microalgae. *Environmental Science and Pollution Research, 23*(22), 22284–22291.

Yang, Y., Tang, S., & Chen, J. P. (2024). Carbon capture and utilization by algae with high concentration CO2 or bicarbonate as carbon source. *Science of the Total Environment, 918*, 170325.

You, X., *et al.* (2021). Integrating acidogenic fermentation and microalgae cultivation of bacterial-algal coupling system for mariculture wastewater treatment. *Bioresource Technology, 320*, 124335.

Zhang, S., Zhang, L., Xu, G., Li, F., & Li, X. (2022). A review on biodiesel production from microalgae: Influencing parameters and recent advanced technologies. *Frontiers in Microbiology, 13*, Article 970028. https://doi.org/10.3389/fmicb.2022.970028

Zhang, Z., *et al.* (2021). Bacterial-algal coupling system for high strength mariculture wastewater treatment: Effect of temperature on nutrient recovery and microalgae cultivation. *Bioresource Technology, 338*, 125574.

3 Impact of Microalgal Biorefineries on Circular Economy and Climate Change Mitigation

Darissa Alves Dutra, Richard Luan Silva Machado, Adriane Terezinha Schneider, Mariany Costa Deprá, Rosangela Rodrigues Dias, Leila Queiroz Zepka, and Eduardo Jacob-Lopes

3.1 INTRODUCTION

The effect of anthropogenic actions in recent years has served as fuel for the greenhouse effect, resulting in persistent changes in Earth's climate patterns. Consequently, climate change now requires contingency plans, and measures necessary to build a future that is less inhospitable to human beings. In this sense, to reduce advances in environmental degradation, a biorefinery emerges as a crucial link in the search for a solid future, boosting the circular economy, to replace the predominant economic model of "extract-produce-use-discard" (Kumar and Verma 2021; Zomer et al., 2024).

The emergence of microalgae biorefinery represents a revolution in the contemporary industrial paradigm, leveraging the potential of biological resources to promote sustainable development. Due to the current scenario, where pressure on fossil resources and environmental demands are intensified due to population expansion in the world, the microalgae biorefinery emerges as an innovative solution (Goswami et al., 2022a; Talebi et al., 2022; Hoang et al., 2023). The biorefinery uses processes and technologies to convert microalgae biomass into a wide variety of products. These microorganisms are ahead of conventional raw materials, helping to mitigate greenhouse gases (GHGs) (Javed et al., 2019; Calijuri et al., 2022).

Combining the potential of a biorefinery and the great resourcefulness of microalgae, the result is a circular economy. This method aims at the continuous reuse of resources, thus finding a powerful ally in the microalgae biorefinery. One of the highlights of the microalgae biorefinery is its power to value industrial waste, using it as a source of nutrients for the cultivation of microalgae, promoting the recovery and reuse of materials, thus reducing environmental impacts (Yadav et al., 2019; Ummalyma et al., 2023). Furthermore, the intrinsic characteristics of microalgae allow the capture of carbon dioxide (CO_2), and the production of multiple products, such as food, biofuels, cosmetics, animal feed, and fertilizers, contributing to innovation in existing market sectors and opportunity for economic investment (Jacob-Lopes et al., 2020a; López-Sánchez et al., 2022; Samoraj et al., 2024).

Realizing the finite nature of conventional resources and their major influence on the progression of climate change, the emphasis is shifting to alternative raw materials with the potential to mitigate environmental stresses and secure renewable natural capital (Jacob-Lopes et al., 2024; Machado et al., 2024). In this vein, this chapter explores how microalgae biorefineries can provide a circular

DOI: 10.1201/9781003685814-3

economy, significantly contribute to the mitigation of climate change, and measure to overcome the obstacles to its implementation.

3.2 MICROALGAE-BASED PRODUCTS PROMOTING A CIRCULAR ECONOMY

The growing energy demand, the intensification of GHG emissions levels, and the grave for fossil resources under construction mean that products from microalgae biomass present themselves as an alternative to the demands of humanity. Biomass is considered a biological resource, which has unquestionably contributed to the development of the bioeconomy and can be applied in the pharmaceutical and food industries (Sarwer et al., 2022). The circular economy is mainly based on the "closed loop" method linked to the microalgae biorefinery (Cheirsilp et al., 2023). Integrating multiple unit operations into the microalgae biorefinery structure contributes to increasing biomass recovery, process efficiency, and profitability, resulting in increased biomass recovery. Therefore, the adoption of bioeconomy and microalgae biomass can minimize waste by treating waste as raw material in closed process cycles, developing bioproducts, and green energy, thus helping to mitigate climate change.

The circular economy approach with a cascade effect encompasses products from biomass, such as polyunsaturated fatty acids (PUFAs), used in the food industry and the production of biofuels, characterizing them for the production of bioproducts with high and low added value, respectively, as demonstrated in Figure 3.1.

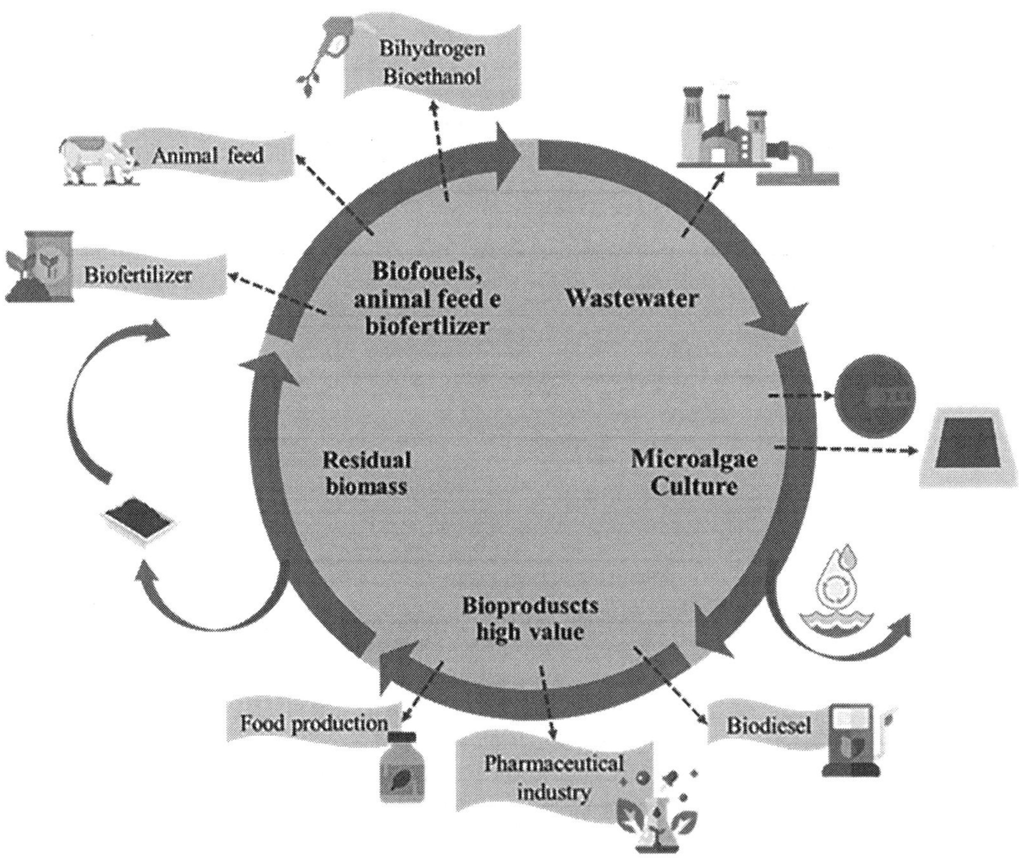

FIGURE 3.1 Holistic circular bioeconomy for closed-loop microalgae biomass conversion.

Several recent studies have explored these cascade processes. For example, Imbimbo et al. (2020) used the French press in a cascade extraction process with *Galdieria phlegrea*. In this study, proteins and phycocyanin were recovered from the supernatant, while carotenoids and lipids were extracted using pressurized liquid extraction and supercritical fluid extraction methods. In another example, Francavilla et al. (2015) extracted lipids from freeze-dried *Dunaliella tertiolecta* using chloroform:methanol and subsequently performed rapid pyrolysis of the residue to produce bio-oil, which was refined for use as biofuel. Lupatini et al. (2017) investigated the extraction of proteins and carbohydrates from defatted *Arthrospira platensis* using Soxhlet extraction, while Ansari et al. (2017) explored several cascade extraction routes with dried *Scenedesmus obliquus*, finding that the protein–lipid–carbohydrate route presented the best recovery of each fraction. In line with the holistic vision of this reuse of waste biomass, Slegers et al. (2020) demonstrated that cascading biomass exploitation improves its utilization by 25%–95%, demonstrating that multi-product biorefinery is more profitable in value chains compared to the production of single products.

Of note, biorefineries play a prominent role in the state of the art, with one of their main challenges being their dependence on fossil energy sources (Khoo et al., 2023; Kumaran et al., 2023). In this sense, biofuels emerge as a beneficial solution for mitigating climate change, which is a victim of the burning of fossil fuels. Among some of the renewable fuels obtained from residual microalgae biomass oil are biohydrogen and bioethanol, considered an alternative for the success of green energy. However, the generation of biofuels from microalgae does not yet offer resilience in the market, due to its high production costs, thus avoiding commercial competition (Jacob-Lopes et al., 2020b). An example of this is the production of biodiesel through microalgae biomass. Some estimates show an average of 2–2.59 USD/L for its production, considering the stages of the process (cultivation, harvesting, biomass treatment, extraction, and transesterification) (Nazloo et al., 2022). An alternative to reduce production costs is the use of wastewater, through bioremediation. These microorganisms are capable of recovering and reusing pollutants present in wastewater, converting them into nutrients. This process not only benefits the treatment of these waters but also generates biomass.

Furthermore, microalgae can be used to produce a variety of bioproducts, while contributing to oxygen production and promoting the circular economy. In this way, not only companies but also educational and research institutions have sought to implement the circular economy. This growing concern offers a promising opportunity to advance microalgae technology in the circular bioeconomy, encompassing not only wastewater treatment but also CO_2 capture and bioproduct production, aiming for sustainable and innovative scalability for the future. In general, the implementation of the circular bioeconomy through microalgae biorefinery plays an important role in reducing climate change, by transforming residual biomass into bioproducts and energy, reducing dependence on fossil resources, which are a significant source of GHGs.

3.3 INTEGRATING MICROALGAE BIOREFINERIES INTO THE CIRCULAR BIOECONOMY CONCEPT

The abrupt increase in CO_2 levels in the atmosphere is undoubtedly one of the main causes of global warming. These high levels result in irreversible long-term changes, such as extreme climate impacts, ocean acidification, and sea rises, and threaten global biodiversity (Onyeaka et al., 2021). It is imperative to take urgent action to mitigate these devastating effects and ensure a sustainable future for future generations.

In this sense, net-zero emissions were presented to balance these anthropogenic emissions. Agreements for the circular bioeconomy to serve as a strategy to meet goals aimed at the United Nations Sustainable Development Goals (SDGs) were drawn up, aiming to optimize resources, minimize waste generation, and create a renewable system, through the cascade circuit (Machado et al., 2024). In this regard, microalgae appear as modern green gold, one of their main characteristics being the efficiency of carbon sequestration and biomass accumulation (Deprá et al., 2024). Unlike terrestrial plants, these photosynthetic microorganisms do not require arable land, presenting superior environmental compatibility. Although microalgae contribute to mitigating climate change and promoting the bioeconomy, it is essential to consider additional variables, such as their upstream and downstream processing, to ensure a holistic approach sustainably (Goswami et al., 2022b).

The upstream processing, cultivation, and growing conditions such as pH, light, temperature, and nutrients are essential. Downstream is the most expensive stage in financial terms and viability, encompassing the harvesting, drying, extraction, and purification of this biomass (Ahmad et al., 2022). In this context, the microalgae biorefinery emerges as a promising and integrated approach to maximize the use of these components present in microalgae and mitigate climate change. Through biotechnological processes, it converts microalgae biomass into a variety of value-added bioproducts such as biofuels, foods, nutritional supplements, cosmetics, and pharmaceuticals. In addition to optimizing production efficiency and maximizing the use of natural resources, the biorefinery contributes significantly to environmental sustainability by reducing CO_2 emissions and waste, promoting a more sustainable production cycle (Okeke et al., 2022).

Predictably, the biorefinery was encouraged by the combination of microalgae and its efficiency in sequestering and capturing CO_2 and its maximum production of microalgal biomass (Wu et al., 2023). Microalgae biorefineries focus on optimizing processes by ensuring optimal conditions such as adequate light sources and sufficient levels of dissolved CO_2 to maximize efficiency. Specifically, in downstream processing, there is the potential to increase profits by avoiding the generation of waste byproducts. This is aligned with the principles of the circular bioeconomy, fully utilizing microalgae biomass, reducing waste, and taking advantage of microalgae to significantly reduce CO_2 emissions into the atmosphere, mitigating climate change (Goswami et al., 2022b).

However, potential bottlenecks in microalgae-based production still lie in their costly processing and the need to improve understanding of integrated microalgae biorefinery (Kumar & Singh, 2019). For the first case, the economic and energetic demerits associated with each step such as cultivation, optimization for target product, harvesting, product extraction, and fuel conversion have been studied separately and potential optimization techniques for universal applications are widespread (Arcigni et al., 2019). Therefore, a consolidated approach involving sequential process optimization is necessary to achieve finite improvements in terms of economics and energy perspectives. El-Dalatony et al. (2019) presented an integrated method to use microalgae biomass efficiently, in which the sequential fermentation of microalgae biomass was carried out in the order of carbohydrates, proteins followed by the transesterification of lipids in *Chlamydomonas mexicana*. After this process, the lipids and glycerol remaining from transesterification can be subjected to additional fermentation. This fermentation resulted in the segmented production of bioethanol (from carbohydrates), higher alcohols (from proteins), biobutanol (from remaining lipids and glycerol), and biodiesel resulting from the transesterification of lipids.

The study by El-Dalatony et al. (2019) demonstrated a conversion efficiency of approximately 89% and an energy recovery of approximately 46%. This indicates that the method developed is promising for the integrated production of various biofuels from microalgae, aiming to use available biomass more efficiently. Other authors have explored different strategies to increase efficiency and sustainability in the production of biofuels from microalgae. González-González et al. (2018) suggested that integrating biodiesel production with biogas in a closed circuit can significantly improve energy, economic efficiency, and environmental safety, especially in relation to the carbon balance. In another study, Carbajal et al. (2020) used *Gluconobacter oxydans* to produce dihydroxyacetone from glycerol byproducts obtained from biodiesel produced with *Scenedesmus dimorphus*.

Based on this, the focus on microalgae biorefineries has been intensified, exploring the byproducts of biodiesel conversions. However, high infrastructure and operation costs in the downstream phase limit the economic viability of biodiesel production compared to traditional fossil fuels (Eldiehy et al., 2021). Despite these challenges, microalgae biorefineries remain a promising solution.

Creating a circular bioeconomy requires high-quality secondary raw materials that can be recycled efficiently. CO_2 bio-sequestration by microalgae cultures is an emerging field in waste management, complementing efforts such as nutrient removal and the production of biogas, hydrogen, and fertilizers through anaerobic digestion of waste microalgae biomass (Vargas-Estrada et al., 2022). Waste management, including wastewater treatment and CO_2 mitigation, can be optimized by combining these techniques. However, to achieve economic scale, more research and development are needed, in addition to attracting a market for value-added products derived from microalgae (Calicioglu & Demirer, 2022).

In the long term, a microalgae biorefinery appears as a promising solution for the environment, offering a commercial alternative with minimal waste generation. The reuse of residual biomass through the cascade method supports the circular bioeconomy and contributes to climate change mitigation. However, to ensure a significant contribution to the net-zero emissions target, it is essential to intensify the development of energy-efficient processing strategies in microalgae biorefineries. In this sense, the viability of selling carbon credits can be an alternative. Due to carbon capture by microalgae, these can generate carbon credits, which can be sold to other companies. Thus, it is possible to establish a virtuous cycle of bioeconomy, which integrates the capture of CO_2 by microalgae and production in biorefineries, promoting the mitigation of climate change and the generation of carbon credits.

3.4 MICROALGAE AND CARBON FOOTPRINT

Indeed, we envision a future where we will achieve sustainability at its highest level. However, in today's globalized world, this still seems to be many decades away. Several studies and technologies still need to be put into play, so that the much-dreamed-of global sustainability stops being just a mere dream (Gills & Morgan, 2022; Javanmardi et al., 2023). In this context, microalgae were a brilliant discovery that brought with it a light at the end of the tunnel for possible mitigation of climate change. The capture of CO_2 through microalgae, in addition to supporting the production of bioproducts, is also a door that opens to the generation of profit through carbon credits – increasingly on the rise – giving light to carbon neutrality in search of a negative carbon footprint (Becker et al., 2020; Dias et al., 2022).

In the current global scenario of GHG emissions, promising initiatives and agreements that aim to achieve carbon neutrality – as early as the next decade – are rapidly approaching their deadline. We are in a critical time pressure where sustainability must assert itself as the predominant solution. To this end, an ambitious goal was set: to achieve production with net-zero carbon emissions (de Lima Barizão et al., 2023). However, for this to be envisioned, processes and products based on microalgae must be analyzed. Promising studies have already reported options for minimizing the carbon footprint throughout the life cycle of products. For example, Dias et al. (2022) in their analysis observed that energy demand is one of the most carbon-emitting points in production. Thus, they concluded that replacing a high-carbon fossil-based electricity grid with a clean, renewable electricity grid would result in carbon-neutral operations.

In this regard, it is important to highlight that microalgae are called biofactories precisely because of their impressive characteristics related to CO_2 fixation. In fact, compared to other terrestrial crops, microalgae achieve 10–50 times greater CO_2 fixation (Schneider et al., 2024). In this sense, the capture and sequestration of carbon carried out by microalgae is, without a doubt, one of the most effective methods currently available in the world (Onyeaka et al., 2021). The CO_2 sequestration mechanism can reach a value of 1.83 kg of CO_2 sequestered from the atmosphere and transformed into 1 kg of biomass, demonstrating its high capacity to assist in mitigating climate change (Arun et al., 2022).

And finally, the integration of CO_2 capture and sequestration with the generation of carbon credits can be a promising program. Carbon credits, in addition to introducing ecological responsibility, mainly in companies, but also in countries, can be a source of income. In turn, carbon credits are used as a currency in carbon markets, being a form of action to reduce GHGs in the atmosphere. Therefore, a carbon credit represents 1 tonne of CO_2 equivalent that is no longer emitted. A simple way to exemplify this theory is when 1 tonne of CO_2 equivalent was not emitted by an entity, it generates a carbon credit that can be sold to entities that have exceeded pre-established CO_2 emissions limits or that voluntarily wish to reduce their emissions toward carbon-neutral production (Schneider et al., 2024; Trouwloon et al., 2023; The World Bank, 2021).

It is clear that microalgae are an essential source of atmospheric carbon mitigation. Although there are still major obstacles to be overcome, the tireless search for a possible sustainable future is undoubtedly worth the effort. Several promising technologies already available on the market involving microalgae still require more robust development, since they are not yet considered economically viable and their environmental impacts are not yet fully clear. The development of methodologies that increase the generation of carbon credits by microalgae and, consequently, help to consolidate zero emissions in production is still necessary. However, strong global collaboration is key to ensuring that the infamous hole in the ozone layer can be completely closed by the year 2050.

3.5 INVESTING IN A BIOREFINERY TO MITIGATING CLIMATE CHANGE

The growing demand for food, driven by population growth, requires the agricultural sector to reinvent itself. To meet growing demand, the agricultural sector urgently needs to seek innovative and sustainable solutions to increase productivity. The adequate presence of nutrients in the soil plays a crucial role in the development and efficiency of agricultural production (Chittora et al., 2020; Kour et al., 2020). Therefore, producers are significantly dependent on fertilizers to meet production needs. In most agricultural areas, conventional methods with chemical fertilizers are a more common response to the sector's challenges (Kumar et al., 2018).

Commonly used nutrients are nitrogen, phosphorus, and potassium (NPK). Despite increasing productivity in the short term, chemical fertilizers can have harmful consequences for the soil, such as acidification, loss of essential nutrients, and microbiota imbalance (Nayak et al., 2019; Sinha et al., 2020; Alam et al., 2024). A new era in agriculture is emerging with the use of biofertilizers derived from microalgae, promoting more sustainable and productive agricultural practices. More than nourishing plants, applying microalgae-based biofertilizers cultivates a promising future for agriculture, where sustainability, productivity, and food security go hand in hand (Ng et al., 2024).

In response to the growing scarcity of natural resources, microalgae emerge as a promising solution for a sustainable circular economy, offering multifaceted biofertilizers (Banu et al., 2020; Moreira et al., 2023). Table 3.1 summarizes the microalgae used to produce biofertilizers, detailing their species, cultivation methods, and the beneficial impacts observed on crops.

Through Table 3.1 it is possible to visualize the feasibility of developing biofertilizers using different species of microalgae. The results of the studies indicate the promise of replacing chemical fertilizers. In addition to increasing productivity, microalgae contribute to improving the soil through structural modifications, introduction of essential biomolecules, and stimulation of microbiological activity (Kalyanasundaram et al., 2020). To affirm the power of microalgae as biofertilizers, studies were developed to eliminate possible doubts and to show the way forward to introduce the necessary agricultural changes. Dineshkumar et al. (2020) highlight that *Chlorella vulgaris* and *A. platensis* strains demonstrated significant positive effects on mung bean growth parameters. The fresh weight of plants increased by between 60.7% and 95% compared to the control, while the weight of ears and kernels also increased by 22% and 90%, respectively, compared to the control. The authors emphasize that the appropriate use of microalgae biofertilizers not only boosts the economic development of sustainable agriculture but also significantly contributes to the reduction of pollution.

This dual advantage is driving the growing interest in incorporating microalgae as active ingredients in biofertilizers and biostimulants. Similarly, Bello et al. (2021) corroborate that

TABLE 3.1
Different Microalgal Biomass as Biofertilizer

Biomass Microalgae	Culture Medium	Production and Application of Biofertilizer/ Biostimulant	Applied Culture	Results Obtained	Reference
Scenedesmus sp.	Domestic wastewater	Application of microalgae biomass to soil after lipid extraction for biodiesel production.	Rice	Increased soil availability of NPK up to 28%, 39%, and 33%, respectively, compared to chemical fertilizer.	Nayak et al. (2019)
Chlorella vulgaris	BG11	Dry biomass powder, along with cow dung, is applied to the soil, or simply microalgae biomass.	Onion	Increase of up to 17.3% of N in soil compared to control. Increase of up to 63.2% in plant height, 17.4% of free amino acids in the post-harvest plant compared to soil control.	Dineshkumar et al. (2020)
Chlorella sp. *Neochloris conjuncta* *Botryococcus braunii*	Basal medium	Liquid effluent is applied to the soil following anaerobic digestion of microalgae biomass with cow dung and silage corn.	Corn	Plant dry weight increased by up to 23% compared to the control.	Ekinci et al. (2019)
Chlorella sp. *Scenedesmus* sp. *Arthrospira platensis* *Synechocystis* sp.	Wastewater from residential treatment plants enriched with human urine	Microalgae extracts (10 g of dried microalgal biomass dissolved in 100 ml of deionized water) were applied by spraying as a liquid.	Tomato	Plant height and chlorophyll content increased by 46%. Germination occurred three days earlier than the control.	Supraja et al. (2020)

Source: Adapted from: Braun and Colla (2023).

microalgae are highly recommended for the formulation of biofertilizers; by carrying out a study, they confirmed that biofertilizers from microalgae promoted the growth of several plant species, such as tomato (*Solanum lycopersicum*), cucumber (*Cucumis sativus*), and pumpkin (*Cucurbita maxima*). A study conducted by Nayak et al. (2019) demonstrated that the use of de-oilified biomass residues from *Scenedesmus* sp. as a biofertilizer in rice (*Oryza sativa*) cultivation resulted in increases in agricultural production and a reduction in the use of chemical fertilizers. This method represents a significant advance toward sustainable agriculture. At the same time, issues such as accelerated urbanization, industrialization, and dependence on fossil fuels also have global impacts. Among these activities, the combustion of fossil fuels for energy generation is the largest source of GHG emissions (Raihan et al., 2022).

Due to the need for change, four generations of biofuels have emerged, with microalgae playing a crucial role in the third generation. Third-generation biofuels offer greater energy density compared to first generation (based on food crops) and second generation (non-food biomass waste), considering the required cultivation area (Dutra et al., 2024). However, aiming to improve the performance of microalgae, the fourth generation uses genetically modified varieties, capable of directly converting CO_2 into fuel. Microalgae stand out for their high lipid concentration of 20%–50%, being capable of producing biodiesel. This characteristic makes them an attractive alternative to conventional biodiesel (Kumawat et al., 2024).

Table 3.2 presents a summary of pretreatment methods, biofuels obtained, yields, and references associated with the study of various species of microalgae for biofuel production.

TABLE 3.2
Biofuels from Microalgae

Microalgae	Pretreatment Method	Biofuel Obtained	Yield	Reference
Scenedesmus sp.	Thermal	Biogas	0.085LCH_4/g COD (chemical oxygen demand)	González-Fernández et al. (2013)
Chlorella vulgaris	Thermochemical method – Gasification	Biohydrogen	81.6%	Liu et al. (2017)
Dunaliella sp.	Thermal	Bioethanol	0.40 g/g^{-1} DCW (dry cell weight)	Karatay et al. (2016)
Arthrospira platensis	Thermal	Produces bio-oil, raw material to produce biodiesel	46.6% weight	Liu et al. (2021)

Table 3.2 demonstrates the versatility of biotechnological applications of microalgae, where different species are capable of producing different types of biofuels. The positive results of the studies presented establish microalgae as a promising source of replacement for conventional fuels. In this context, Nazloo et al. (2022) compared the use of fresh water in biodiesel production, highlighting that 591 liters of fresh water are needed to produce one liter of biodiesel from the microalgae *C. vulgaris*, while the production of soy biodiesel requires 33 times more fresh water per liter of biodiesel. In line, Mehrabadi et al. (2016) demonstrated that cultivating microalgae biomass in wastewater can significantly reduce biodiesel production costs. This approach contributes not only to environmental sustainability by reusing wastewater but also to the economic viability of producing microalgae-based biofuels.

There are no simulations or even experiments in the literature that combine the production of biofertilizers with biofuels. Braun and Colla (2023) propose an approach to integrate both processes in a biorefinery, suggesting the prior extraction of proteins from biomass. These proteins can be targeted for the synthesis of biofertilizers, while the carbohydrates remaining in the biomass can be used for the production of bioethanol through hydrolysis and subsequent fermentation (Menegotto et al., 2020; Halim et al., 2022). The process in biorefineries involves the sequential extraction of lipids, proteins, and carbohydrates. This order is crucial due to the harsh conditions required for each step, such as high temperatures and low pH, which can impact the properties of the other components of the microalgal biomass (Braun & Colla, 2023).

Integrating biofuels and biofertilizers into microalgae biorefineries not only improves resource efficiency but also promotes a robust circular economy by replacing conventional chemicals and reducing GHG emissions associated with the production of synthetic fertilizers and fossil fuels. These biorefineries emerge as a promising solution for a more sustainable and climate-resilient economy. Given the global urgency for sustainable alternatives, especially in the face of climate change and threats to food security, investing in microalgae platforms for bioproducts can ensure an alternative source of essential proteins and nutrients (Talebi et al., 2022; Ahmad & Ashraf, 2023). This choice represents a crucial step toward a prosperous future, where the sustainable production of biofuels, biochemicals, food, and other products is fundamental to food security and environmental preservation.

3.6 CONCLUSION

Growing awareness of ecological impacts is driving a global transformation, elevating environmental issues to the same level as the economic concerns of the world's main powers. The traditional dichotomy between development and environmental preservation is being replaced by a new paradigm that seeks to harmonize these two aspects. Investing in a biorefinery is not just a strategic choice, but a key to future environmental and energy security. In addition to offering a viable alternative to indispensable conventional raw materials, the microalgae biorefinery promotes a closed production cycle, minimizing waste and valuing byproducts. As we move toward a more conscious and responsible future, investing in technologies like this is not just strategic, but an urgent necessity. This transition not only contributes to the preservation of our planet's limited natural resources but also strengthens energy security by diversifying energy sources and raw materials. Therefore, the microalgae biorefinery represents not only the key to the environmental and energy security of the future but also a unique opportunity to lead the global transition to a zero-waste, low-environmental-impact development model.

DECLARATION

The authors have not used any generative AI tools during the writing of this manuscript.

REFERENCES

Ahmad, A., & Ashraf, S. S. (2023). Sustainable food and feed sources from microalgae: Food security and the circular bioeconomy. *Algal Research*, 74, 103185.

Ahmad, S., Iqbal, K., Kothari, R., Singh, H. M., Sari, A., & Tyagi, V. V. (2022). A critical overview of upstream cultivation and downstream processing of algae-based biofuels: Opportunity, technological barriers and future perspective. *Journal of Biotechnology*, 351, 74–98.

Alam, S. K., Li, P., & Fida, M. (2024). Groundwater nitrate pollution due to excessive use of N-fertilizers in rural areas of Bangladesh: Pollution status, health risk, source contribution, and future impacts. *Exposure and Health*, 16(1), 159–182.

Ansari, F. A., Shriwastav, A., Gupta, S. K., Rawat, I., & Bux, F. (2017). Exploration of microalgae biorefinery by optimizing sequential extraction of major metabolites from *Scenedesmus obliquus*. *Industrial & Engineering Chemistry Research*, 56(12), 3407–3412.

Arcigni, F., Friso, R., Collu, M., & Venturini, M. (2019). Harmonized and systematic assessment of microalgae energy potential for biodiesel production. *Renewable and Sustainable Energy Reviews*, 101, 614–624.

Arun, J., Raghu, R., Hanif, S. S. M., Thilak, P. G., Sridhar, D., Nirmala, N., ..., & Pugazhendhi, A. (2022). A comparative review on photo and mixotrophic mode of algae cultivation: Thermochemical processing of biomass, necessity of bio-oil upgrading, challenges and future roadmaps. *Applied Energy*, 325, 119808.

Banu, J. R., Kavitha, S., Gunasekaran, M., & Kumar, G. (2020). Microalgae based biorefinery promoting circular bioeconomy-techno economic and life-cycle analysis. *Bioresource Technology*, 302, 122822.

Becker, S., Bouzdine-Chameeva, T., & Jaegler, A. (2020). The carbon neutrality principle: A case study in the French spirits sector. *Journal of Cleaner Production*, 274, 122739.

Bello, A. S., Saadaoui, I., & Ben-Hamadou, R. (2021). "Beyond the source of bioenergy": Microalgae in modern agriculture as a biostimulant, biofertilizer, and anti-abiotic stress. *Agronomy*, 11(8), 1610.

Braun, J. C., & Colla, L. M. (2023). Use of microalgae for the development of biofertilizers and biostimulants. *BioEnergy Research*, 16(1), 289–310.

Calicioglu, O., & Demirer, G. N. (2022). Role of microalgae in circular economy. In *Integrated Wastewater Management and Valorization using Algal Cultures* (pp. 1–12). Elsevier.

Calijuri, M. L., Silva, T. A., Magalhães, I. B., de Paula Pereira, A. S. A., Marangon, B. B., de Assis, L. R., & Lorentz, J. F. (2022). Bioproducts from microalgae biomass: Technology, sustainability, challenges and opportunities. *Chemosphere*, 305, 135508.

Carbajal, E. M. T., Hernández, E. M., Linares, L. F., Maldonado, E. N., & Ballesteros, R. L. (2020). Techno-economic analysis of *Scenedesmus dimorphus* microalgae biorefinery scenarios for biodiesel production and glycerol valorization. *Bioresource Technology Reports*, 12, 100605.

Cheirsilp, B., Maneechote, W., Srinuanpan, S., & Angelidaki, I. (2023). Microalgae as tools for bio-circular-green economy: Zero-waste approaches for sustainable production and biorefineries of microalgal biomass. *Bioresource Technology*, 387, 129620.

Chittora, D., Meena, M., Barupal, T., Swapnil, P., & Sharma, K. (2020). Cyanobacteria as a source of biofertilizers for sustainable agriculture. *Biochemistry and Biophysics Reports*, 22, 100737.

de Lima Barizão, A. C., de Oliveira Gomes, L. E., Brandão, L. L., Sampaio, I. C. F., de Moura, I. V. L., Gonçalves, R. F., ..., & Cassini, S. T. (2023). Microalgae as tertiary wastewater treatment: Energy production, carbon neutrality, and high-value products. *Algal Research*, 72, 103113.

Deprá, M. C., Dias, R. R., Zepka, L. Q., & Jacob-Lopes, E. (2024). Algae for environmental sustainability: Trends and future outlook. In *Algae as a Natural Solution for Challenges in Water-Food-Energy Nexus: Toward Carbon Neutrality* (pp. 861–869). Springer Nature Singapore.

Dias, R. R., Deprá, M. C., Zepka, L. Q., & Jacob-Lopes, E. (2022). Roadmap to net-zero carbon emissions in commercial microalgae-based products: Environmental sustainability and carbon offset costs. *Journal of Applied Phycology*, 34(3), 1255–1268.

Dineshkumar, R., Subramanian, J., Arumugam, A., Ahamed Rasheeq, A., & Sampathkumar, P. (2020). Exploring the microalgae biofertilizer effect on onion cultivation by field experiment. *Waste and Biomass Valorization*, 11, 77–87.

Dutra, D. A., Schneider, A. T., Dias, R. R., Zepka, L. Q., & Jacob-Lopes, E. (2024). Microalgae-based diesel: A historical perspective to future directions. In *Wastewater Exploitation: From Microbiological Activity to Energy* (pp. 241–252). Springer Nature Switzerland.

Ekinci, K., Erdal, I., Uysal, Ö., Uysal, F. Ö., Tunce, H., & Doğan, A. (2019). Anaerobic digestion of three microalgae biomasses and assessment of digestates as biofertilizer for plant growth. *Environmental Progress & Sustainable Energy*, 38(3), e13024.

El-Dalatony, M. M., Salama, E. S., Kurade, M. B., Kim, K. Y., Govindwar, S. P., Kim, J. R., ..., & Jeon, B. H. (2019). Whole conversion of microalgal biomass into biofuels through successive high-throughput fermentation. *Chemical Engineering Journal*, 360, 797–805.

Eldiehy, K. S., Das, V., Borah, D., Mandal, M., & Deka, D. (2021). Strategic approaches to enhance lipid accumulation in microalgae for bioprospecting. In *Algal Biorefinery* (pp. 50–84). Routledge.

Francavilla, M., Kamaterou, P., Intini, S., Monteleone, M., & Zabaniotou, A. (2015). Cascading microalgae biorefinery: Fast pyrolysis of *Dunaliella tertiolecta* lipid extracted-residue. *Algal Research*, 11, 184–193.

Gills, B., & Morgan, J. (2022). Global climate emergency: After COP24, climate science, urgency, and the threat to humanity. In *Economics and Climate Emergency* (pp. 253–270). Routledge.

González-Fernández, B. Sialve, N. Bernet, J.P. Steyer, (2013). Effect of organic loading rate on anaerobic digestion of thermally pretreated Scenedesmus sp. biomass, *Bioresource Technology*, 129, 219–223,

González-González, L. M., Correa, D. F., Ryan, S., Jensen, P. D., Pratt, S., & Schenk, P. M. (2018). Integrated biodiesel and biogas production from microalgae: Towards a sustainable closed loop through nutrient recycling. *Renewable and Sustainable Energy Reviews*, 82, 1137–1148.

Goswami, L., Kayalvizhi, R., Dikshit, P. K., Sherpa, K. C., Roy, S., Kushwaha, A., ..., & Rajak, R. C. (2022a). A critical review on prospects of bio-refinery products from second and third generation biomasses. *Chemical Engineering Journal*, 448, 137677.

Goswami, R. K., Agrawal, K., Upadhyaya, H. M., Gupta, V. K., & Verma, P. (2022b). Microalgae conversion to alternative energy, operating environment and economic footprint: An influential approach towards energy conversion, and management. *Energy Conversion and Management*, 269, 116118.

Halim, R., Papachristou, I., Chen, G. Q., Deng, H., Frey, W., Posten, C., & Silve, A. (2022). The effect of cell disruption on the extraction of oil and protein from concentrated microalgae slurries. *Bioresource Technology*, 346, 126597.

Hoang, A. T., Sirohi, R., Pandey, A., Nižetić, S., Lam, S. S., Chen, W. H., ..., & Pham, V. V. (2023). Biofuel production from microalgae: Challenges and chances. *Phytochemistry Reviews*, 22(4), 1089–1126.

Imbimbo, P., Bueno, M., D'Elia, L., Pollio, A., Ibañez, E., Olivieri, G., & Monti, D. M. (2020). Green compressed fluid technologies to extract antioxidants and lipids from *Galdieria phlegrea* in a biorefinery approach. *ACS Sustainable Chemistry & Engineering*, 8(7), 2939–2947.

Jacob-Lopes, E., Maroneze, M. M., Queiroz, M. I., & Zepka, L. Q. (Eds.). (2020a). *Handbook of Microalgae-Based Processes and Products: Fundamentals and Advances in Energy, Food, Feed, Fertilizer, and Bioactive Compounds*. Academic Press.

Jacob-Lopes, E., Santos, A. B., Severo, I. A., Deprá, M. C., Maroneze, M. M., & Zepka, L. Q. (2020b). Dual production of bioenergy in heterotrophic cultures of cyanobacteria: Process performance, carbon balance, biofuel quality and sustainability metrics. *Biomass and Bioenergy*, 142, 105756.

Javanmardi, E., Liu, S., & Xie, N. (2023). Exploring the challenges to sustainable development from the perspective of grey systems theory. *Systems*, 11(2), 70.

Javed, F., Aslam, M., Rashid, N., Shamair, Z., Khan, A. L., Yasin, M., ..., & Bazmi, A. A. (2019). Microalgae-based biofuels, resource recovery and wastewater treatment: a pathway towards sustainable biorefinery. *Fuel*, 255, 115826.

Kalyanasundaram, G. T., Ramasamy, A., Rakesh, S., & Subburamu, K. (2020). Microalgae and cyanobacteria: Role and applications in agriculture. *Applied Algal Biotechnology*. Nova Science Publishers.

Karatay, S. E., Erdoğan, M., Dönmez, S., & Dönmez, G. (2016). Experimental investigations on bioethanol production from halophilic microalgal biomass. *Ecological Engineering, 95*, 266 270. https://doi.org/10.1016/j.ecoleng.2016.06.079

Khoo, K. S., Ahmad, I., Chew, K. W., Iwamoto, K., Bhatnagar, A., & Show, P. L. (2023). Enhanced microalgal lipid production for biofuel using different strategies including genetic modification of microalgae: A review. *Progress in Energy and Combustion Science*, 96, 101071.

Kour, D., Rana, K. L., Yadav, A. N., Yadav, N., Kumar, M., Kumar, V., ..., & Saxena, A. K. (2020). Microbial biofertilizers: Bioresources and eco-friendly technologies for agricultural and environmental sustainability. *Biocatalysis and Agricultural Biotechnology*, 23, 101487.

Kumar, B., & Verma, P. (2021). Biomass-based biorefineries: an important architype towards a circular economy. *Fuel*, 288, 119622.

Kumar, D., & Singh, B. (2019). Algal biorefinery: An integrated approach for sustainable biodiesel production. *Biomass and Bioenergy*, 131, 105398.

Kumar, K. A., Swain, D. K., & Bhadoria, P. B. S. (2018). Split application of organic nutrient improved productivity, nutritional quality and economics of rice-chickpea cropping system in lateritic soil. *Field Crops Research*, 223, 125–136.

Kumaran, M., Palanisamy, K. M., Bhuyar, P., Maniam, G. P., Rahim, M. H. A., & Govindan, N. (2023). Agriculture of microalgae *Chlorella vulgaris* for polyunsaturated fatty acids (PUFAs) production employing palm oil mill effluents (POME) for future food, wastewater, and energy nexus. *Energy Nexus*, 9, 100169.

Kumawat, G., Vyas, P., Choudhary, S., & Meena, M. (2024). Microalgal biodiesel as a sustainable and green energy alternative: A metabolomic approach. *Biomass and Bioenergy*, 186, 107257.

Liu, B., Wang, Z., & Feng, L. (2021). Effects of reaction parameter on catalytic hydrothermal liquefaction of microalgae into hydrocarbon rich bio-oil. *Journal of the Energy Institute*, 94, 22–28.

Liu, G., Liao, Y., Wu, Y., Ma, X., & Chen, L. (2017). Characteristics of microalgae gasification through chemical looping in the presence of steam. *International Journal of Hydrogen Energy*, 42(36), 22730–22742.

López-Sánchez, A., Silva-Gálvez, A. L., Aguilar-Juárez, Ó., Senés-Guerrero, C., Orozco-Nunnelly, D. A., Carrillo-Nieves, D., & Gradilla-Hernández, M. S. (2022). Microalgae-based livestock wastewater treatment (MbWT) as a circular bioeconomy approach: Enhancement of biomass productivity, pollutant removal and high-value compound production. *Journal of Environmental Management*, 308, 114612.

Lupatini, A. L., de Oliveira Bispo, L., Colla, L. M., Costa, J. A. V., Canan, C., & Colla, E. (2017). Protein and carbohydrate extraction from *S. platensis* biomass by ultrasound and mechanical agitation. *Food Research International*, 99, 1028–1035.

Machado, R. L. S., Dutra, D. A., Schneider, A. T., Dias, R. R., Zepka, L. Q., & Jacob-Lopes, E. (2024). Recent advances in algal nexus for circular economy. In *Algae as a Natural Solution for Challenges in Water-Food-Energy Nexus: Toward Carbon Neutrality* (pp. 61–77). Springer

Mehrabadi, A., Craggs, R., & Farid, M. M. (2016). Biodiesel production potential of wastewater treatment high rate algal pond biomass. *Bioresource Technology*, 221, 222–233.

Menegotto, A. L. L., Fernandes, I. A., Colla, L. M., Duarte, J., Andrade, M. Z., Abirached, C., ..., & Valduga, E. (2020). Thermic and techno-functional properties of *Arthrospira platensis* protein fractions obtained by membrane separation process. *Journal of Applied Phycology*, 32(6), 3885–3900.

Moreira, J. B., Santos, T. D., Duarte, J. H., Bezerra, P. Q. M., de Morais, M. G., & Costa, J. A. V. (2023). Role of microalgae in circular bioeconomy: From waste treatment to biofuel production. *Clean Technologies and Environmental Policy*, 25(2), 427–437.

Nayak, M., Swain, D. K., & Sen, R. (2019). Strategic valorization of de-oiled microalgal biomass waste as biofertilizer for sustainable and improved agriculture of rice (*Oryza sativa* L.) crop. *Science of the Total Environment*, 682, 475–484.

Nazloo, E. K., Moheimani, N. R., & Ennaceri, H. (2022). Biodiesel production from wet microalgae: Progress and challenges. *Algal Research*, 68, 102902.

Ng, Z. Y., Ajeng, A. A., Cheah, W. Y., Ng, E. P., Abdullah, R., & Ling, T. C. (2024). Towards circular economy: Potential of microalgae–bacterial-based biofertilizer on plants. *Journal of Environmental Management*, 349, 119445.

Okeke, E. S., Ejeromedoghene, O., Okoye, C. O., Ezeorba, T. P. C., Nyaruaba, R., Ikechukwu, C. K., ..., & Orege, J. I. (2022). Microalgae biorefinery: an integrated route for the sustainable production of high-value-added products. *Energy Conversion and Management: X*, 16, 100323.

Onyeaka, H., Miri, T., Obileke, K., Hart, A., Anumudu, C., & Al-Sharify, Z. T. (2021). Minimizing carbon footprint via microalgae as a biological capture. *Carbon Capture Science & Technology*, 1, 100007.

Raihan, A., Begum, R. A., Nizam, M., Said, M., & Pereira, J. J. (2022). Dynamic impacts of energy use, agricultural land expansion, and deforestation on CO_2 emissions in Malaysia. *Environmental and Ecological Statistics*, 29(3), 477–507.

Samoraj, M., Çalış, D., Trzaska, K., Mironiuk, M., & Chojnacka, K. (2024). Advancements in algal biorefineries for sustainable agriculture: Biofuels, high-value products, and environmental solutions. *Biocatalysis and Agricultural Biotechnology*, 58, 103224.

Sarwer, A., Hamed, S. M., Osman, A. I., Jamil, F., Al-Muhtaseb, A. A. H., Alhajeri, N. S., & Rooney, D. W. (2022). Algal biomass valorization for biofuel production and carbon sequestration: A review. *Environmental Chemistry Letters*, 20(5), 2797–2851.

Schneider, A. T., Dias, R. R., Deprá, M. C., Zepka, L. Q., & Jacob-Lopes, E. (2024). Microalgae and its potential to generate carbon credits. *Algae Mediated Bioremediation: Industrial Prospectives*, 2, 539–547.

Sinha, D., & Tandon, P. K. (2020). An overview of nitrogen, phosphorus and potassium: Key players of nutrition process in plants. In *Sustainable Solutions for Elemental Deficiency and Excess in Crop Plants* (pp. 85–117).Springer.

Slegers, P. M., Olivieri, G., Breitmayer, E., Sijtsma, L., Eppink, M. H., Wijffels, R. H., & Reith, J. H. (2020). Design of value chains for microalgal biorefinery at industrial scale: Process integration and techno-economic analysis. *Frontiers in Bioengineering and Biotechnology*, 8, 550758.

Supraja, K. V., Behera, B., & Balasubramanian, P. (2020). Efficacy of microalgal extracts as biostimulants through seed treatment and foliar spray for tomato cultivation. *Industrial Crops and Products*, 151, 112453.

Talebi, S., Edalatpour, A., & Tavakoli, O. (2022). Biorrefinaria de algas: uma solução potencial para o nexo alimento–energia–água–meio ambiente. *Sustainable Energy & Fuels*, 6(11), 2623–2664.

The World Bank (2021). State and Trends of Carbon Pricing 2021, World Bank, Washington, DC.

Trouwloon, D., Streck, C., Chagas, T., & Martinus, G. (2023). Understanding the use of carbon credits by companies: A review of the defining elements of corporate climate claims. *Global Challenges*, 7(4), 2200158.

Ummalyma, S. B., Sirohi, R., Udayan, A., Yadav, P., Raj, A., Sim, S. J., & Pandey, A. (2023). Sustainable microalgal biomass production in food industry wastewater for low-cost biorefinery products: A review. *Phytochemistry Reviews*, 22(4), 969–991.

Vargas-Estrada, L., Longoria, A., Arenas, E., Moreira, J., Okoye, P. U., Bustos-Terrones, Y., & Sebastian, P. J. (2022). A review on current trends in biogas production from microalgae biomass and microalgae waste by anaerobic digestion and co-digestion. *BioEnergy Research*, 15(1), 77–92.

Wu, G., Tham, P. E., Chew, K. W., Munawaroh, H. S. H., Tan, I. S., Wan, W. A. A. Q. I., ..., & Show, P. L. (2023). Net zero emission in circular bioeconomy from microalgae biochar production: A renewed possibility. *Bioresource Technology*, 388, 129748.

Yadav, G., Dash, S. K., & Sen, R. (2019). A biorefinery for valorization of industrial waste-water and flue gas by microalgae for waste mitigation, carbon-dioxide sequestration and algal biomass production. *Science of the Total Environment*, 688, 129–135.

Zomer, T., McAloone, T., & Pigosso, D. (2024). Categorization of manufacturing companies' readiness profiles for the transition to the circular economy: A multidimensional cluster analysis. *Journal of Industrial Ecology*, 28(2), 277–288.

4 Road Map from Microalgae Biomass to Biorefinery Prospects and Challenges

Tesfaye Kassaw Bedru, Workisa Bacha Garuma, R. Baskaran, Mani Jayakumar, and T. Selvaganapathy

4.1 INTRODUCTION

The traditional linear patterns of production and consumption often lead to massive waste and resource depletion; thus, the circular economy has lately gained a lot of interest as a possible replacement. This uses up a lot of resources and is usually quite inefficient. Reducing waste, maximizing reuse of resources, and incorporating renewable energy sources are the tenets of the circular economy (Prieto-Sandoval et al., 2018).More and more focus is being placed on microalgae biomass as a renewable resource within the framework of the circular economy because of its potential in biorefinery processes (Rajesh Banu et al., 2020).

Microalgae outperform traditional biological feedstock in a number of ways, including photosynthetic efficiency, adaptability to non-farming environments, and freshwater requirements. Microalgae are a great choice for sustainable production due to their diverse features, which make them perfect for making products such as biofuels, bioplastics, and nutrients (Chanana et al., 2023). Greenhouse gas emissions can be mitigated and carbon storage enhanced by the absorption of atmospheric carbon dioxide by microalgae (Politaeva et al., 2023).

There is a lot of room for improvement in microalgae-based biorefineries before they see widespread commercial application, despite their promising future. The economic viability of large-scale algal production, the energy-intensive processes required for harvesting and processing, and the development of effective technology to extract high-value products from microalgae biomass are all examples of such challenges (Yadav & Sen, 2018).Continued research is paving the way for advancements in microalgae-based biorefineries, which could revolutionize microalgae's status in the circular economy, despite these limitations.

As the need for renewable energy and sustainable resources continues to rise around the world, microalgae are expected to play an increasingly important role in biorefineries. Reducing dependency on fossil fuels and promoting a more sustainable bio-based economy could be achieved through the integration of microalgae into frameworks for the circular economy.

4.1.1 Importance of Sustainable and Renewable Energy

To satisfy the increasing global need for energy in a sustainable manner, a renewable energy source that is less harmful to the environment is urgently required. By utilizing natural processes to produce electricity, renewable energy sources such as solar, wind, hydropower, and bioenergy provide a solution, lessen reliance on non-renewable resources, and decrease carbon footprints (Shuba & Kifle, 2018).Issues like climate change, resource loss, and environmental degradation are more critical now than they have ever been. Sustainable and renewable energy sources have become increasingly

DOI: 10.1201/9781003685814-4

important due to these difficulties. Here, the prospect of bioenergy produced by microalgae is extremely attractive. Large quantities of biomass may be produced by microalgae with great photosynthetic efficiency and the capacity to thrive in many settings. This biomass can then be transformed into biofuels like biodiesel and bioethanol. Microalgae are unique among plants in that they can absorb and store carbon dioxide without requiring extensive cultivation; this property contributes to lowering atmospheric CO_2 levels and slowing the rate of climate change (Singh et al., 2023).

It is in line with the concepts of the circular economy to incorporate microalgae into renewable energy strategies. A technique that improves environmental sustainability and economic feasibility is the conversion of waste biomass into bioenergy and other valuable goods (Ali et al., 2022). Improving microalgae efficiency, increasing biomass production, and decreasing energy costs in harvesting and processing are all technical and economic hurdles that must be overcome before the renewable energy potential of microalgae can be fully realized. Despite these challenges, microalgae are getting closer to large-scale, practical solutions for sustainable energy production thanks to ongoing research and technical advancements (Khan et al., 2018). One way to ensure a cleaner and more adaptable energy future is to develop and utilize sustainable energy sources, such microalgae bioenergy.

4.1.2 Potential of Fast-Growing, Lipid-Rich Microalgae as a Biofuel Feedstock

Finding long-term replacements for fossil fuels has become more of a priority as the world moves away from reliance on these resources. One of the most intriguing prospects among the various biofuel feedstocks is fast-growing, lipid-rich microalgae, which have a high lipid content and produce biomass quickly. The photosynthetic capacity of microalgae is well-documented, and some species of this type of single-celled plant can even store lipids that might be converted into biodiesel. With fossil fuel sources dwindling and environmental sustainability becoming an increasingly pressing issue, these traits make them a promising candidate for biofuel production (Gaurav et al., 2024; Van Lal et al., 2023).

One major advantage of employing microalgae as a biofuel feedstock is their ability to produce substantial biomass yields in a relatively short period of time. When compared to traditional biofuel crops like corn or soybeans, microalgae are better for the environment because they don't demand as much freshwater or farmland. Growing microalgae in saltwater or wastewater increases its sustainability potential even further. By fixing carbon dioxide through photosynthetic processes, microalgae can also aid in reducing emissions of greenhouse gases (Vasistha et al., 2020).

Microalgae have the potential to be a biofuel feedstock due to their high lipid content. Particularly in times of nutritional stress, several species of microalgae can accumulate substantial amounts of lipids. Biodiesel can be made from these lipids when they are collected. Research on lipid productivity and biofuel conversion efficiency has shown particularly promising results from lipid-rich species such as *Nannochloropsis*, *Chlorella*, and *Botryococcus braunii* (Sivaramakrishnan et al., 2022).. These microalgae have the potential to increase biofuel production and meet the world's energy demands since they can produce significantly more lipids per hectare than traditional oil crops (Shiong et al., 2023).

Despite the promising future of lipid-rich microalgae, there are a number of challenges to be solved. Growing, collecting, and extracting lipids from large quantities of algae is still prohibitively expensive, which greatly delays commercialization. Additionally, energy-intensive processes are sometimes required to convert algal biomass into biofuel, which might reduce the system's overall energy balance (Gaurav et al., 2024; Wang et al., 2024).. The focus of current research is on improving farming practices, creating more cost-effective harvesting and processing technologies, and refining genetic engineering tactics to increase lipid yields (Jothibasu et al., 2021).

The economic feasibility of producing biofuels from microalgae has been significantly enhanced by recent advancements in bioreactor design and growing techniques. Systems such as ponds and closed-loop bioreactors have been utilized. More output with less water and energy used is possible with its open design. Along with biofuels, other useful biological products are also produced, including medicines and bioactive chemicals. This can assist in reducing the expenses (Farzaneh & Ahmad, 2022).

4.2 SIGNIFICANCE OF MICROALGAE-BASED BIOREFINERY IN CIRCULAR ECONOMY

4.2.1 Biorefinery

In recent years, microalgae-based biorefineries have gained significant interest as a potential long-term answer to pressing global issues like energy insecurity, climate change, and resource depletion through the production of high-value products and renewable energy. High growth rates, non-agricultural site suitability, and the capacity to utilize wastewater and CO_2 from businesses as feedstock are just a few of the ways microalgae outperform conventional feedstock. This helps to lower emissions of greenhouse gases and supports the ideas of the circular economy (Olguín et al., 2022).

Microalgae are perfect for biorefineries because of their adaptability, which can extract several uses for a single biomass feedstock. A wide variety of goods are produced from these processes, including biofuels like biodiesel and bioethanol as well as valuable substances used in the nutraceutical, cosmetic, and pharmaceutical industries, such as omega-3 fatty acids, pigments, proteins, and bioactive molecules. Microalgae biorefinery facilities can be made more efficient and financially viable by integrating transformation technologies such as hydrothermal liquefaction (HTL), anaerobic digestion (AD), fermentation, and others (Narayanan, 2024).

Additionally, microalgae-based biorefinery aids sustainable development by opening up new avenues for rural development and employment creation, particularly in inhospitable regions for traditional farming. Bioreactors equipped with LED lighting and real-time monitoring are examples of cutting-edge farming systems that boost biomass yield while decreasing water and energy use. It still has a ways to go before it can be considered commercially viable, due to issues including high manufacturing costs and scaling issues that necessitate ongoing research and technological innovation (Kuo et al., 2021; Mahmood et al., 2023).

4.2.2 Circular Economy

More and more people are realizing that microalgae are essential to advancing the principles of the circular economy, which stress the need of reducing waste, increasing resource efficiency, and creating sustainable production cycles. Microalgae thrive in non-arable land using nutrient-rich waste streams, including wastewater, and grow rapidly; these traits make them ideal for circular bioeconomy models. To close nutrient loops and reduce reliance on synthetic fertilizers, new research suggests that wastewater treatment facilities can employ microalgae-based systems to collect nutrients and produce biomass (Fuentes-Grünewald et al., 2020; Ezhumalai et al., 2024; Stiles et al., 2018).. In addition, microalgae can be cultivated using CO_2 emissions from industries; this helps to reduce emissions of greenhouse gases and store carbon, which is in line with the goals of the circular economy (Onyeaka et al., 2021).

Animal feed, biofuels, bioplastics, and high-value bio-chemicals including omega-3 fatty acids and antioxidants are just a few of the many possible products of microalgae biomass. This multi-product technique exemplifies the principles of the circular economy by maximizing resource use and decreasing waste (Sarma et al., 2021).. It does this by ensuring that every percent of the biomass is utilized efficiently. Incorporating AD to produce biogas from residual biomass is just one example of how biorefinery design is evolving to lead to closed-loop systems that boost sustainability through the generation of renewable energy and the reduction of trash.

4.3 MICROALGAE CULTIVATION AND HARVESTING

4.3.1 Overview of Microalgae Cultivation Methods

There are many different approaches to microalgae farming, which can be categorized into open and closed systems. Each system has its own set of advantages and disadvantages. The ease of usage and cheap operating expenses (OPEX) make open pond systems like raceway ponds and high-rate algae ponds (HRAPs) popular. The low cost and adaptability of raceway pond systems make them attractive, but they also pose risks of pollution, are vulnerable to uncontrolled environmental conditions, and produce significantly lower yields than closed systems (Yadav & Sen, 2018).. Raceway ponds are shallow, oval-shaped pools that use a paddle wheel to move water. HRAPs are a kind of raceway ponds that are engineered to have an ideal water-to-light ratio for maximum efficiency. Their limitations in environmental monitoring and contamination remain unchanged, despite their widespread application in biofuel production and wastewater treatment (Dudek et al., 2012).Closed systems, also known as photo-bioreactors (PBRs), offer a more regulated setting for growing microalgae, which boosts output while decreasing contamination chances. One costly option is the tubular PBR, which has recently grown in popularity owing to its capacity to increase photosynthetic efficiency by ensuring uniform light exposure across the culture (Liu et al., 2013). However, this setup requires a substantial initial investment and ongoing maintenance. Another practical design that makes the most of biomass production and allows for optimal light penetration is the flat-panel PBR. Its high surface-to-volume ratio makes it a powerful tool. However, biofouling and overheating are potential issues with these systems that could limit their utility in the future (Tan et al., 2018).Closed systems can also be achieved through connected farming, which involves growing microalgae on a solid substrate. This method simplifies harvesting and produces a high biomass density. Light penetration is a potential limitation and the approach may necessitate costly substrates, nevertheless these benefits (Liu et al., 2013).

Another alternative that does not include enclosures is the cultivation of microalgae on surfaces that are not suspended in a solution or in biofilms. These technologies are appealing for extensive use because of their reputation for minimal water consumption and simplicity of harvesting. But problems like biofouling can arise, and better anti-fouling methods are required to keep operations afloat (Katarzyna et al., 2015).Growing microalgae on a big scale has never been easier or more economical than using wastewater-based growth, which uses nutrient-rich effluent as a substrate. Through phytoremediation, this approach aids in environmental cleanup while simultaneously reducing cultivation expenses and improving nutrient recovery. Nevertheless, the efficacy is highly dependent on the reliability and standard of the wastewater that is used (Taylor et al., 2015).

The most up-to-date research in microalgae farming aims to improve these processes by using inexpensive nutrient sources, genetically modified strains, and combining farming with biorefinery to create a sustainable and prosperous economy. For example, new approaches are being developed to capture carbon dioxide (CO_2) from the air using microalgae, which will significantly help in the fight against climate change. One example is the production of biofuels and other bio-products from microalgae biomass, which may be used in both direct and indirect ways to enhance carbon capture (Penloglou et al., 2024). By enhancing the effectiveness of PBRs for microalgae biomass yields, researchers have also investigated optimal development under various environmental circumstances (Farooq, 2022). Although there are advantages and disadvantages to each microalgae cultivation method, new technologies are constantly improving their performance and scalability, opening up exciting possibilities for biofuel generation, environmental remediation, and the creation of valuable bio-products.

4.3.2 Photo-Bioreactors

Light, temperature, food supply, carbon dioxide levels, etc. can all be precisely controlled in PBRs, which are specialized production systems designed to maximize the growth of microalgae. In order to make microalgae cultivation more efficient and cheaper, new designs and procedures have been created (Chanquia & Kara, 2022). A recent advancement in PBR technology is the creation of tubular photo reactors. These reactors offer microalgae more light efficiency due to their higher surface-to-volume ratios. These setups are perfect for high-density algae farms because they evenly distribute light and efficiently exchange gases. Reactor design and operation must undergo ongoing development to address issues such as biofouling, oxygen build-up, and temperature control (Abdur et al., 2024).

A further noteworthy advancement is the use of flat-panel PBRs. When compared to conventional open pool systems, these reactors provide more efficient use of space and more light penetration. Recent research has shown that their modular architecture makes them ideal for large-scale production because of the adaptability and flexibility they provide to various environmental conditions. However, the high costs of building and maintenance still prevent their broad implementation (Oliva et al., 2023).

There has been a lot of recent study focused on the possibility of incorporating PBRs into wastewater treatment systems. In addition to providing an efficient means of wastewater cleanup, this technology encourages the growth of microalgae, which supply vital nutrients. Using PBRs to treat wastewater can drastically cut down on phosphorus and nitrogen levels while simultaneously producing biomass that has several potential uses, including biofuels (Dammak et al., 2023).

Using modern materials and coatings to boost PBR efficiency is another area that has recently been investigated. Light transmission, fouling, and reactor lifespan are all areas that stand to benefit from these advancements. To keep the lighting within the reactors at an optimal level while reducing maintenance costs, for instance, new anti-reflective coatings and self-cleaning surfaces might be utilized (Wei et al., 2023).In addition, PBR management using Artificial intelligence (AI) and machine learning (ML) may optimize performance measures including light intensity, temperature, and nutrient supply, leading to far faster microalga growth rates with much less energy use. In order to maximize the performance of PBRs, automated control systems are developed and growth rate predictions for algae are made using AI and ML techniques.

Hybrid PBRs that combine phototrophic and mixotrophic culture modes have also demonstrated potential following these improvements. Particularly in regions with erratic light supply, these hybrid systems can optimize the light and dark cycles to increase biomass output and yield. This opens up a lot of room for microalgae culture, which makes PBR a viable option for increasing production on a big scale from an economic perspective (Abdur Razzak et al., 2024).

New developments and studies in PBR technology have great promise for the large- and small-scale cultivation of microalgae. But, PBRs still have a ways to go before they can be considered a viable option for microalgae production in the long run. Problems like high initial and ongoing costs and the need for complex control systems still need to be ironed out.

4.3.2.1 Open Ponds

Since they are both adaptable and inexpensive to run, open ponds have long been a go-to method for growing microalgae. Some examples of these systems are circular ponds, raceway ponds, and unstirred ponds. Because they are typically constructed outdoors to soak up natural sunlight, open ponds are an excellent option for industrial-scale production. On the other hand, new research has shown both the benefits and drawbacks of open pond farming (Alberto et al., 2014).

The reduced construction and operating costs of open pond systems in comparison to closed PBR systems are a key advantage of these systems. Because of their reduced price, microalgae are an attractive option for applications requiring vast quantities of biomass, such as biofuel production and wastewater treatment. Wastewater is just one of many nutrients that may be used in open ponds to treat waste and recycle nutrients, which in turn helps keep costs down and promotes sustainable development (Rossi et al., 2023).

In spite of all these benefits, open pools nevertheless encounter a lot of problems, the most common of which being pollution and environmental restrictions. Microalgae productivity can be negatively impacted by exposed structures because of temperature, light, and contamination from undesired microorganisms You et al., 2023).In addition, open ponds can be difficult to keep in an ideal environment for microalgae growth due to their sensitivity to changes in temperature, rainfall, and sunshine. These considerations may necessitate supplementary environmental management strategies or restrict their use to areas with relatively stable climates (Novovesk et al., 2023).

New research has looked at potential solutions to these problems. To make the process more sustainable and cost-effective, you can combine open ponds with wastewater treatment. This will boost the availability of nutrients and allow for the retreat of wastewater, which is great for the environment (Alavianghavanini et al., 2024). Modified pond designs and surfaces are one example of how new materials and designs can enhance light penetration and decrease contamination risk.

In general, open ponds are still useful for growing microalgae because of their adaptability and cheap cost, but further study is required to overcome their drawbacks. New developments in pond architecture, ecological administration, and the incorporation of other eco-friendly methods are necessary to expand their potential uses.

4.3.3 Harvesting Technologies

Different microalgae harvesting methods including centrifugation, flocculation, and filtration are necessary for efficient biomass recovery and have varying costs and energy requirements.

4.3.4 Centrifugation

Because of its efficiency and effectiveness in recovering biomass, centrifugation has become a popular and reliable technique for harvesting microalgae. This technique separates biomass from culture media with recoveries of approximately 95%–100% by using centrifugal force to hasten the sedimentation of microalgae cells (Gaurav et al., 2024).

The pharmaceutical, nutraceutical, and cosmetic sectors, among others, place a premium on the purity of the biomass they harvest, making centrifugation an ideal process for these applications. Powerful mechanical equipment are required to produce the centrifugal force; however, this process is coupled with considerable energy consumption. When it comes to producing inexpensive goods like biofuels, centrifugation is among the most costly harvesting technologies for large-scale harvesting (Silva et al., 2023).

A number of methods to lessen centrifugation's financial and ecological footprint have been the subject of recent research. Recent developments in continuous-flow centrifuge technology, for instance, have demonstrated the potential to lower operational expenses by decreasing energy consumption without sacrificing stability. It has been suggested to enhance overall efficiency and reduce energy consumption of the centrifugation system by combining centrifugation with other harvesting methods, such as flocculation or filtration by pre-concentrating microalgae biomass (Banu et al., 2024). The goal of these hybrid approaches is to gather microalgae in a more sustainable and economical way by combining the best features of several technologies.

Still, centrifugation's operational expenses remain an obstacle to its widespread usage in microalgae collection, despite the method's dependability, scalability, and capacity to generate high-purity biomass. Reducing the energy consumption and costs of centrifugation while maintaining or improving its efficiency in recovering biomass is anticipated to be the focus of future research and technical development.

4.3.5 Flocculation

The low cost and adaptability of flocculation make it a viable option for microalgae collecting. The microalgae cells gather into big clumps during this phase, and then they may be readily removed from the growing medium. Chemicals like aluminum sulfate, natural flocculants like chitosan, and bio flocculants made from microbes are some of the things that can be used to flocculate it. One major benefit of flocculation over other harvesting processes, including centrifugation, is the reduced energy need (Banu et al., 2024).

Modern flocculation methods have evolved to make the process more efficient and environmentally friendly. For instance, bio-based flocculants have gained acceptance as a greener substitute for traditional chemical flocculants. By decreasing the environmental effect of chemical residues, these bio-flocculants provide sustainable solutions; they are often generated from natural polymers or microbial by-products (Saranya et al., 2022).In addition, new hybrid techniques that combine flocculation with other harvesting methods have shown promise for dissolved air flotation, which could lead to cost savings and increased efficiency Zhou et al., 2022).

Optimizing flocculation for various microalgae species and growth conditions continues to be a challenge, despite recent developments. Achieving efficient flocculation requires precise control of factors including pH, ionic strength, and flocculant content. Additionally, additional research is needed to determine whether flocculants affect the quality of biomass that is harvested, particularly for use in the food and pharmaceutical sectors (Banu et al., 2024).

Research into flocculation's efficiency, sustainability, and adaptability to different production scales is continuous because of the technology's essential role in microalgae harvesting. Combining it with other harvesting techniques might increase its usefulness in the expanding microalgae market.

4.3.5.1 Filtration

Because of its capacity to manage huge volumes of growth medium, filtration is an essential technique for microalgae harvesting, particularly when the biomass it produces is of high purity. A semi-permeable membrane is used to contain the algal cells and let water and tiny particles flow through while a microalgae suspension is passed through. Scalability, user-friendliness, and the absence of chemical additions that could taint the biomass are some of the benefits of this method. Problems including membrane fouling, high operational costs, and energy consumption from pressure demands continue to significantly limit its application in large-scale microalgae growth (Daramola et al., 2023).

Improved membrane materials and techniques have been the focus of recent filtering technology advancements aimed at reducing fouling and increasing lifetime. To combat biofouling and prolong membrane service life, new technologies have emerged, such as anti-fouling coatings and nanomaterials. To illustrate the point, there is hope that nano-clay-modified membranes can improve the effectiveness of the filtering procedures under consideration and decrease fouling during microalgae harvesting (Lakshmi et al., 2023).To further improve overall efficiency, it has been suggested to combine filtering with additional harvesting techniques such flocculation. This would reduce the initial microalgae cell density and minimize fouling (Banu et al., 2024).

When it comes to collecting tiny algae species, membrane filtration methods like microfiltration and ultrafiltration really shine in terms of cell density and purity. Nevertheless, the method's economic feasibility is sometimes called into question because of the high expenses of membrane replacement and maintenance, as well as the regularity with which pollution like fouling must be addressed. Research is being focused on developing environmentally friendly filtering methods that can reduce energy consumption and membrane fouling without sacrificing recovery rates. The feasibility of using filtration to produce microalgae biomass will be enhanced as a result (Santos et al., 2024).

4.3.6 Biomass Recovery and Improving Biomass Efficiency

The economic viability and environmental footprint of microalgae biorefineries are directly impacted by biomass recovery and biomass improvement in microalgae farming, two crucial aspects of sustainable production. Recent advances have maximized biomass yield and optimized recovery processes through the use of novel farming techniques, process integration, and nutrient recycling, among other measures. Microalgae culture's incorporation into wastewater treatment systems boosts nutrient recovery and biomass yields by providing a constant supply of crucial nutrients like nitrogen and phosphorus, which are typically scarce in traditional media. This has been shown to increase nutrient recovery and biomass production. In addition, there has been encouraging progress in biofilm-based growth systems toward optimizing harvesting operations and reducing water usage; these developments bode well for the overall efficiency of biomass and the reduction of operational expenses (Ramanna et al., 2021).

Recent research has also highlighted the need of illumination optimization for increasing biomass production and microalgae growth rate. Modern advancements in reactor design, such as new PBR layouts and materials that disperse light more evenly and decrease biofouling, have substantially raised biomass outputs (Chernova et al., 2023). Microalgae strains that have been enhanced in photosynthetic efficiency and have better growth characteristics in different environmental conditions have been created by genetic engineering and strain improvement (Zhang et al., 2023).

Beyond these technical developments, a new strategy for pre-treatment of microalgae biomass using environmentally friendly solvents has shown promise for the recovery of important components including lipids and proteins. Improved product extraction efficiency with less harmful chemical use is one way these methods contribute to a more environmentally friendly biorefinery process (Choi et al., 2022). To fully utilize microalgae for biofuels, bio-products, and environmentally sustainable uses, there must be ongoing progress in improving biomass recovery and efficiency (Figure 4.1 and Table 4.1).

4.4 PROCESSING TECHNIQUES FOR ENERGY AND VALUE-ADDED PRODUCTS FROM MICROALGAE BIOMASS

Because of its rapid growth rates, CO_2 trapping capabilities, and biochemical composition (lipids, proteins, carbs, and pigments), microalgae biomass is a promising raw material for renewable energy and value-added product production. In order to produce biofuels such as biodiesel, bioethanol,

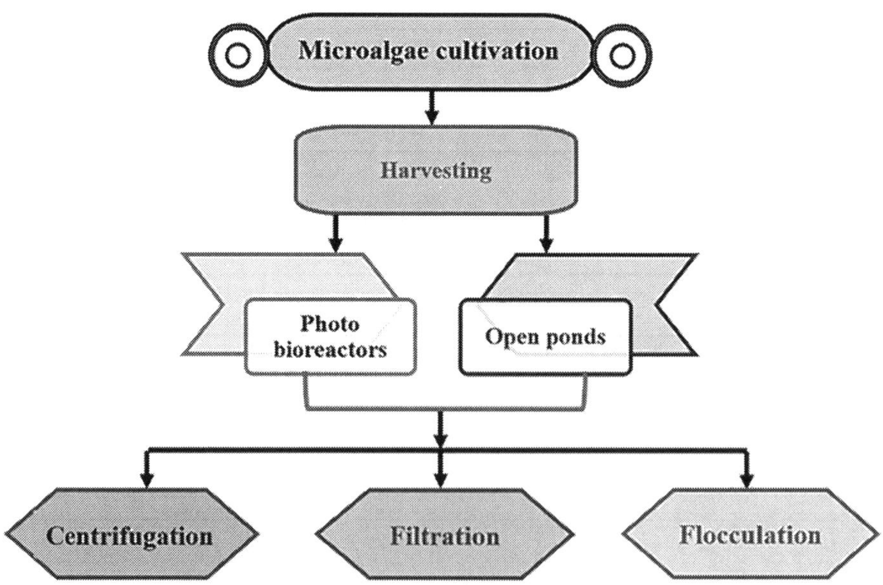

FIGURE 4.1 Microalgae cultivation and harvesting techniques.

TABLE 4.1
Microalgae Harvesting Technologies Merits, Demerits, and Challenges

S. No	Harvesting Technology	Merits	Demerits	Challenges	References
1.	Centrifugation	Highly efficient with rapid separation and high recovery rates; effective for small cells and low-density suspensions	Energy-intensive and costly, limiting large-scale use	Reducing energy consumption and cost for large-scale applications	(Barros et al., 2015; Christenson and Sims, 2011)
2.	Flocculation	Cost-effective, simple, and low energy requirement; suitable for large-scale application	Chemical flocculants can contaminate biomass, affecting further processing	Developing environmentally friendly bio-flocculants to ensure biomass purity	(Salim et al., 2011; Roselet et al., 2021)
3.	Filtration	Energy-efficient, capable of continuous operation, suitable for large-scale	Prone to membrane fouling, leading to increased maintenance and reduced efficiency	Overcoming fouling issues and improving membrane durability for long-term use	(Zhao et al., 2013; Show et al., 2015)

and biogas, processing methods have evolved to focus on recovering valuable compounds such as omega-3 fatty acids, antioxidants, and pigments. HTL, pyrolysis, and transesterification are just a few of the biofuel production and quality techniques that have been the subject of substantial research. However, there is hope for extracting bioactive chemicals of high purity through the use of innovative extraction techniques such as microwave-assisted extraction and supercritical fluid extraction. Waste may be minimized and economic and environmental benefits enhanced by incorporating these processing processes into a biorefinery. This allows for the full biomass of microalgae to be utilized efficiently (Rawat et al., 2013; Khan et al., 2018).

The transformation of microalgae biomass into beneficial bioenergy products relies heavily on biochemical reactions including fermentation and AD. An abundant renewable energy source, AD involves microbes decomposing organic materials in an oxygen-free environment to produce biogas, which mostly consists of methane and carbon dioxide. The method is highly effective at producing biogas from the microalgae's protein- and carbohydrate-rich byproducts, and it also reduces waste and recovers nutrients (Vigani et al., 2015).

To make bio-based fuels like bioethanol, fermentation transforms the sugars in microalgae biomass into these fuels by utilizing certain kinds of microbes. To maximize production yields and efficiencies, this approach is best applied to microalgae feedstock that are carbohydrate rich, as well as to microalgae species that have undergone improvements in metabolic engineering and fermentation technologies [67]. The combination of AD and fermentation in microalgae biorefinery makes microalgae a renewable energy source that is both economically viable and environmentally friendly.

4.4.1 Anaerobic Digestion

The biochemical process known as AD uses microorganisms operating in an oxygen-free environment to transform microalgae biomass into biogas, which mostly consists of carbon dioxide and methane. The capacity of this technique to generate renewable energy while also managing microalga biomass sustainably has attracted a lot of attention. The high protein and carbohydrate content of microalgae makes them perfect for AD, where anaerobic bacteria may break them down into simpler chemicals and eventually create methane (Ward et al., 2014).

AD has recently made strides in improving a number of parameters to boost biogas output and process stability. It is becoming more common to use pre-treatment methods like heat, chemicals, or enzymes to break down the tough cell walls of microalgae. These walls contain complex polysaccharides like cellulose and hemicellulose, which microorganisms can't digest. The rate of methane production is boosted because these pre-treatments make organic matter more bioavailable (Passos et al., 2014; Mahdy et al., 2015).The carbon-nitrogen ratio is balanced, ammonia retention is reduced, and biogas production is improved (Wirth et al., 2015)when microalgae are digested alongside other organic starts like food wastes or sewage sludge.

The efficiency of microalgae digestion has been enhanced by developments in reactor design, particularly high-rate anaerobic digesters like the anaerobic membrane bioreactor (AnMBRs) and the up-flow anaerobic sludge blanket reactor (UASB). So, the rate of methane production goes up (Ehimen et al., 2011).More valuable products, such as biological fertilizer, can be recovered from digestion when AD is integrated with other processes in the biorefining framework. This increases the overall economic and environmental sustainability of microalgae use (Schenk et al., 2008).

Nevertheless, there are still obstacles to overcome, such as microalgae's high ammonia concentration, which might hinder methanogenesis, and the necessity for efficient pre-treatment techniques to degrade algal cell walls. Optimal working circumstances are the target of ongoing research efforts to resolve these challenges. One potential way to transform biomass into renewable energy while also reducing waste is to establish close partnerships with microbes and to investigate innovative pre-treatment procedures.

Converting carbohydrate components like starch and cellulose into biofuels and useful compounds like bioethanol, butanol, and organic acids is the main goal of fermentation, a crucial biochemical transformation process of microalga biomass. Microalgae that are abundant in carbohydrates are perfect for fermentation because their high sugar content can be converted efficiently by microbes. Recent studies have looked into developing genetically modified strains, improving fermentation conditions for greater efficiency, and producing bio-ethanol from microalgae. Scientists have discovered that certain strains of bacteria and yeast can alter the metabolic pathway to produce fewer by-products, leading to improved sugar consumption and higher ethanol yield (Harun et al., 2010).

Enzymatic hydrolysis, acid hydrolysis, and steam explosion are common pre-treatment procedures used to convert complex polysaccharides found in microalgae cell walls into sugars that may be fermented, thus increasing the efficiency of fermentation. Various techniques hydrolysis technology has advanced recently, allowing for the employment of individualized enzyme cocktails to accelerate the conversion of algal biomass to sugars, which in turn facilitates subsequent fermentation processes (Ho et al., 2013). The simultaneous extraction of high-value co-products like pigments, proteins, and lipids is made possible through the integration of fermentation inside a biorefinery framework. Because of this, it is now more economically viable to produce biofuels from microalgae (Borowitzka, 2013).

Maximizing production while minimizing costs is the problem of fermentation for microalgae biomass conversion. Fermentation parameters, including pH, temperature, and nutrient delivery, must be optimized. Additionally, fixing issues brought on by inhibitory compounds, including phenolics produced during biomass pre-treatment, is crucial to achieving effective fermentation operations. To improve the fermentation process's overall robustness, it is recommended to investigate sophisticated detoxification procedures and adaptive evolution tactics to increase microbial resistance to these inhibitors Chen et al., 2013).Notwithstanding these obstacles, fermentation appears to be a viable option for producing sustainable energy and valuable byproducts from microalgae biomass.

4.4.2 Thermochemical Conversion

An essential way to transform microalgae biomass into renewable energy and products with added value is through thermochemical conversion, which includes processes like gasification, pyrolysis, and HTL. By subjecting the algae to high temperatures and, in certain cases, pressure, these processes decompose the microalgae's complex organic components into simpler substances like charcoal, syngas, and bio-oil. Although pyrolysis and gasification are effective methods for producing biofuels and syngas, HTL is more appropriate for processing wet microalgae since it does not necessitate costly drying processes. Recent advances in catalyst research, reactor design, and process optimization have substantially enhanced the efficiency and product yields of these processes, making them more suitable for biofuel production on a wide scale (Rana et al., 2022).

Thermochemical gasification of microalgal biomass into syngas, mostly composed of methane, carbon monoxide, and hydrogen, is an attractive and potentially useful technique. Microalgae undergo partial oxidation at elevated temperatures in this procedure (often between 800 and 1000 degrees Celsius) with a controlled quantity of air or oxygen present. Divide things into their individual parts; it will be easier. Recent studies have aimed to optimize gasification parameters including feedstock preparation, temperature, and pressure in order to increase the yield of syngas and its quality for subsequent conversion into chemicals or liquid fuels (Zhu, 2015).

Microalgae provide a special problem for gasification due to their high moisture content. This often requires an energy-intensive drying phase. But there is a way to convert wet microalgae without drying them first: supercritical water gasification (SCWG). In an aqueous medium, this process operates at high pressures and temperatures. This boosts gasification efficiency and drastically cuts down on energy input (Guo et al., 2015).Furthermore, it has been established that improvements in catalytic gasification, which uses catalysts like iron or nickel, boost reaction rates and decrease tar formation, producing cleaner syngas and better hydrogen yields (Udomsap & Shams, 2019).

Co-gasification, in which microalgae are gasified with another sort of biomass, such as agricultural waste or municipal solid waste. This approach boosts gasification efficiency by maximizing syngas quality, increasing energy recovery, and keeping the carbon-to-nitrogen ratio in check (Soratana et al., 2012). The evolution of complex gasification reactors, like entrained flow gasifiers and fluidized beds, has enhanced process efficiency and scalability while facilitating improved heat and mass transfer.

Despite these advancements, microalgae gasification remains unfeasible from an economic standpoint, mostly due to the high initial and ongoing expenses associated with the process. Optimizing reactor designs, creating more cost-effective catalysts, and integrating gasification with other conversion processes are all areas that need further investigation in order to achieve higher efficiencies and increase the economic viability of microalgae gasification.

4.4.3 Pyrolysis

Pyrolysis is a thermochemical process that breaks down microalgae biomass into useful biofuels and chemicals by heating them in the absence of oxygen. Typically, between 300 and 700 degrees Celsius, this process results in the production of bio-oil, syngas, and biochar. Each type is used in the energy and materials sector. Recent studies have concentrated on optimizing the pyrolysis parameters, such as temperature, heating rate, and the use of catalysts to enhance the yield and quality of bio-oil. It is regarded a viable alternative to petroleum-based fuels, with special attention being made to catalytic pyrolysis for its capacity to improve the breakdown of long-chain hydrocarbons acquired which raises the caloric value of the bio-oil and reduces the oxygen content (Abdullah et al., 2021).

In addition, advances in reactor design such as fluidized bed reactors are being developed to allow greater heat transfer and mixing. It allows for more efficient conversion of microalgae biomass into pyrolysis products with other microalgae biomass pyrolysis ingredients, such as agricultural waste or plastic waste. It has shown promise in improving bio-oil quality and increasing overall energy recovery (Nam et al., 2022).The quality of bio-oil can still be significantly impacted by issues like high water and nitrogen content in microalgae. The efficiency and economics of the pyrolysis process are being improved by pre-treatment techniques such as dewatering, drying, and nutrient optimization

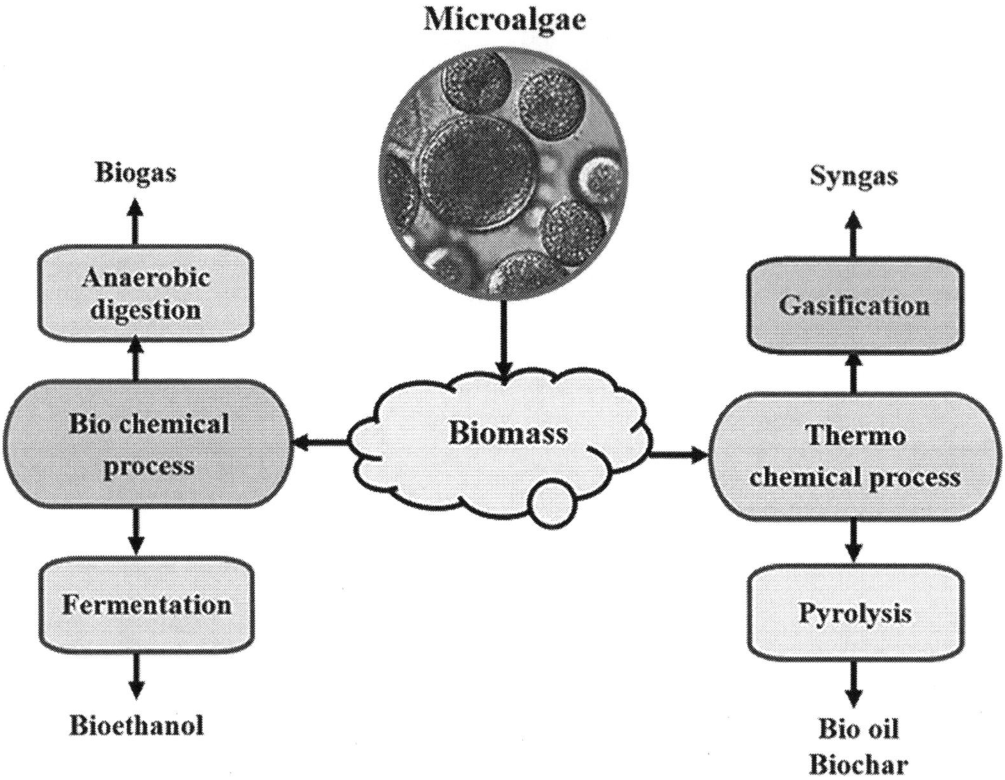

FIGURE 4.2 Processing techniques for energy and value-added products from microalgae biomass.

during cultivation (Prado et al., 2020).The aforementioned developments suggest that pyrolysis is a viable approach for the sustainable production of energy from microalgae; nonetheless, additional investigation is necessary to optimize its economic prospects (Figure 4.2 and Table 4.2).

4.5 MICROALGAE BIOREFINERY

Microalgae biorefineries are a viable and sustainable technique to produce biofuels, chemicals, and a wide range of high-value products from microalgae biomass. This is different from general biorefinery that relies on terrestrial crops; microalgae have many advantages, such as their rapid growth rate, high photosynthesis efficiency, and the ability to grow non-arable land using non-potable water. Because of these features, microalgae are a promising feedstock for the integrated production of value-added products like omega-3 fatty acids, antioxidants, pigments, and bioplastics, as well as biofuels (such as biodiesel, bioethanol, and biogas) (Rani et al., 2020).

Recent advances in microalgae biorefinery technology have focused on developing comprehensive processes that use fractions of the entire microalgae biomass to maximize economic returns and reduce waste. Including boosting the efficiency of farming systems to improve fat, protein, and carbohydrate production as well as the application of new extraction conversion procedures to recover beneficial substances and biofuels. It also comprises integrated processes that incorporate biochemical and thermochemical conversion, such as AD and HTL, and that have shown the ability to boost energy recovery and lower overall processing costs (Nagarajan et al., 2023).Moreover, the utilization of new innovative solvents and enzymatic techniques is being researched for effective extraction (Xu et al., 2021).

TABLE 4.2
Various Processing Technology, Pros, Cons, and Challenges

S. No	Processing Technology	Pros	Cons	Challenges	References
4.1.	**Biochemical conversion**				
4.1.1.	Anaerobic digestion	Effective for converting organic material to biogas (methane) and digestate	Slow process and requires a large infrastructure	Optimizing microbial activity and reducing process time	(Sialve et al., 2009)
4.1.2.	Fermentation	Produces bioethanol and other bio-based products	Limited to specific types of algae; high cost of enzymes	Improving fermentation efficiency and lowering enzyme costs	(Ho et al., 2013)
4.2.	**Thermochemical conversion**				
4.2.1.	Gasification	Converts biomass into syngas, which can be used for energy production	High temperatures required, energy-intensive	Reducing energy input while maintaining efficiency	(Goyal et al., 2008)
4.2.2.	Pyrolysis	Produces bio-oil, biochar, and syngas at lower temperatures than gasification	Requires high-quality biomass input	Managing the production of consistent-quality bio-oil	(Chen et al., 2015)

Despite these developments, there are still challenges with scalability, high capital and OPEX, and changing biomass composition. The continuing study intends to solve these issues by developing novel biorefinery models that harness advancements in cultivation processing and product recovery to realize the promise of microalgae as a diverse feedstock for the bio-economy.

4.5.1 Biorefinery Models Self-Sustainable

The self-sustaining microalgae biorefineries model is designed to enhance resource efficiency and decrease waste. This is in keeping with the ideas of circular bio-economics. These models integrate a number of conversion processes to completely use fractions of the total microalgae biomass to produce a range of products, such as biofuels, high-value chemicals, and nutraceuticals, while also recycling nutrients and energy within the system. Recent improvements in recycled self-sustainable biorefinery models stress the need of closed-loop systems whereby products such as carbon dioxide, wastewater, and residual biomass are absorbed back into the production cycle. These systems minimize environmental effect and reduce external inputs (Wang et al., 2018; Rawat et al., 2013).

One possible option is to combine microalgae cultivation with wastewater treatment and carbon dioxide collection. The model eliminates the need for synthetic fertilizers by growing microalgae in nutrient-rich wastewater. Additionally, it boosts algae production while capturing carbon dioxide emissions from industrial processes. Biogas, bio-oil, and bioactive compounds can be extracted from the biomass using integrated biorefinery processes such pyrolysis, AD, and enzymatic hydrolysis. This guarantees the efficient utilization of all fractions (Rajesh Banu et al., 2021). Optimizing growing conditions, reducing overall energy usage, and exhibiting potential to improve production can be achieved employing modern cultivation systems, such as bioreactors integrated with LED lights and real-time monitoring (Chisti, 2007; Gangadhar et al., 2023).

Microalgae biorefinery presents both economic and environmental issues; these self-sustaining models find solutions by making processes more efficient and using less outside resources. The high starting costs and technological complexity make it difficult to scale these models from lab to industrial scale. In order to be widely used and financially effective, self-sustaining microalgae biorefineries must keep innovating in reactor design, process integration, and resource recovery.

4.5.2 Microalgae Biorefineries Sustainability Assessment

To determine if a microalgae biorefinery is sustainable, it is necessary to conduct a thorough evaluation utilizing multiple frameworks, including the life cycle assessment (AoLC). In order to reduce emissions, energy consumption, and resource consumption, the AoLC framework examines the environmental impacts associated with the microalgae biorefinery process at every stage, from cultivation to product conversion (Singh et al., 2019; Amer et al., 2011; Lardon et al., 2009).Microalgae biorefineries can be better understood and optimized for cost-effectiveness with the help of economic analysis technical (EAT), which focuses on these viability issues by analyzing capital and OPEX, technological performance, and market dynamics (Tariq et al., 2022; Show et al., 2017). Economic analysis social (EAS) assesses the social ramifications, such as employment creation, community benefits, and societal acceptance, to guarantee that the biorefinery promotes social sustainability and impacts neighboring communities Chia et al., 2018).As a whole, these models explain microalgae biorefineries' long-term viability and can guide the development of greener, more cost-effective methods of producing biofuel, and other bio-products.

4.5.3 Assessment of Life Cycle (AoLC)

The AoLC assesses the environmental impact of all stages of production and the transition to biofuels and bio-products, from farm to processing and final use. Microalgae have advantages over conventional biofuels, such as higher yields per hectare and the ability to grow in non-agricultural areas but high energy consumption is challenged, especially during farming. Open pond cultivation is less expensive but requires more water, whereas PBRs require more energy for CO_2 supply, lighting, and temperature control despite being more controlled. Carbon footprints can be decreased and nutrients can be supplied by combining microalgae production with wastewater treatment or CO_2 collection from industrial emissions (Lardon et al., 2009).Combining coagulation with membrane filtration has the potential to reduce energy use. Conversion processes such as HTL can handle wet biomass and it helps reduce the energy-intensive drying process, while co-products and nutrients can be recycled to improve overall sustainability (Amer et al., 2011).By combining these processes and renewable energy sources to increase efficiency in microalgae, biorefinery can increase environmental and economic viability. This makes it a promising avenue for sustainable biofuel production (Ruiz et al., 2016; Patil et al., 2009). (Figure 4.3).

4.5.4 Economic Analysis Technical (EAT)

The EAT of microalgae biorefineries focuses on evaluating the technical aspects of economic feasibility, including capital expenditures (CAPEX), OPEX, process efficiency, and market competitiveness of various technologies used in the production of biofuels and bio-products from microalgae. A key component of this analysis is the costs associated with the farming system, harvesting and dewatering technologies, and downstream processing techniques. According to recent research, there is still a significant financial barrier to the sustainable production of biofuels from microalgae biorefineries despite their promising potential (Kumar et al., 2010; Khoo et al., 2011). These barriers include the high cultivation costs, especially in PBRs, and the energy-intensive harvesting and drying processes. Research focuses on optimizing cultivating strategies to reduce CAPEX and OPEX to improve economic viability. For example, integrating microalgae production with wastewater treatment can significantly reduce costs. It provides a nutrient-rich growing medium and reduces the need for synthetic fertilizers. Additionally, co-location with an industrial plant allows it to capture CO_2 emissions to increase growth and it also improves the cost-effectiveness of microalgae cultures (Zhu, 2015). Advances in low-energy harvesting techniques such as flocculation and electrocoagulation and the development of innovative dewatering technology have also been shown to reduce operating costs associated with biomass recovery (Salim et al., 2011).Additionally, the efficiency of

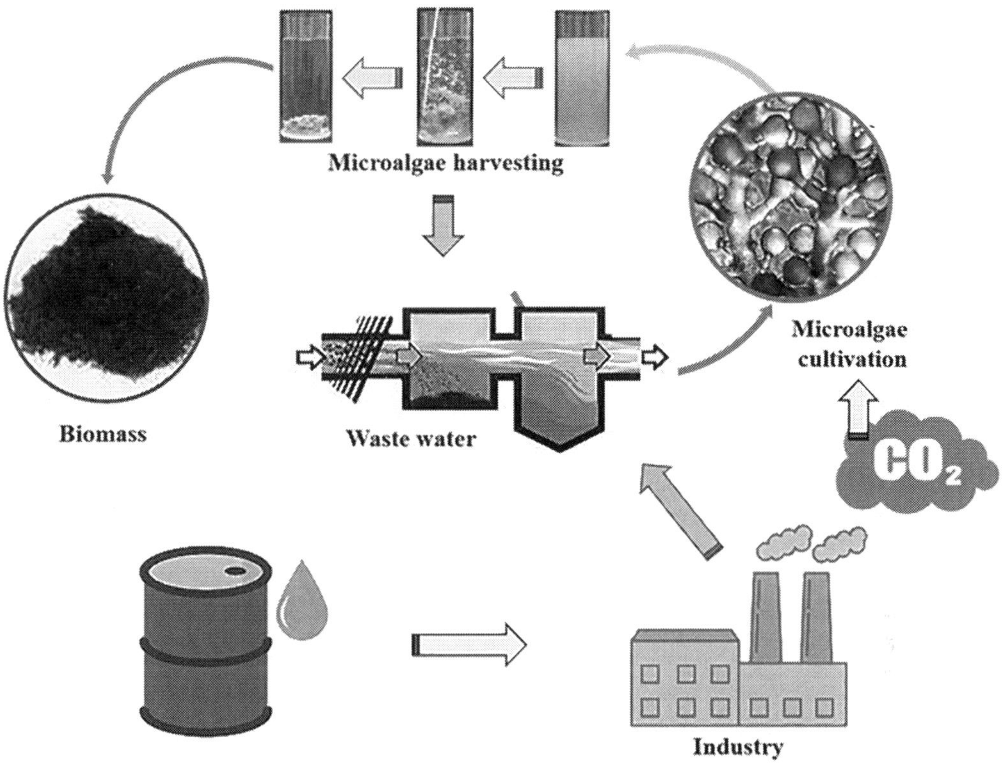

FIGURE 4.3 Assessment of life cycle biofuel from microalgae biomass.

conversion processes such as HTL and pyrolysis can be improved, which can directly process wet biomass without the need for energy-intensive drying. It can greatly reduce energy consumption and costs. The biorefinery model's overall economic viability can be improved by utilizing co-products like biochar and extracted high-value compounds (such as proteins and pigments) to create new revenue streams (Zhou et al., 2010).However, there are difficulties in preserving product quality and consistency, optimizing the process, and lowering equipment costs when bringing these technologies from the laboratory to the commercial scale. Future economic analysis will focus on the integration of technological innovation. Process optimization for joint product evaluation and exploring market opportunities to make microalgae biorefinery more competitive in the global bio-economy.

4.5.5 Economic Analysis Social (EAS)

The primary objective of their EAS is to assess the growing and operating microalgae biorefineries' societal benefits and costs. Benefits to public health, energy security, rural development, social inclusion, and employment are some of the aspects that are taken into account in this research. From cultivation and harvesting through processing and distribution, microalgae biorefineries have the ability to generate direct and indirect job opportunities, which might greatly benefit the local economy. New research shows that rural areas might benefit economically from growing microalgae. In regions where farmland and water are scarce, it can help bring people together and offer a living wage for farmers and laborers (Chia et al., 2018).

Additionally, microalgae biorefineries may lessen dependence on fossil fuels and increase diversity of energy sources, both of which contribute to better energy security. An increase in energy independence can be achieved by areas through the conversion of microalgae into biofuels. This helps keep society stable and resilient in the face of unpredictable fuel prices around the world. Better nutrition and health outcomes for local communities are additional social benefits of co-producing high-value products including proteins, pigments, and omega-3 fatty acids (Kumar et al., 2010; Singh et al., 2019).

However, there are still obstacles to overcome challenges with cost scalability, public acceptance, and significant initial investment. To tackle these issues, we need to raise people's consciousness about the positive effects microalgae biorefineries have on society and the environment and rally behind laws that encourage the study, creation, and implementation of environmentally friendly technology. Social acceptability and equitable benefit distribution can both be enhanced by involving local stakeholders in decision-making processes (Saratale et al., 2021).

By integrating environmental and social concerns, microalgae biorefinery has the potential to be an effective instrument for sustainable development.

4.5.6 Open Loop and Closed Loop Approach

The efficiency, sustainability, and economic feasibility of microalgae biorefineries are affected by whether they are categorized as closed-loop or open-loop systems, which are determined by the principles of design and operation. Microalgae biorefineries employ an open-loop technique that uses biomass to a partial extent while discarding some by-products and waste streams. This approach frequently leads to inefficiencies and environmental impact. On the other hand, a closed-loop system is based on the circular economy's tenets and seeks to recycle and repurpose all waste streams and by-products in order to maximize resource use (Chia et al., 2018).

In order to create biofuels and other essential products, the open-loop biorefinery will grow, store, and process microalgae. Despite its low entry barrier and relative ease of implementation, this method isn't perfect when it comes to recycling residual biomass or other by-products. Environmental issues, such as eutrophication and the emission of greenhouse gases, can arise from the disposal of waste, such as surplus nutrients or used biomass. Despite their potential usefulness in pilot projects, recent research has shown that open-loop systems are inefficient and expensive to run on a large scale since they don't recover energy or nutrients (Xu et al., 2021; Saratale et al., 2021).

In contrast, closed-loop biorefineries are designed to make the most of all available resources, recycle CO_2, H_2O, and nutrients back into the manufacturing process; for instance, the biomass residue following biofuel extraction can be utilized as a substrate for AD to produce biogas. After the AD process, the biorefinery can potentially obtain bio-fertilizer and energy. Reducing waste and dependency on external sources is the goal of this technique. Protecting the environment and ensuring the long-term viability of biorefineries, the latest innovations in closed-loop technology have the potential to integrate processes such as nutrient recycling, wastewater treatment, and carbon dioxide sequestration, ultimately leading to zero net emissions and increased profitability (Saratale et al., 2021; Sirohi et al., 2022).

Ultimately, a closed-loop system is more suited to achieve sustainability objectives and adhere to the principles of the circular economy, even though an open-loop system may provide easier implementation and cheaper initial expenses. Considerations of economic, environmental, and regulatory factors should be considered while deciding between these methods.

4.6 CHALLENGES AND PROSPECTS

4.6.1 Challenges

The sustainable production of biofuels and high-value goods could be achieved by microalgae biorefineries, but there are numerous obstacles that must be overcome before they can be economically viable and implemented on a large scale. One of the major restrictions, especially when employing a PBR, is the high production cost of microalgae cultivation. Although PBR is necessary for controlled growth conditions, it is expensive both initially and in the long run. The process of biomass concentration involves energy-intensive steps including harvesting and dewatering, which adds to the overall cost (Laurens et al., 2017).

There are a lot of problems with downstream processing as well. HTL, transesterification, and lipid extraction are a few of the difficult and expensive processes needed to transform microalgae biomass into biofuels and other goods. Improving these processes to increase yields with less energy and chemicals used still requires significant research. Recent studies have shown that cutting-edge methods, such as supercritical fluid extraction and microwave-assisted extraction, can boost productivity; however, these methods may necessitate further cost and scale reductions before they become economically viable (Laurens et al., 2017; Zuorro et al., 2022).In addition, the different species of microalgae have different biochemical compositions, which makes it difficult to standardize biorefinery processes. The species determines the unique requirements for optimal growth and the yields of compounds like lipids and proteins. Developing a cultivation and conversion procedure that works for everyone is quite challenging.

Finally, the commercialization of microalgae biorefinery is hindered by marketing and regulatory barriers. The lack of consistent regulations for microalgae products, along with the unpredictable nature of the biofuels and bio-products markets, discourages investment and slows down their widespread use. Making plans for policies that encourage sustainable practices is one approach to fixing these problems. Raising awareness via fostering partnerships between corporations and academic institutions and creating cutting-edge technological solutions (Thiruvenkadam et al., 2015).

4.6.2 Prospects

Sustainable energy generation, environmental protection, and resource utilization are three of the world's most pressing problems, and microalgae biorefineries could be the answer. The versatility of microalgae makes them a promising feedstock for the synthesis of biofuels, bio-chemicals, and other valuable products. Because of its incapacity to compete with food crops, its quick growth rate, and the fact that it can grow on non-arable terrain with non-potable water, microalgae provide significant benefits over traditional biomass sources (Borowitzka, 2013).

More efficient farming and harvesting methods have been the focus of recent technical developments aimed at lowering the production costs of microalgae biorefineries. One new development is a closed bioreactor system that uses LED lights and automatic monitoring to increase microalgae production while decreasing energy use (Chisti, 2007).Furthermore, by incorporating microalgae production into wastewater treatment and collecting carbon dioxide from industrial greenhouse gas emissions, we can reduce pollution, sequester carbon, and provide a nutrient-rich environment for algae growth raising the microalgae biorefinery's sustainability profile (Zhu, 2015).

Microalgae biorefineries are economically viable because they adopt a multi-product strategy that makes the most of available resources and generates income from a variety of sources. This method utilizes a single biomass source to extract bioactive chemicals, proteins, pigments, and biofuels (Ho et al., 2013). Expensive initial capital expenses, labor-intensive harvesting methods, and uneven biomass composition are some of the remaining concerns that require attention. Use the microalgae biorefinery to its maximum potential process improvement must be prioritized to resolve these difficulties.

4.7 CONCLUSION

In this chapter, we lay out a detailed plan for biorefining microalgae biomass and adding it to the circular economy. It begins by stressing the significance of renewable energy sources and continues by showcasing the possibilities of lipid-rich microalgae, which grow quickly, as a feedstock for biofuels. From PBRs and open ponds to flocculation, centrifugation, and filtration, this chapter covers it all when it comes to microalgae-based biorefinery and its role in the circular economy. It delves into important methods for processing microalgae biomass into useful byproducts like biodiesel, biogas, biofuels, and ethanol. This chapter also includes a model of a self-sustaining biorefinery that uses economic evaluation and life cycle analysis to determine sustainability. At last, the chapter delves into the pros and cons of microalgae biorefinery. It draws attention to potential avenues for future innovation in this field.

DECLARATION

The authors have not used any generative AI tools during the writing of this manuscript.

REFERENCES

Abdur Razzak, S., Bahar, K., Islam, K. M. O., Haniffa, A. K., Faruque, M. O., Hossain, S. M. Z., et al. (2024). Microalgae cultivation in photobioreactors: Sustainable solutions for a greener future. *Green Chemical Engineering, 5*(4), 418–439. https://doi.org/10.1016/j.gce.2023.10.004

Abdur, S., Bahar, K., Islam, K. M. O., Khaleel, A., Omar, M., Hossain, S. M. Z., et al. (2024). Green chemical engineering microalgae cultivation in photobioreactors: Sustainable solutions for a greener future. *Green Chemical Engineering, 5*(4), 418–439. https://doi.org/10.1016/j.gce.2023.10.004

Abomohra, A. E. F., & Almutairi, A. W. (2020). A close-loop integrated approach for microalgae cultivation and efficient utilization of agar-free seaweed residues for enhanced biofuel recovery. *Bioresource Technology, 317*, 124027. https://doi.org/10.1016/j.biortech.2020.124027

Ahmed, S. F., Inayat, A., & Kusumo, F. (2021). Microalgae biomass as a sustainable source for biofuel, biochemical and biobased value-added products: An integrated biorefinery concept. *Fuel, 307*, 121782.

Alavianghavanini, A., Shayesteh, H., Bahri, P. A., Vadiveloo, A., & Moheimani, N. R. (2024). Science of the total environment microalgae cultivation for treating agricultural effluent and producing value-added products. *Science of The Total Environment, 912*, 169369. https://doi.org/10.1016/j.scitotenv.2023.169369

Alberto, J., Costa, V., & de Morais, M. G. (2014). An Open Pond System for Microalgal Cultivation. In *Biofuels from Algae* (pp. 1–22). Elsevier B.V. http://dx.doi.org/10.1016/B978-0-444-59558-4.00001-2

Ali, S. S., Mastropetros, S. G., Schagerl, M., Sakarika, M., Elsamahy, T., El-Sheekh, M., et al. (2022). Recent advances in wastewater microalgae-based biofuels production: A state-of-the-art review. *Energy Reports, 8*, 13253–13280. https://doi.org/10.1016/j.egyr.2022.09.143

Augusta, M., Silvello, D. C., Gasparotto, G. A., & Goldbeck, R. (2023). Enzymatic hydrolysis of carbohydrate-rich *Chlorella vulgaris* for third-generation bioethanol production by cellulase-recombinant yeast. *Clean Chemical Engineering, 6*, 100111.

Babich, O., Ivanova, S., Michaud, P., Budenkova, E., Kashirskikh, E., Anokhova, V., et al. (2024). Fermentation of micro- and macroalgae as a way to produce value-added products. *Biotechnology Reports, 41*, e00827. https://doi.org/10.1016/j.btre.2023.e00827

Barros, A. I., Gonçalves, A. L., Simões, M., & Pires, J. C. M. (2015). Harvesting techniques applied to microalgae: A review. *Renewable and Sustainable Energy Reviews, 41*, 1489–1500.

Batan, L. Y., Graff, G. D., & Bradley, T. H. (2016). Techno-economic and Monte Carlo probabilistic analysis of microalgae biofuel production system. *Bioresource Technology, 219*, 45–52. http://dx.doi.org/10.1016/j.biortech.2016.07.085

Bernard, O. (2009). Policy analysis life-cycle assessment of biodiesel production from microalgae. *Environmental Science & Technology, 43*(17), 6475–6481.

Bilad, M. R., Arafat, H. A., & Vankelecom, I. F. J. (2014). Membrane technology in microalgae cultivation and harvesting: A review. *Biotechnology Advances, 32*, 1283–1300. http://dx.doi.org/10.1016/j.biotechadv.2014.07.008

Bridgwater, A. V. (2012). Review of fast pyrolysis of biomass and product upgrading. *Biomass and Bioenergy, 38*, 68–94. http://dx.doi.org/10.1016/j.biombioe.2011.01.048

Brown, R. C. (2011). Introduction to Thermochemical Processing of Biomass into Fuels, Chemicals, and Power. In *Thermochemical Processing of Biomass: Conversion into Fuels, Chemicals and Power* (pp. 1–12). Wiley.

Casanova, L. M., Brantes, L., Mendes, B., Souza, T. De, Bernardo, R., Joao, R. R., et al. (2023). Development of microalgae biodiesel: Current status and perspectives. *Microorganisms, 11*, 34.

Chanana, I., Kaur, P., Kumar, L., Kumar, P., & Kulshreshtha, S. (2023). Advancements in microalgal biorefinery technologies and their economic analysis and positioning in energy resource market. *Fermentation, 9*, 202.

Chanquia, S. N., & Kara, S. (2022). Photobioreactors for cultivation and synthesis: Specifications, challenges, and perspectives. *Engineering in Life Sciences, 22*(12), 712–724.

Chen, W., Lin, B., Huang, M., & Chang, J. (2014). Thermochemical conversion of microalgal biomass into biofuels: A review. *Bioresource Technology, 184*, 314–327. http://dx.doi.org/10.1016/j.biortech.2014.11.050

Chiaramonti, D., Prussi, M., Buffi, M., Maria, A., & Pari, L. (2017). Review and experimental study on pyrolysis and hydrothermal liquefaction of microalgae for biofuel production. *Applied Energy, 185*, 963–972. http://dx.doi.org/10.1016/j.apenergy.2015.12.001

Collet, P., Spinelli, D., Lardon, L., Steyer, J., & Bernard, O. (2014). Life-Cycle Assessment of Microalgal-Based Biofuels. In *Biofuels from Algae* (pp. 287–312). Elsevier.

Dammak, I., Fersi, M., Hachicha, R., & Abdelkafi, S. (2023). Current insights into growing microalgae for municipal wastewater treatment and biomass generation. *Resources, 12*, 119.

Daneshvar, E., Wicker, R. J., Show, P. L., & Bhatnagar, A. (2022). Biologically-mediated carbon capture and utilization by microalgae towards sustainable CO2 biofixation and biomass valorization – A review. *Chemical Engineering Journal, 427*, 130884. https://doi.org/10.1016/j.cej.2021.130884

de Morais, E. G., Sampaio, I. C. F., Gonzalez-Flo, E., Ferrer, I., Uggetti, E., & García, J. (2023). Microalgae harvesting for wastewater treatment and resources recovery: A review. *New Biotechnology, 78*, 84–94. https://doi.org/10.1016/j.nbt.2023.10.002

Deepa, P., & Sowndhararajan, K. (2023). A review of the harvesting techniques of microalgae. *Water, 15*, 3074.

Duan, P., Li, S., Jiao, J., Wang, F., & Xu, Y. (2018). Supercritical water gasification of microalgae over a two-component catalyst mixture. *Science of The Total Environment, 630*, 243–253. https://doi.org/10.1016/j.scitotenv.2018.02.226

Dudek, M., Grala, A., Dębowski, M., Zieliński, M., Krzemieniewski, M., Dudek, M., et al. (2012). Investigations into the use of algae for energy production have been carried. *Polish Journal of Natural Sciences, 27*(2), 151–164.

Ekinci, K., Erdal, I., Uysal, Ö., Uysal, F. Ö., Tunce, H., & Doğan, A. (2019). Anaerobic digestion of three microalgae biomasses and assessment of digestates as biofertilizer for plant growth. *Environmental Progress & Sustainable Energy, 38*(4), 131–139.

Ennaceri, H., Ishika, T., Mkpuma, V. O., & Moheimani, N. R. (2023). Microalgal biofilms: Towards a sustainable biomass production. *Algal Research, 72*, 103124. https://doi.org/10.1016/j.algal.2023.103124

Ennaceri, H., Mkpuma, V. O., & Moheimani, N. R. (2023). Nano-clay modified membranes: A promising green strategy for microalgal antifouling filtration. *Science of The Total Environment, 902*, 166479. https://doi.org/10.1016/j.scitotenv.2023.166479

Ezhumalai, G., Arun, M., Manavalan, A., Rajkumar, R., & Heese, K. (2024). A holistic approach to circular bioeconomy through the sustainable utilization of microalgal biomass for biofuel and other value added products. *Microbial Ecology, 87*(61). https://doi.org/10.1007/s00248-024-02376-1

Fal, S., Smouni, A., & El H. (2023). Integrated microalgae-based biorefinery for wastewater treatment, industrial CO2 sequestration and microalgal biomass valorization: A circular bioeconomy approach. *Environmental Advances, 12*, 100365. https://doi.org/10.1016/j.envadv.2023.100365

Fal, S., Smouni, A., & El H. (2023). Integrated microalgae-based biorefinery for wastewater treatment, industrial CO2 sequestration and microalgal biomass valorization: A circular bioeconomy approach. *Environmental Advances, 12*, 100365. https://doi.org/10.1016/j.envadv.2023.100365

Farooq, W. (2022). Maximizing energy content and CO2 bio-fixation efficiency of an indigenous isolated microalga *Parachlorella kessleri* HY-6 through nutrient optimization and water recycling during cultivation. *Frontiers in Bioengineering and Biotechnology, 9*, 804608.

Farzaneh, S., & Ahmad, K. (2022). Recent advances in the bioapplication of microalgaederived biochemical metabolites and development trends of photobioreactorbased culture systems. *3 Biotech, 12*, 260. https://doi.org/10.1007/s13205-022-03327-8

Francisco, J., Ribes, J., & Seco, A. (2024). Towards optimisation of microalgae cultivation through monitoring and control in membrane photobioreactor systems. *Water, 16*, 155.

Fuentes-Grünewald, C., Gayo-Peláez, J. I., Ndovela, V., Wood, E., Kapoore, R. V., Llewellyn, C. A., et al. (2020). Towards a circular economy: A novel microalgal two-step growth approach to treat excess nutrients from digestate and to produce biomass for animal feed. *Bioresource Technology, 320*, 124349. https://doi.org/10.1016/j.biortech.2020.124349

Gaur, S., Kaur, M., Kalra, R., Rene, E. R., & Goel, M. (2024). Application of microbial resources in biorefineries: Current trend and future prospects. *Heliyon, 10*(8), e28615. https://doi.org/10.1016/j.heliyon.2024.e28615

Gaurav, K., Neeti, K., & Singh, R. (2024). Microalgae-based biodiesel production and its challenges and future opportunities: A review. *Green Technologies and Sustainability, 2*(1), 100060. https://doi.org/10.1016/j.grets.2023.100060

Gaurav, K., Neeti, K., & Singh, R. (2024). Microalgae-based biodiesel production and its challenges and future opportunities: A review green technologies and sustainability microalgae-based biodiesel production and its challenges and future opportunities: A review. *Green Technologies and Sustainability, 2*(1), 100060. https://doi.org/10.1016/j.grets.2023.100060

González-Delgado, Á. D., Álvarez-Cordero, A. S., & De Ávila-Alvis, Y. P. (2022). Environmental evaluation of microalgae-based biodiesel production via in-situ transesterification. *Chemical Engineering Transactions, 92*, 571–576.

Gul, A., Hruza, J., & Yalcinkaya, F. (2021). Fouling and chemical cleaning of microfiltration membranes: A mini-review. *Polymers (Basel), 13*, 846.

Gurreri, L., Calanni, M., Luciano, A., Lima, S., Scargiali, F., Fino, D., et al. (2024). Environmental sustainability of microalgae-based production systems: Roadmap and challenges towards the industrial implementation. *Sustainable Chemistry and Pharmacy, 35*, 101191. https://doi.org/10.1016/j.scp.2023.101191

Hasan, M. M., Mo, M., Uddin, M. N., Kabir, Z., Badruddin, I. A., & Khan, T. M. Y. (2024). Insights into anaerobic digestion of microalgal biomass for enhanced energy recovery. *Energy Research, 12*, 1355686.

Huang, K. X., Vadiveloo, A., Zhou, J. L., Yang, L., & Chen, D. Z. G. F. (2023). Integrated culture and harvest systems for improved microalgal biomass production and wastewater treatment. *Bioresource Technology, 376*, 128941.

John, R. P., Anisha, G. S., Nampoothiri, K. M., & Pandey, A. (2011). Micro and macroalgal biomass: A renewable source for bioethanol. *Bioresource Technology, 102*(1), 186–193. http://dx.doi.org/10.1016/j.biortech.2010.06.139

Jothibasu, K., Dhar, D. W., & Rakesh, S. (2021). Biomass and bioenergy recent developments in microalgal genome editing for enhancing lipid accumulation and biofuel recovery. *Biomass and Bioenergy, 150*, 106093. https://doi.org/10.1016/j.biombioe.2021.106093

Katarzyna, L., Sai, G., & Singh, O. A. (2015). Non-enclosure methods for non-suspended microalgae cultivation: Literature review and research needs. *Renewable and Sustainable Energy Reviews, 42*, 1418–1427. http://dx.doi.org/10.1016/j.rser.2014.11.029

Khan, M. I., Shin, J. H., & Kim, J. D. (2018). The promising future of microalgae: Current status, challenges, and optimization of a sustainable and renewable industry for biofuels, feed, and other products. *Microbial Cell Factories, 17*(36), 1–21. https://doi.org/10.1186/s12934-018-0879-x

Kim, K., Hourfar, F., Halim, A., Abdul, B., Rizwan, M., Almansoori, A., et al. (2023). Importance of microalgae and municipal waste in bioenergy products hierarchy—Integration of biorefineries for microalgae and municipal waste processing: A review. *Energies, 16*, 6361.

Krishna, A., & Show, P. (2019). Bio-processing of algal bio-refinery: A review on current advances and future perspectives. *Bioengineered, 10*(1), 574–592. https://doi.org/10.1080/21655979.2019.1679697

Krohn, I., Menanteau-Ledouble, S., Hageskal, G., Astafyeva, Y., Jouannais, P., Nielsen, J. L., et al. (2022). Health benefits of microalgae and their microbiomes. *Microbial Biotechnology, 15*(7), 1966–1983.

Kuo, C., Sun, Y., Lin, C., Lin, C., & Wu, H. (2021). Cultivation and biorefinery of microalgae (Chlorella sp.) for producing biofuels and other byproducts: A review. *Sustainability, 13*, 13480.

Kuppan, P., Sudharsanam, A., Venkateswarlu, K., & Megharaj, M. (2023). Solar technology–closed loop synergy facilitates low-carbon circular bioeconomy in microalgal wastewater treatment. *npj Clean Water, 6*, 43.

Lee, J., Kwon, E. E., & Park, Y. (2020). Recent advances in the catalytic pyrolysis of microalgae. *Catalysis Today, 355*, 263–271. https://doi.org/10.1016/j.cattod.2019.03.010

Leong, H. Y., Chang, C. K., Khoo, K. S., Chew, K. W., Chia, S. R., Lim, J. W., et al. (2021). Waste biorefinery towards a sustainable circular bioeconomy: A solution to global issues. *Biotechnology for Biofuels, 14*, 87. https://doi.org/10.1186/s13068-021-01939-5

Li, X., Wang, Z., He, Y., Wang, Y., Wang, S., Zheng, Z., et al. (2024). A comprehensive review of the strategies to improve anaerobic digestion: Their mechanism and digestion performance. *Methane, 3*, 227–256.

Liu, T., Wang, J., Hu, Q., Cheng, P., Ji, B., Liu, J., et al. (2013). Bioresource technology attached cultivation technology of microalgae for efficient biomass feedstock production. *Bioresource Technology, 127*, 216–222. http://dx.doi.org/10.1016/j.biortech.2012.09.100

Liu, Z., Hao, N., Hou, Y., Wang, Q., Liu, Q., Yan, S., & Chen, F. Z. L. (2023). Technologies for harvesting the microalgae for industrial applications: Current trends and perspectives. *Bioresource Technology, 357*, Article 127630.

Mahmood, T., Hussain, N., Shahbaz, A., Mulla, S. I., Iqbal, H. M. N., & Bilal, M. (2023). Sustainable production of biofuels from the algaederived biomass. *Bioprocess and Biosystems Engineering, 46*(8), 1077–1097. https://doi.org/10.1007/s00449-022-02796-8

Mishra, S., Roy, M., & Mohanty, K. (2019). Microalgal bioenergy production under zero-waste biorefinery approach: Recent advances and future perspectives. *Bioresource Technology, 292*, 122008. https://doi.org/10.1016/j.biortech.2019.122008

Mohsenpour, S. F., Hennige, S., Willoughby, N., Adeloye, A., & Gutierrez, T. (2021). Integrating micro-algae into wastewater treatment: A review. *Science of The Total Environment, 752*, 142168. https://doi.org/10.1016/j.scitotenv.2020.142168

Molina Grima, E., Belarbi, E. H., Acién Fernández, F. G., Robles Medina, A., & Chisti, Y. (2003). Recovery of microalgal biomass and metabolites: Process options and economics. *Biotechnology Advances, 20*, 491–515.

Moshood, T. D., Nawanir, G., & Mahmud, F. (2021). Microalgae biofuels production: A systematic review on socioeconomic prospects of microalgae biofuels and policy implications. *Environmental Challenges, 5*, 100207. https://doi.org/10.1016/j.envc.2021.100207

Muñoz-P, K. M., Robledo-Narvaez, P. N., & Edgar, V. (2023). Engineering the Metabolic Landscape of Microorganisms for Lignocellulosic Conversion. *Microorganisms, 11*, 2197.

Najjar, Y. S. H., & Abu-Shamleh, A. (2020). Harvesting of microalgae by centrifugation for biodiesel production: A review. *Algal Research, 51*, 102046. https://doi.org/10.1016/j.algal.2020.102046

Narayanan, M. (2024). Biorefinery products from algal biomass by advanced biotechnological and hydrothermal liquefaction approaches. *Discover Applied Sciences, 6*(146). https://doi.org/10.1007/s42452-024-05777-6

Novovesk, L., Nielsen, S. L., Tufan, O., Haznedaroglu, B. Z., Fazi, S., Robbens, J., et al. (2023). Overview and challenges of large-scale cultivation of photosynthetic microalgae and cyanobacteria. *Marine Drugs, 21*, 445.

Okeke, E. S., Ejeromedoghene, O., Okoye, C. O., Ezeorba, T. P. C., Nyaruaba, R., Ikechukwu, C. K., et al. (2022). Microalgae biorefinery: An integrated route for the sustainable production of high-value-added products. *Energy Conversion and Management X, 16*, 100323. https://doi.org/10.1016/j.ecmx.2022.100323

Olabi, A. G., Alami, A. H., Alasad, S., Aljaghoub, H., Sayed, E. T., Shehata, N., et al. (2022). Emerging technologies for enhancing microalgae biofuel production: Recent progress, barriers, and limitations. *Fermentation, 8*, 649.

Olguín, E. J., Gloria, S., Arias-Olguín, I., Melo, F. J., González, R. E., Cruz, L., et al. (2022). Microalgae-based biorefineries: Challenges and future trends to produce carbohydrate enriched biomass, high-added value products and bioactive compounds. *Biology (Basel), 11*, 1146.

Oliva, G., Gino, M., Buonerba, A., Hasan, S. W., Belgiorno, V., Naddeo, V., et al. (2023). Case studies in chemical and environmental engineering carbon capture and utilization in waste to energy approach by leading-edge algal photo-bioreactors: The influence of the illumination wavelength. *Case Studies in Chemical and Environmental Engineering, 7*, 100348. https://doi.org/10.1016/j.cscee.2023.100348

Omokaro, G. O., & Nafula, Z. S. (2023). Microalgae – Harnessing environmental and nature benefits – A comprehensive review. *American Journal of Agricultural Science Engineering Technology, 7*(3), 47–56.

Onyeaka, H., Miri, T., Obileke, K., Hart, A., Anumudu, C., & Al-Sharify, Z. T. (2021). Minimizing carbon footprint via microalgae as a biological capture. *Carbon Capture Science and Technology, 1*, 100007.

Pan, Z., Sun, X., Huang, Y., Liang, T., Lu, J., Zhang, L., et al. (2024). Anaerobic co-digestion of food waste and microalgae at variable mixing ratios: Enhanced performance, kinetic analysis, and microbial community dynamics investigation. *Applied Sciences, 14*, 4387.

Penloglou, G., Pavlou, A., & Kiparissides, C. (2024). Recent Advancements in photo-bioreactors for microalgae cultivation: A brief overview. *Processes, 12*, 1104.

Pereira, A. S. A. P., Silva, T. A. D., Magalhães, I. B., Ferreira, J., Braga, M. Q., Lorentz, J. F., Assemany, P. P., & Couto, E. A. D. C. M. (2024). Biocompounds from wastewater-grown microalgae: A review of emerging cultivation and harvesting technologies. *Science of The Total Environment, 920*, 170918.

Politaeva, N., Ilin, I., Velmozhina, K., & Shinkevich, P. (2023). Carbon dioxide utilization using Chlorella microalgae. *Environments, 10*, 109.

Prieto-Sandoval, V., Jaca, C., & Ormazabal, M. (2018). Towards a consensus on the circular economy. *Journal of Cleaner Production, 179*, 605–615. https://doi.org/10.1016/j.jclepro.2017.12.224

Qin, S., Wang, K., Gao, F., Ge, B., Cui, H., & Li, W. (2023). Biotechnologies for bulk production of microalgal biomass: From mass cultivation to dried biomass acquisition. *Biotechnology for Biofuels and Bioproducts, 16*, 131. https://doi.org/10.1186/s13068-023-02382-4

Raheem, A., Ji, G., Memon, A., Sivasangar, S., Wang, W., & Zhao, M. (2018). Catalytic gasification of algal biomass for hydrogen-rich gas production: Parametric optimization via central composite design. *Energy Conversion and Management Journal, 158*, 235–245.

Raj, T., Morya, R., Chandrasekhar, K., Kumar, D., Soam, S., Kumar, R., & Patel, A. K. K. S. (2023). Microalgae biomass deconstruction using green solvents: Challenges and future opportunities. *Bioresource Technology, 69*(128429), 128429. https://doi.org/10.1038/s41598-023-29883-4

Rajesh Banu, J., Preethi, Kavitha, S., Gunasekaran, M., & Kumar, G. (2020). Microalgae based biorefinery promoting circular bioeconomy—Techno economic and life-cycle analysis. *Bioresource Technology, 302*, 122822. https://doi.org/10.1016/j.biortech.2020.122822

Rao, N. R. H., Beyer, V. P., Henderson, R. K., Thielemans, W., & M. K. (2023). Microalgae harvesting using flocculation and dissolved air flotation: Selecting the right vessel for lab-scale experiments. *Bioresource Technology, 374*, 128786. https://doi.org/10.1016/j.biortech.2020.122998

Rodriguez, C., Alaswad, A., Mooney, J., Prescott, T., & Olabi, A. G. (2015). Pre-treatment techniques used for anaerobic digestion of algae. *Fuel Processing Technology, 138*, 765–779. http://dx.doi.org/10.1016/j.fuproc.2015.06.027

Rossi, S., Mantovani, M., Marazzi, F., Bellucci, M., Casagli, F., Mezzanotte, V., et al. (2023). Microalgal cultivation on digestate: Process efficiency and economics. *Chemical Engineering Journal, 460*, 141753. https://doi.org/10.1016/j.cej.2023.141753

Rumin, J., de Oliveira Junior, R. G., Bérard, J. B., & Picot, L. (2021). Improving microalgae research and marketing in the European Atlantic Area: Analysis of major gaps and barriers limiting sector development. *Marine Drugs, 19*, 319.

Salim, S., Bosma, R., Vermuë, M. H., & Wijffels, R. H. (2011). Harvesting of microalgae by bio-flocculation. *Journal of Applied Phycology, 23*(5), 849–855.

Samoraj, M., Çalış, D., Trzaska, K., Mironiuk, M., & Chojnacka, K. (2024). Advancements in algal biorefineries for sustainable agriculture: Biofuels, high-value products, and environmental solutions. *Biocatalysis and Agricultural Biotechnology, 58*, 103224. https://doi.org/10.1016/j.bcab.2024.103224

Sankaran, R., Andres, R., Cruz, P., Pakalapati, H., Show, P. L., Ling, T. C., et al. (2019). Recent advances in the pretreatment of microalgal and lignocellulosic biomass: A comprehensive review. *Bioresource Technology, 298*, 122476. https://doi.org/10.1016/j.biortech.2019.122476

Sarma, S., Sharma, S., Rudakiya, D., Upadhyay, J., Rathod, V., & Patel, A. (2021). Valorization of microalgae biomass into bioproducts promoting circular bioeconomy: A holistic approach of bioremediation and biorefinery. *3 Biotech, 11*(8), 1–29. https://doi.org/10.1007/s13205-021-02911-8

Shiong, K., Ahmad, I., Wayne, K., Iwamoto, K., & Bhatnagar, A. (2023). Enhanced microalgal lipid production for biofuel using different strategies including genetic modification of microalgae: A review. *Progress in Energy and Combustion Science, 96*, 101071. https://doi.org/10.1016/j.pecs.2023.101071

Shuba, E. S., & Kifle, D. (2018). Microalgae to biofuels: 'Promising' alternative and renewable energy, review. *Renewable and Sustainable Energy Reviews, 81*, 743–755.

Singh, I., Pandey, A., Shangdiar, S., Rai, P. K., Kumar, A., Amesho, K. T. T., et al. (2023). Towards sustainable energy: Harnessing microalgae biofuels for a greener future. *Sustainability, 15*, 14029.

Sivaramakrishnan, R., Suresh, S., Kanwal, S., & Ramadoss, G. (2022). Microalgal biorefinery concepts' developments for biofuel and bioproducts: Current perspective and bottlenecks. *International Journal of Molecular Sciences, 23*, 2623.

Stiles, W. A. V., Styles, D., Chapman, S. P., Esteves, S., Melville, L., Silkina, A., et al. (2018). Using microalgae in the circular economy to valorise anaerobic digestate: Challenges and opportunities. *Bioresource Technology, 267*, 732–742. https://doi.org/10.1016/j.biortech.2018.07.100

Sundaram, T., Rajendran, S., Gnanasekaran, L., Rachmadona, N., Jiang, J. J., Khoo, K. S., et al. (2023). Bioengineering strategies of microalgae biomass for biofuel production: Recent advancement and insight. *Bioengineered, 14*(1), 2252228. https://doi.org/10.1080/21655979.2023.2252228

Tan, J. S., Lee, S. Y., Chew, K. W., Lam, M. K., Lim, J. W., Ho, S. H., et al. (2020). A review on microalgae cultivation and harvesting, and their biomass extraction processing using ionic liquids. *Bioengineered, 11*(1), 116–129. https://doi.org/10.1080/21655979.2020.1711626

Tan, X. B., Lam, M. K., Uemura, Y., Lim, J. W., Wong, C. Y., & Lee, K. T. (2018). Cultivation of microalgae for biodiesel production: A review on upstream and downstream processing. *Chinese Journal of Chemical Engineering, 26*, 17–30.

Taylor, P., Pacheco, M. M., Hoeltz, M., Moraes, M. S. A., & Schneider, R. C. S. (2015). Microalgae: Cultivation techniques and wastewater phycoremediation. *Journal of Environmental Science and Health, Part A, 50*(6), 585–601.

Torres, A., Padrino, S., Brito, A., & Díaz, L. (2023). Biogas production from anaerobic digestion of solid microalgae residues generated on different processes of microalgaetobiofuel production. *Biomass Conversion and Biorefinery, 13*, 4659–4672. https://doi.org/10.1007/s13399-021-01898-9

Van Lal, M., Ruatpuia, J. V. L., Ao, S., Chumpi, A., Belur, K., & Lalthazuala, S. (2023). Microalgae as a sustainable feedstock for biodiesel and other production industries: Prospects and challenges. *Energy Nexus, 12*, 100255. https://doi.org/10.1016/j.nexus.2023.100255

Vandamme, D., Foubert, I., & Muylaert, K. (2013). Flocculation as a low-cost method for harvesting microalgae for bulk biomass production. *Trends in Biotechnology, 31*(4), 233–239. http://dx.doi.org/10.1016/j.tibtech.2012.12.005

Vasistha, S., Khanra, A., Clifford, M., & Rai, M. P. (2020). Current advances in microalgae harvesting and lipid extraction processes for improved biodiesel production: A review. *Renewable and Sustainable Energy Reviews, 137*, 110498. https://doi.org/10.1016/j.rser.2020.110498

Veerabadhran, M. (2021). Anaerobic digestion of microalgal biomass for bioenergy production, removal of nutrients and microcystin: Current status. *Journal of Applied Microbiology, 131*, 1639–1651.

Venkata Mohan, S., Hemalatha, M., Chakraborty, D., Chatterjee, S., Ranadheer, P., & Kona, R. (2020). Algal biorefinery models with self-sustainable closed loop approach: Trends and prospective for blue-bioeconomy. *Bioresource Technology, 295*, 122128. https://doi.org/10.1016/j.biortech.2019.122128

Vuppaladadiyam, A. K., Prinsen, P., Raheem, A., Luque, R., & Zhao, M. (2018). Sustainability analysis of microalgae production systems: A review on resource with unexploited high-value reserves. *Environmental Science & Technology, 52*(24), 14031–14049.

Wang, M., Ye, X., Bi, H., & Shen, Z. (2024). Microalgae biofuels: Illuminating the path to a sustainable future amidst challenges and opportunities. *Biotechnology for Biofuels and Bioproducts, 17*(10). https://doi.org/10.1186/s13068-024-02461-0

Ward, A. J., Lewis, D. M., & Green, F. B. (2014). Anaerobic digestion of algae biomass: A review. *Algal Research, 5*(1), 204–214. http://dx.doi.org/10.1016/j.algal.2014.02.001

Wei, Y., Wu, Q., Meng, H., Zhang, Y., & Cao, C. (2023). Recent advances in photocatalytic self-cleaning performances of TiO2-based building materials. *RSC Advances, 13*(30), 20584–20597.

Wiatrowski, M., Klein, B. C., Davis, R. W., Quiroz-Arita, C., Tan, E. C. D., Hunt, R. W., et al. (2022). Techno-economic assessment for the production of algal fuels and value-added products: Opportunities for high-protein microalgae conversion. *Biotechnology for Biofuels and Bioproducts, 15*(8). https://doi.org/10.1186/s13068-021-02098-3

Yadav, G., & Sen, R. (2018). Sustainability of Microalgal Biorefinery: Scope, Challenges, and Opportunities. In *Green Energy and Technology* (pp. 335–351). MDPI.

You, N., Deng, S., Wang, C., Ngo, H. H., Wang, X., Yu, H., et al. (2023). Review and opinions on the research, development and application of microalgae culture technologies for resource recovery from wastewater. *Water, 15*, 1192.

Zabochnicka, M., Krzywonos, M., Romanowska-Duda, Z., Szufa, S., Darkalt, A., & Mubashar, M. (2022). Algal biomass utilization toward circular economy. *Life, 12*, 1480.

Zhang, B., & Wei, Z. (2024). Flocculation with intermittent dosing for enhanced microalgae harvesting. *Separation and Purification Technology, 330*, 125445.

Zhao, Z., Muylaert, K., & Vankelecom, I. F. J. (2023). Applying membrane technology in microalgae industry: A comprehensive review. *Renewable and Sustainable Energy Reviews, 172*, 113041.

Ziyue, T., Chen, W., Chen, Y., Yang, H., & Chen, H. (2018). Co-pyrolysis of microalgae and plastic: Characteristics and interaction effects. *Bioresource Technology, 274*, 145–152. https://doi.org/10.1016/j.biortech.2018.11.083

5 Role of Brown Algae in Climate Change Mitigation

Fayaz A. Malla, Suhaib A. Bandh, Afaan A. Malla, and Showkat Rashid

5.1 INTRODUCTION

Coastal and intertidal zones around the globe are home to brown algae, which are part of the Phaeophyceae family of marine macroalgae. Fucoxanthin, a brown pigment found in these algae, allows them to efficiently photosynthesize in marine conditions with low light levels by hiding the green hue of chlorophyll (Lammerant *et al.*, 2023). The rapid growth, large carbon sequestration capacities, and possible biofuel generation of species such as *Ascophyllum nodosum* and *Macrocystis pyrifera* have attracted a lot of research interest. Brown algae play a crucial role in reducing the impact of climate change by their ability to adapt and endure. They can act as carbon sinks and provide bioeconomy applications (Sultana *et al.*, 2023, Malla *et al.*, 2024). An important part of marine ecosystem regulation is the presence of kelp forests, which are made up of brown algae. According to Dominguez Almela *et al.* (2023), these ecosystems trap as much as 50 times more carbon dioxide than land-based forests. In addition to providing a haven for a wide variety of marine animals, kelp forests boost marine production and increase biodiversity. To add to the stabilization of marine ecosystems in response to climate change, brown algae play a role in nutrient cycling and act as a buffer against the acidification of the ocean (Zhang *et al.*, 2024).

The objective of this chapter is to investigate brown algae's ecological, industrial, and economic functions in order to determine their potential to reduce the effects of climate change. Capabilities for carbon sequestration, bioeconomy contributions, and sustainable development implications are all part of this discussion. Evaluating new policy frameworks and technological developments that might improve the efficiency of brown algae in combating climate change is also within the purview of this project.

5.2 BIOLOGICAL AND ECOLOGICAL CHARACTERISTICS OF BROWN ALGAE

5.2.1 Taxonomy and Distribution

Multicellular marine macroalgae commonly found in coastal habitats around the world, brown algae (Phaeophyceae) are a subclass of the phylum Ochrophyta. Fucoxanthin, a brown pigment, is their defining characteristic; it helps them absorb light for photosynthesis, even in low-light underwater settings (Lammerant *et al.*, 2023). About 2,000 species of brown algae are categorized into orders such as Fucales (including *Fucus vesiculosus*), Laminariales (including *M. pyrifera*), and Dictyotales (including *Dictyota dichotoma*). Brown algae predominate in benthic areas in cooler seas, where they form extensive kelp forests. A vast variety of marine creatures rely on these environments for food and shelter, making them essential to biodiversity. Most notably, *M. pyrifera*, the

DOI: 10.1201/9781003685814-5

biggest brown algae, creates enormous underwater forests that have a major role in the dynamics of carbon in the ocean (Dominguez Almela *et al.*, 2023).

5.2.2 Photosynthetic Efficiency and Carbon Sequestration

The distinctive light-harvesting systems of brown algae allow them to perform photosynthesis with remarkable efficiency. These algae may survive in low-light situations, including murky water or deeper water, because fucoxanthin increases light absorption (Zhang *et al.*, 2024). One of their key roles in oceanic carbon sequestration is photosynthesis, which converts atmospheric CO_2 into organic compounds. According to Sultana *et al.* (2023), kelp forests may sequester 600–1,200 g C/m^2/year, demonstrating the incredible carbon sequestration potential of brown algae. Brown algae, in contrast to plants on land, store a large amount of their organic carbon in the deep ocean, where it stays for generations. Brown algae play a crucial part in reducing atmospheric CO_2 levels and so mitigating climate change. Their "blue carbon" storage capability is a prime example of this.

5.2.3 Adaptations to Marine Environments

Brown algae have adapted in many ways, both structurally and physiologically, to survive and even thrive in the harsh and ever-changing sea habitat. The formation of holdfasts, which function as anchors for the algae to cling to substrates and endure powerful ocean currents, is an important adaptation. Furthermore, numerous species possess air bladders and pliable stipes, which aid in buoyancy and maximize light absorption for photosynthesis. Additionally, brown algae can withstand changes in salinity and nutrient variability, and their quick growth is supported by their efficient nutrient uptake systems. found that certain species can grow as much as 50 cm every day, which helps them dominate marine ecosystems. In addition, they are resistant to environmental stresses like herbivory and desiccation because their cell walls are abundant with polysaccharides like alginate, which serve as structural support.

5.3 BROWN ALGAE AS A NATURAL CARBON SINK

5.3.1 Carbon Fixation and Storage Mechanisms

The effective carbon fixation and storage mechanisms of brown algae make them an essential component of natural carbon sinks. These marine macroalgae take in carbon dioxide from the air and turn it into organic carbon through a process called photosynthesis. Brown algae are able to improve light absorption through the pigment fucoxanthin, enabling photosynthesis even in marine conditions with low light levels. *M. pyrifera*, sometimes known as gigantic kelp, is one of these efficient species that may fix as much as 1,965 g C/m^2/year (Stahl *et al.*, 2024). Brown algae fix CO_2 and then release it as either dissolved organic carbon (DOC), particulate organic carbon (POC), or biomass. Long-term carbon storage is helped by POC as it descends to deeper ocean layers, although DOC can be stubborn and stays in marine systems for a long time. Because they increase oceanic carbon storage and decrease atmospheric CO_2 concentrations, these activities are essential for climate change mitigation (Hellige *et al.*, 2024).

5.3.2 Role in Blue Carbon Ecosystems

Blue carbon ecosystems, which encompass salt marshes, seagrass beds, and mangrove forests, rely heavily on brown algae. Despite their long-standing exclusion from blue carbon assessments, macroalgae are now known to have a substantial role in carbon sequestration. According to Raven (2018), these creatures transfer carbon from their bodies to sediments in coastal and deep-sea environments, where it is preserved for eons. In coastal vegetated ecosystems, for instance, over

20%–60% of the carbon is contained in macroalgal detritus, making it a significant addition to sedimentary carbon reserves. According to Kwan *et al.* (2022), macroalgae such as Ulva and Sargassum have the ability to store more carbon than seagrasses do above ground, reaching 450 Mg C/year in Singapore. In addition, brown algae help reduce ocean acidification (OA) by releasing oxygen and absorbing carbon dioxide, which makes the water a better environment for marine life that relies on calcification. Their ecological and climatic relevance is underscored by their dual role as carbon sinks and ecosystem stabilizers (Krause-Jensen *et al.*, 2018).

5.3.3 Quantitative Assessment of Carbon Sequestration

To incorporate brown algae into global climate mitigation measures, it is vital to quantify their carbon sequestration potential. Global macroalgal primary productivity is estimated to be more than 1 Pg C/year, with a considerable amount of this carbon being exported to deep-sea sediments (Ross *et al.*, 2022). Despite widespread belief to the contrary, macroalgae play a significant part in carbon fixation in the Antarctic, accounting for 0.9%–2.8% of the world's total. Their complex communities, which may be found at depths of more than 70 m, play an essential role in polar ecosystems' carbon cycling and sequestration processes (Tait *et al.*, 2024). The benefits of seaweed farming in terms of carbon sequestration can also be measured. In South Korea, for example, the seaweed farming business produces biomass, releases DOC, and buries POC, all of which contribute to a substantial amount of CO_2. Its scalability for climate mitigation is highlighted by the anticipated carbon sequestration capacity of this industry, which is around 16 Tg C/year (Kim, 2024).

5.4 MITIGATION OF OCEAN ACIDIFICATION

5.4.1 The Carbonate-Bicarbonate Buffer System

To stabilize ocean pH and reduce acidification, the carbonate-bicarbonate buffer system plays an essential role in the global carbon cycle. Carbonic acid (H_2CO_3) is formed when atmospheric CO_3 dissolves in saltwater. This acid then dissociates into bicarbonate ions (HCO_3^-) and hydrogen ions (H^+). OA occurs when this process causes the pH of the ocean to fall. By acting as buffers, bicarbonate and carbonate ions (CO_3^{3+}) mitigate the consequences of elevated acidity (Comeau *et al.*, 2013). Corals and mollusks rely on this buffering capacity to enable their calcification, which is essential for marine ecosystem maintenance. Indirectly, brown algae help this system out by making photosynthesis easier. According to Cornwall *et al.* (2024), their metabolic activities lower CO_2 concentrations in the area, which makes the carbonate buffer less stressed and improves its ability to neutralize acidity.

Furthermore, brown algae perform an essential function by converting Dissolved Inorganic Carbon (DIC) into organic carbon, which can then be transported to lower levels of the ocean. Indirectly supporting the carbonate equilibrium and reducing the consequences of OA, this sequestration removes carbon from surface waters (Johnson *et al.*, 2014).

5.4.2 Brown Algae and pH Regulation in Marine Systems

The photosynthetic and metabolic activities of brown algae have a substantial impact on the dynamics of pH in aquatic habitats. As a consequence of taking CO_2 in for photosynthesis, their local concentrations of hydrogen ions drop, leading to a detectable rise in pH. Researchers have discovered that calcifying creatures in acidifying waters can find sanctuary in dense macroalgal environments like kelp forests, where the pH can rise by as much as 0.2 units during the day (Saderne & Wahl, 2013). A common intertidal brown alga, *Fucus serratus*, has been found to lower pCO_2 levels in its immediate vicinity, which helps to lessen the acidification effects on epibiotic communities that it is a part of. This includes calcifying species like *Electra pilosa* (Noisette *et al.*, 2022). Upwelling events send CO_2-rich, low-pH waters to the surface, making this localized buffering capacity more crucial in certain places. In addition, the carbonate chemistry is affected by the organic molecules released by brown algae, which include polysaccharides and phenolic substances. Further support for pH regulation can be provided by these compounds, which can boost the development of stable bicarbonate complexes. Such interactions have been associated with enhanced resistance to OA in ecosystems where Fucus species predominate, such as the Baltic Sea (Gao *et al.*, 2016).

5.5 POTENTIAL FOR BIOENERGY AND BIOMATERIALS

5.5.1 Brown Algae as a Feedstock for Biofuels

The high biomass yield and abundant carbohydrate content of brown algae—especially alginate, laminarin, and mannitol—make them a promising renewable feedstock for biofuel production (Kruger *et al.*, 2022). Fermentation and other bioprocesses can transform these carbs into biodiesel, bioethanol, and biogas. Brown algae outpace terrestrial crops in terms of growth rate, land requirement, and freshwater dependency; they thrive in saltwater environments. According to Laurens *et al.* (2017), gigantic kelp, scientifically known as *M. pyrifera*, is one of the most productive marine bioresources, capable of producing up to 100 metric tons of wet biomass per hectare each year. Additionally, according to Rathour *et al.* (2023), integrated biorefinery models show that the economic feasibility can be greatly improved by combining biofuel production with co-products such as biofertilizers and bioplastics. Enzymatic hydrolysis of brown algal polysaccharides has been refined by recent advances in fermentation technology, leading to improved bioethanol yields. As a bonus, methane yields surpass 200 mL/g volatile solids when algal biomass is anaerobically digested for biogas production (Khan *et al.*, 2021).

5.5.2 Bioplastics and Sustainable Alternatives

An environmentally preferable substitute for plastics made from petroleum, brown algae have recently gained attention as a potential source for long-term bioplastics. The film-forming capabilities and structural flexibility of alginate, a polysaccharide derived from brown algae, make it an essential component of biodegradable bioplastics (Berger *et al.*, 2023). By replacing traditional plastics with bioplastics made from algae, we can lessen our impact on the environment by using these materials in packaging, medical equipment, and agricultural films. For example, according to life cycle assessments, bioplastics made from algae can cut carbon emissions from conventional plastics by as much as half (Almomani *et al.*, 2023). Bioplastics' mechanical qualities and water resistance have been further improved by innovative methods, such as combining alginate with other natural polymers, increasing their use. According to Omorogi *et al.* (2024), bioplastics made from algae are becoming more popular in Europe's circular bioeconomy since they help achieve carbon neutrality and sustainable manufacturing goals.

5.5.3 Economic Viability and Technological Challenges

Producing biofuels and bioplastics on a massive scale from brown algae is an intriguing but technically and financially challenging prospect. A major obstacle to commercialization is the high cost of cultivation, harvesting, and processing. For example, according to Yadav *et al.* (2022), the costs of producing gasoline from algal biorefineries surpass $5/gallon, making them uncompetitive in the absence of subsidies. Cutting expenses also requires making the most of biomass yield and extracting carbohydrates. New approaches are being explored, including genetic engineering to increase the growth rates and carbohydrate yields of algae and the creation of more efficient techniques of enzymatic hydrolysis (Laurens *et al.*, 2017).

There are logistical hurdles associated with scaling up production infrastructure as well. Offshore growing methods for algae must be able to endure the severe marine environment and guarantee sustainable nutrient management, both of which demand large initial investments (Kruger *et al.*, 2022). Nevertheless, carbon credits and renewable energy subsidies are examples of governmental incentives that might greatly contribute to the expansion of this sector.

5.6 NUTRIENT CYCLING AND ECOSYSTEM SERVICES

5.6.1 Role in Nutrient Uptake and Water Purification

Marine environments rely heavily on brown algae for nitrogen cycling and water filtration. Agricultural runoff and urban garbage discharge create coastal eutrophication, which can be controlled by their ability to absorb and use surplus nutrients like nitrogen and phosphorus. Research has shown that certain species, such as *F. vesiculosus* and *M. pyrifera*, are quite good at absorbing nutrients. For example, Rodrigues-Filho *et al.* (2023) found that these species may assimilate up to 70 μmol/g of dry weight per day of nitrogen. Algal blooms are prevented, and hazardous chemicals are reduced, leading to enhanced water quality, thanks to this nutrient intake. Brown algae, when added to fish farming systems, reduce nutrient contamination in areas where aquaculture is common. One example is the use of macroalgae as biofilters in integrated multi-trophic aquaculture (IMTA) systems, which effectively reduce nitrogen levels in the water column (Liu *et al.*, 2020). Water bodies affected by hypoxia can be re-oxygenated with the help of brown algae, which produce oxygen through photosynthesis. In areas prone to eutrophication, in particular, their function as a natural purifier helps to keep ecosystems balanced and provides a boost to biodiversity.

5.6.2 Support for Marine Biodiversity

Brown algae grow in thick mats that provide shelter for many aquatic creatures, including fish, crustaceans, and microbes. The kelp forests play a crucial role in sustaining local biodiversity by offering refuge, breeding grounds, and feeding locations. Rockfish, lobsters, and more than a thousand species of fish, including those with commercial significance, can be found in only one kelp forest (Atkinson *et al.*, 2023). Brown algae not only provide physical cover, but they also drive primary productivity, which contributes to marine biodiversity. Their ability to produce organic carbon through rapid photosynthetic processes is fundamental to the stability of the food chain. In coastal ecosystems, nitrogen cycling and microbial populations are supported by dissolved organic matter, which is released when algae decompose (Gubelit & Grossart, 2020). Brown algae also help biodiversity in a roundabout way by lowering the energy of waves and stabilizing sediment. Seagrasses and other benthic creatures benefit from this, which promotes ecosystem resilience. Recognized as biodiversity hotspots, coastal lagoons and intertidal zones dominated by brown algae provide important ecosystem services to human communities through their biological activities (Apostolaki *et al.*, 2020).

5.7 BROWN ALGAE CULTIVATION AND ITS IMPACTS

5.7.1 Sustainable Aquaculture Practices

Sustainable aquaculture relies heavily on the farming of brown algae like *M. pyrifera* and *Laminaria japonica*. Because it doesn't necessitate fertilizers, freshwater, or farmland, seaweed farming leaves less of an ecological imprint than other types of aquacultures (Jagtap & Meena, 2022). Some species of brown algae may grow 5–10 cm/day, which makes them very efficient at producing biomass and cycling nutrients (Khan *et al.*, 2024). As an additional component of sustainable aquaculture, IMTA involves the cultivation of brown algae in conjunction with fish and shellfish. According to Tan *et al.* (2023), this method improves water quality, decreases the environmental impact of fish farming, and helps with nutrient pollution by absorbing excess nitrogen and phosphorus. In addition to improving ocean quality, algae farms serve as carbon sinks by removing CO_2 from the air.

5.7.2 Scaling Cultivation for Carbon Mitigation

Growing brown algae on a larger scale could help reduce greenhouse gas emissions and slow global warming. An estimated \$11.8 billion is generated from the annual production of about 35.8 million tons of seaweed by farmers around the world (Sultana *et al.*, 2023), making up 97% of the total seaweed production. Recent developments in seaweed farming infrastructure in South Korea have shown that large-scale carbon sequestration activities are feasible; these operations have the ability to collect up to 16 Tg CO_2/year (Kim, 2024). Sequestration of carbon through POC and DOC pathways is the primary emphasis of large-scale agricultural operations. Effectively lowering atmospheric CO_2 levels, brown algae farms store carbon in deep-sea sediments and export biomass to less accessible places. Optimizing scalability, however, requires addressing site-specific issues, including nutrient competition and ecological trade-offs.

5.7.3 Socio-Economic Benefits

Brown algae farming has far-reaching social and economic effects, helping with food security, jobs, and community development, among other things. Because of their high nutrient and bioactive component content, brown algae are being used more and more in pharmaceuticals, biofertilizers, animal feed, and as a sustainable food source (Omoto *et al.*, 2024). Seaweed farming is a major economic driver and source of income for more than 6 million people in Southeast Asian countries (Rehman *et al.*, 2024). Women gain agency in the business as a whole, but especially in developing countries where they play an essential role in growing and processing. According to Sultana *et al.* (2023), seaweed farming has the ability to alleviate poverty and promote gender equality in nations such as the Philippines and Indonesia. One area where seaweed farming is starting to make a splash is in the climate mitigation market, where carbon credits can be sold. The importance of brown algae as a resource for sustainable development is highlighted by the way environmental and economic policies are integrated.

5.8 APPLICATIONS IN CLIMATE POLICY FRAMEWORKS

5.8.1 Incorporating Brown Algae into Global Climate Goals

One method to get zero emissions is to include brown algae in global climate plans. The fast growth and carbon sequestration capabilities of brown algae make them compatible with the goals of agreements like the Paris Agreement. According to Sultana *et al.* (2023), brown algae have the capacity to sequester around 40 tons of CO_2/year/ac. It is possible that countries may achieve their emissions reduction goals if brown algae production were included in Nationally Determined Contributions (NDCs). As an example, South Korea is utilizing its existing infrastructure to maximize carbon collection and encourage sustainable habits through seaweed aquaculture, which is part of their climate mitigation strategy (Kim, 2024). Brown algae can help reduce emissions, save marine life, and improve biodiversity, all of which are in line with the United Nations Sustainable Development Goals (SDGs). This is especially true of SDG 13 (Climate Action) and SDG 14 (Life below Water).

5.8.2 Synergies with Carbon Trading and Offset Programs

One potential offset method that could be included in carbon trading markets is brown algae aquaculture. According to Hu *et al.* (2024), nations and companies can make large-scale cultivation of algae more financially viable by generating carbon credits, which are based on the quantification of algae farms' carbon sequestration capacity. The possibility for algae-based solutions to fill legislative gaps in carbon offset programs is illustrated by China's attempts to include ocean carbon sinks into emissions trading markets. It will be crucial to tackle institutional obstacles, including ambiguous property rights and regulatory frameworks, in order to scale these projects (Racine *et al.*, 2021). Seaweed farming as part of carbon offset programs can help with two fronts at once: reducing emissions and cleaning up polluted waterways by bio-extracting nutrients. As an example, studies conducted in the Gulf of Mexico show that seaweed farming can accomplish two goals at once: decreasing coastal eutrophication and producing carbon offsets (Lu *et al.*, 2023).

5.8.3 Policy Recommendations

To maximize the climate benefits of brown algae, specific policy measures must be adopted:

1. **Incentivize Cultivation**: Governments should provide subsidies or tax incentives for seaweed farming, recognizing its dual role in carbon sequestration and ecosystem restoration.
2. **Establish Carbon Accounting Standards**: Uniform methods for quantifying carbon sequestration in seaweed systems are essential for integration into carbon markets (Park *et al.*, 2021).
3. **Promote International Collaboration**: Cross-border partnerships can advance research and share best practices, particularly in regions with underdeveloped seaweed industries.
4. **Integrate into Existing Frameworks**: Aligning brown algae initiatives with international mechanisms like the Kyoto Protocol's Clean Development Mechanism (CDM) or the International Civil Aviation Organization's (ICAO's) CORSIA can broaden their impact (Painuly, 2001).
5. **Foster Public–Private Partnerships**: Engaging private sector stakeholders can accelerate technological innovations and drive investments in seaweed aquaculture.

5.9 CONCLUSION

Carbon sequestration, OA, and biodiversity enhancement are three of the most pressing climate change mitigation issues, and brown algae are at the forefront of finding answers to these problems. Their important function in nutrient cycling and restoration of marine ecosystems, as well as their effectiveness as carbon sinks (they may absorb up to 40 tons of CO_2/ha/year), are crucial pieces of information. The economic and environmental benefits of these substances are highlighted by their incorporation into sustainable aquaculture, as well as their potential for bioenergy and bioplastics. Carbon trading systems and global climate goals both include brown algae, which shows how important these algae could be for climate policy. Incentives for brown algae production, carbon accounting rules, and the incorporation of algae-based solutions into global climate plans should be policymakers' top priorities. Concurrently, scientists need to figure out how to make carbon quantification methods more accurate, find new uses for products made from algae, and solve problems with scaling. By working together on a global scale, particularly through public–private partnerships, we can hasten progress and make brown algae a pillar of climate resilience and sustainable development. To make the most of brown algae's ability to fight climate change, innovative research and focused policy actions are needed.

DECLARATION

The authors have not used any generative AI tools during the writing of this manuscript.

REFERENCES

Almomani, F., Abdelbar, A., & Ghanimeh, S. (2023). A review of the recent advancement of bioconversion of carbon dioxide to added value products: A state of the art. *Sustainability*. https://app.dimensions.ai/details/publication/pub.1160344234

Apostolaki, S., Akinsete, E., Koundouri, P., & Samartzis, P. (2020). Freshwater: The importance of freshwater for providing ecosystem services. *Elsevier*. https://app.dimensions.ai/details/publication/pub.1123794656

Atkinson, C. L., Hopper, G. W., Kreeger, D. A., Lopez, J. W., Maine, A. N., Sansom, B. J., Schwalb, A., & Vaughn, C. C. (2023). Gains and gaps in knowledge surrounding freshwater mollusk ecosystem services. *Freshwater Mollusk Biology and Conservation*. https://app.dimensions.ai/details/publication/pub.1156447622

Berger, N. J., Masri, M. A., Brück, T., Garbe, D., Pfeifer, C., & Lindorfer, J. (2023). Utilizing a CHP power plant's energy and CO_2 emissions for the manufacture of affordable and carbon neutral algae bioplastic for re-useable packaging. *Industrial & Engineering Chemistry Research*. https://app.dimensions.ai/details/publication/pub.1157580782

Comeau, S., Carpenter, R. C., & Edmunds, P. J. (2013). Coral reef calcifiers buffer their response to ocean acidification using both bicarbonate and carbonate. *Proceedings of the Royal Society B*. https://app.dimensions.ai/details/publication/pub.1037903183

Cornwall, C. E., Comeau, S., & Harvey, B. P. (2024). Are physiological and ecosystem-level tipping points caused by ocean acidification? A critical evaluation. *Earth System Dynamics*. https://app.dimensions.ai/details/publication/pub.1172370578

Dominguez Almela, V., Tompkins, E. L., Dash, J., & Tonon, T. (2023). Brown algae invasions and bloom events need routine monitoring for effective adaptation. *Environmental Research Letters*. https://app.dimensions.ai/details/publication/pub.1166541419

Gao, G., Liu, Y., Li, X., Feng, Z., & Xu, J. (2016). An ocean acidification acclimatized green tide alga is robust to changes of seawater carbon chemistry but vulnerable to light stress. *PLoS One*. https://app.dimensions.ai/details/publication/pub.1067095501

Gubelit, Y. I., & Grossart, H.-P. (2020). New methods, new concepts: What can be applied to freshwater periphyton? *Frontiers in Microbiology*. https://app.dimensions.ai/details/publication/pub.1128684084

Hellige, I., Mundanatt, A. A., Massing, J. C., & Hehemann, J.-H. (2024). Roots of coastal plants stabilize carbon fixed by marine algae. *bioRxiv*. https://app.dimensions.ai/details/publication/pub.1182961993

Hu, X., Shan, H., & Zhang, Q. (2024). Institutional obstacles and countermeasures to improve the Chinese ocean carbon sink trading market. *Sustainability*. https://app.dimensions.ai/details/publication/pub.1173533140

Jagtap, A. S., & Meena, S. N. (2022). Chapter 23 Seaweed farming: A perspective of sustainable agriculture and socio-economic development. *Elsevier*. https://app.dimensions.ai/details/publication/pub.1144686287

Johnson, M. D., Price, N. N., & Smith, J. E. (2014). Contrasting effects of ocean acidification on tropical fleshy and calcareous algae. *PeerJ*. https://app.dimensions.ai/details/publication/pub.1029842331

Khan, N., Sudhakar, K., & Mamat, R. (2021). Role of biofuels in energy transition, green economy and carbon neutrality. *Sustainability*. https://app.dimensions.ai/details/publication/pub.1143152359

Khan, N., Sudhakar, K., & Mamat, R. (2024). Macroalgae farming for sustainable future: Navigating opportunities and driving innovation. *Heliyon*. https://app.dimensions.ai/details/publication/pub.1169894010

Kim, K. Y. (2024). Harnessing seaweed farming for climate mitigation in South Korea: Evaluating carbon dioxide removal potential and future research directions. *Algae*. https://app.dimensions.ai/details/publication/pub.1183763007

Krause-Jensen, D., Lavery, P., Serrano, O., Marbà, N., Masque, P., & Duarte, C. M. (2018). Sequestration of macroalgal carbon: The elephant in the Blue Carbon room. *Biology Letters*. https://app.dimensions.ai/details/publication/pub.1104994752

Kruger, J. S., Wiatrowski, M., Davis, R. E., Dong, T., Knoshaug, E. P., Nagle, N. J., Laurens, L. M. L., & Pienkos, P. T. (2022). Enabling production of algal biofuels by techno-economic optimization of co-product suites. *Frontiers in Chemical Engineering*. https://app.dimensions.ai/details/publication/pub.1144629298

Kwan, V., Fong, J., Ng, C. S. L., & Huang, D. (2022). Temporal and spatial dynamics of tropical macroalgal contributions to blue carbon. *The Science of the Total Environment*. https://app.dimensions.ai/details/publication/pub.1146080098

Lammerant, R., Norkko, A., & Gustafsson, C. (2023). A functional perspective on the factors underpinning biomass-bound carbon stocks in coastal macrophyte communities. *Marine Environmental Research*. https://app.dimensions.ai/details/publication/pub.1166527246

Laurens, L. M. L., Markham, J., Templeton, D. W., Christensen, E. D., Van Wychen, S., Vadelius, E. W., Chen-Glasser, M., Dong, T., Davis, R., & Pienkos, P. T. (2017). Development of algae biorefinery concepts for biofuels and bioproducts; a perspective on process-compatible products and their impact on cost-reduction. *Energy & Environmental Science*. https://app.dimensions.ai/details/publication/pub.1090576868

Liu, Q., Zhang, Y., Wu, H., Liu, F., Peng, W., Zhang, X., Chang, F., Xie, P., & Zhang, H. (2020). A review and perspective of eDNA application to eutrophication and HAB control in freshwater and marine ecosystems. *Microorganisms*. https://app.dimensions.ai/details/publication/pub.1125728266

Lu, J., Burton, J., Garzon-Garcia, A., Jackson, C., Newham, M., Bloesch, P., Ramsay, I., Rogers, J., Griffith, M., & Saeck, E. (2023). Scientific challenges and biophysical knowledge gaps for implementing nutrient offset projects. *Journal of Environmental Management*. https://app.dimensions.ai/details/publication/pub.1157224366

Malla, F. A., Malla, A. A., Bandh, S. A., Sofi, N. A., & Ahmed, M. (2024). The Role of Brown Algae in Global Warming Mitigation. In: Kurniawan, T. A., Anouzla, A. (eds) *Algae as a Natural Solution for Challenges in Water-Food-Energy Nexus: Environmental Science and Engineering*. Springer, Singapore. https://doi.org/10.1007/978-981-97-2371-3_28

Noisette, F., Pansch, C., Wall, M., Wahl, M., & Hurd, C. L. (2022). Role of hydrodynamics in shaping chemical habitats and modulating the responses of coastal benthic systems to ocean global change. *Global Change Biology*. https://app.dimensions.ai/details/publication/pub.1146338794

Omoto, R., Uehara, M., Seki, D., & Kinjo, M. (2024). Supply chain-based coral conservation: The case of Mozuku seaweed farming in Onna Village, Okinawa. *Sustainability*. https://app.dimensions.ai/details/publication/pub.1170182995

Painuly, J. P. (2001). The Kyoto Protocol, emissions trading and the CDM: An analysis from developing countries perspective. *The Energy Journal*. https://app.dimensions.ai/details/publication/pub.1072973626

Park, J.-S., Shin, S. K., Wu, H., Yarish, C., Yoo, H. I., & Kim, J. K. (2021). Evaluation of nutrient bioextraction by seaweed and shellfish aquaculture in Korea. *Journal of the World Aquaculture Society*. https://app.dimensions.ai/details/publication/pub.1140210028

Racine, P., Marley, A., Froehlich, H. E., Gaines, S. D., Ladner, I., MacAdam-Somer, I., & Bradley, D. (2021). A case for seaweed aquaculture inclusion in U.S. nutrient pollution management. *Marine Policy*. https://app.dimensions.ai/details/publication/pub.1137228202

Rathour, R. K., Behl, M., Dhashmana, K., Sakhuja, D., Ghai, H., Sharma, N., Meena, K. R., Bhatt, A. K., & Bhatia, R. K. (2023). Non-food crops derived lignocellulose biorefinery for sustainable production of biomaterials, biochemicals, and bioenergy: A review on trends and techniques. *Industrial Crops and Products*. https://app.dimensions.ai/details/publication/pub.1163062063

Raven, J. (2018). Blue carbon: Past, present and future, with emphasis on macroalgae. *Biology Letters*. https://app.dimensions.ai/details/publication/pub.1107380096

Rehman, Z. U., Punnaanan, K., Nootong, K., & In-na, P. (2024). The status of algal decarbonization in Southeast Asian region: A review. *Environmental Research Letters*. https://app.dimensions.ai/details/publication/pub.1174973416

Rodrigues-Filho, J. L., Macêdo, R. L., Sarmento, H., Pimenta, V. R. A., Alonso, C., Teixeira, C. R., Pagliosa, P. R., Netto, S. A., Santos, N. C. L., Daura-Jorge, F. G., Rocha, O., Horta, P., Branco, J. O., Sartor, R., Muller, J., & Cionek, V. M. (2023). From ecological functions to ecosystem services: Linking coastal lagoons biodiversity with human well-being. *Hydrobiologia*. https://app.dimensions.ai/details/publication/pub.1156105180

Ross, F., Tarbuck, P., & Macreadie, P. I. (2022). Seaweed afforestation at large-scales exclusively for carbon sequestration: Critical assessment of risks, viability and the state of knowledge. *Frontiers in Marine Science*. https://app.dimensions.ai/details/publication/pub.1152866760

Saderne, V., & Wahl, M. (2013). Differential responses of calcifying and non-calcifying epibionts of a brown macroalga to present-day and future upwelling pCO_2. *PLoS One*. https://app.dimensions.ai/details/publication/pub.1051158322

Stahl, F., Kappas, L., Uhl, F., Oppelt, N., & Bischof, K. (2024). Feasibility study for kelp afforestation in the German Bight: Habitat availability and light requirements of *Laminaria hyperborea*. *Journal of Sea Research*. https://app.dimensions.ai/details/publication/pub.1172646057

Sultana, F., Wahab, A., Nahiduzzaman, M., Mohiuddin, M. Z. I., Shakil, A., Mamun, A. A., Rahman Khan, S., Wong, L. L., & Asaduzzaman Md . (2023). Seaweed farming for food and nutritional security, climate change mitigation and adaptation, and women empowerment: A review. *Aquaculture and Fisheries*. https://app.dimensions.ai/details/publication/pub.1151850066

Tait, L. W., Chin, C., Nelson, W., George, S., Marriott, P., O'Driscoll, R. L., Lamare, M., Mills, V. S., & Cummings, V. J. (2024). Deep-living and diverse Antarctic seaweeds as potentially important contributors to global carbon fixation. *Communications Earth & Environment*. https://app.dimensions.ai/details/publication/pub.1170758224

Tan, H., Das, S. K., Zainee, N. F. A., Yana, R., & Rozaimi, M. (2023). Ocean acidification and aquacultured seaweeds: Progress and knowledge gaps. *Journal of Marine Science and Engineering*. https://app.dimensions.ai/details/publication/pub.1154244918

Yadav, K., Vasistha, S., Nawkarkar, P., Kumar, S., & Rai, M. P. (2022). Algal biorefinery culminating multiple value-added products: Recent advances, emerging trends, opportunities, and challenges. *3 Biotech*. https://app.dimensions.ai/details/publication/pub.1150460817

Zhang, J., Cran, M., Gao, L., Xie, Z., & Gray, S. (2024). Contribution of seaweed farming to the mitigation of greenhouse gas emissions and microplastics pollution. *Algal Research*. https://app.dimensions.ai/details/publication/pub.1173973413

6 Achieving Net Zero by Algal Biorefineries

Mailin Misson, Grace Joy Wei Lie Chin, Suryani Saallah, Joshophine Pei Ning Kong, Ignatia Justine, Syafiqah Syazwani Jaffar, Vun Yee Thien, Kenneth Francis Rodrigues, Nyuk Ling Ma, and Wilson Thau Lym Yong

6.1 INTRODUCTION

Climate change is one of the greatest challenges of our time, affecting ecosystems, economies, and communities worldwide. To address this urgent issue, we must reduce greenhouse gas emissions, enhance carbon absorption, and transition to sustainable energy sources. One of the key strategies is achieving net-zero emissions, and another is adopting circular economy principles. This chapter explores these ideas with a focus on algal biorefineries, which offer a sustainable way to produce biofuels and other valuable products. By examining how algae can help mitigate climate change and build resilience, we aim to highlight their potential in driving the shift toward a more sustainable future.

6.2 OVERVIEW OF CLIMATE CHANGE AND NET-ZERO GOALS

6.2.1 Definition of Net Zero

Net zero is the equilibrium between the amount of greenhouse gases emitted and the amount sequestered or removed from the atmosphere within a specified timeframe (Fankhauser et al., 2022). Achieving net zero necessitates substantial reductions in current emissions and the active removal of greenhouse gases through diverse strategies, such as carbon capture and storage (CCS) (Gür, 2022), reforestation (Liu et al., 2023), and various carbon offset programs, including hydroelectric projects and renewable energy installations like solar and wind power plants (Sayed et al., 2021). Net zero addresses the fundamental cause of climate change, which is the excessive greenhouse gas emissions, and is, therefore, critical for mitigating its impacts. Attaining net zero requires a multi-sectoral approach, incorporating significant advancements in technology, behavioral shifts, and robust policy frameworks across energy, transportation, industry, and agriculture sectors (Arent et al., 2022). These measures ensure that emissions produced are counterbalanced by equivalent removals, thereby preventing further accumulation of greenhouse gases in the atmosphere.

The concept of net zero gained significant attention following the Paris Agreement in 2015, where various nations pledged to limit global temperature rise to well below 2°C above pre-industrial levels, with aspirations to cap it at 1.5°C (Allen et al., 2022). This global commitment underscores the necessity of transitioning to net-zero emissions to avoid the most severe consequences of climate change. By adopting net-zero targets, countries, corporations, and organizations are implementing strategies to reduce emissions while investing in both natural and technological carbon sequestration methods. The ultimate aim of net zero is to stabilize the global climate system, preventing further environmental degradation, and ensure a sustainable, resilient future for all (Fankhauser et al., 2022).

DOI: 10.1201/9781003685814-6

6.2.2 Importance of Achieving Net Zero in the Context of Global Climate Change

Achieving net-zero emissions is fundamental to addressing the underlying drivers of global climate change, primarily the accumulation of greenhouse gases like carbon dioxide in the atmosphere (Obobisa, 2022). These gases intensify global warming, leading to severe environmental disruptions such as extreme weather events (including heavy rainfall and heatwaves) (AghaKouchak et al., 2020), rising sea levels (Horton et al., 2020), and significant impacts on biodiversity (Taylor et al., 2021) and human communities (Mora et al., 2018). Reaching net zero can halt the rise of these gases, stabilizing global temperatures and preventing further escalation into an irreversible climate crisis. This stabilization is crucial for maintaining ecological balance and safeguarding vital natural systems.

Furthermore, the pursuit of net-zero emissions drives the widespread adoption of innovative technologies and sustainable practices across various sectors, including energy, transportation, and industry. This transition is essential for fostering economic resilience and environmental sustainability (Tang et al., 2023). Investments in renewable energy sources, energy efficiency, and carbon capture technologies not only reduce emissions but also stimulate economic growth by creating green jobs and advancing technological innovation. Achieving net zero also promotes global cooperation and the implementation of cohesive policies that enhance environmental stewardship (Barbhuiya et al., 2024). Thus, the commitment to net zero serves as a cornerstone for ensuring a sustainable future, strengthening global efforts to combat climate change, and building resilient societies.

6.3 ROLE OF ALGAL BIOREFINERIES IN CIRCULAR ECONOMY

6.3.1 Concept of Circular Economy

The circular economy is an advanced economic model designed to minimize waste and maximize resource efficiency through practices such as sharing, leasing, reusing, repairing, refurbishing, and recycling materials and products (Iacovidou et al., 2021). Unlike the traditional linear economy, which operates on a 'take, make, dispose' approach, the circular economy (Figure 6.1) emphasizes

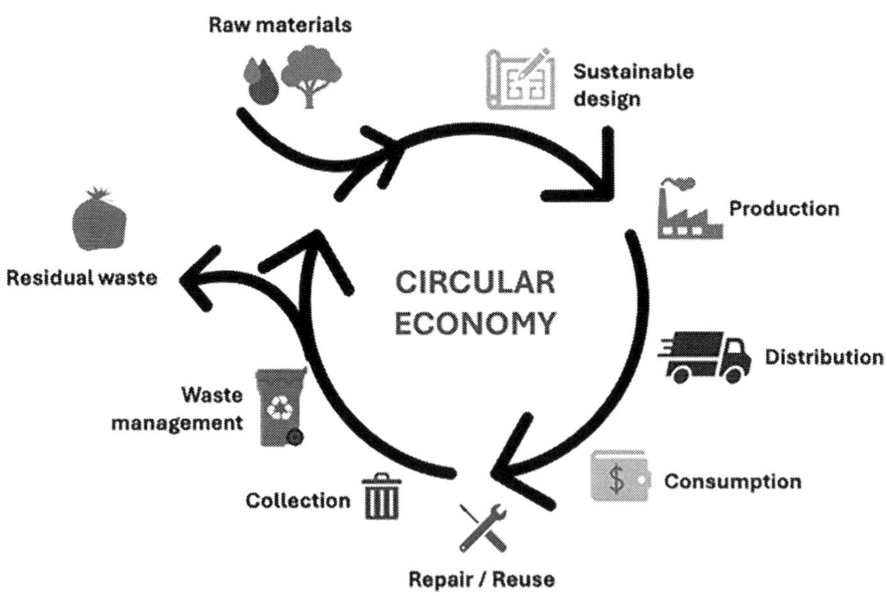

FIGURE 6.1 Circular economy model showcasing the sustainability of raw materials through closed-loop processes that minimize waste and maximize resource efficiency.

sustainability and closed-loop systems (Rashid & Malik, 2023). Its core principles include the elimination of waste and pollution, maintaining products and materials in circulation, and the regeneration of natural systems. This model aims to develop a sustainable economic system that yields benefits for businesses, society, and the environment by promoting the reuse, refurbishment, and recycling of materials and products. Increasingly, the circular economy is recognized as vital for achieving long-term environmental sustainability and ensuring resource security (Velenturf & Purnell, 2021).

6.3.2 The Role of Algal Biorefineries in Promoting Sustainability and Climate Change Resilience

Algal biorefineries exemplify an innovative approach within the circular economy framework, presenting significant environmental and economic advantages. Algae, encompassing both microalgae and macroalgae, can be grown on non-arable land and in diverse aquatic environments, making them a highly adaptable and sustainable feedstock for producing valuable commodities (Chin et al., 2023). These biorefineries transform algal biomass into a variety of products, including biofuels, bioplastics, and biochemicals, which reduces dependence on fossil fuels and lowers greenhouse gas emissions (Arora et al., 2020). Moreover, algae's capacity to absorb substantial amounts of CO_2 through photosynthesis helps in carbon sequestration, contributing to climate change mitigation (Prasad et al., 2021).

Integrating algal biorefineries with wastewater treatment processes exemplifies an eco-friendly and innovative approach. This integration utilizes nutrients from waste streams to promote algal growth while simultaneously purifying water, aligning with circular economy principles by closing nutrient loops and minimizing waste (Justine et al., 2023). Microalgae play a pivotal role in wastewater treatment through bioaccumulation, biodegradation, and biosorption (Hariz & Takriff, 2017). Additionally, microalgae's ability to sequester carbon dioxide via photosynthesis and absorb nutrients like nitrogen and phosphorus at high rates enhances water quality. This process yields valuable biomass for biofuels and bioproducts, promoting both economic growth and environmental sustainability (Mustafa et al., 2021).

Beyond energy production, algal biorefineries demonstrate versatility in producing animal feed, fertilizers, pharmaceuticals, and cosmetics, making algae an attractive feedstock for biorefinery processes (Thanigaivel et al., 2022). Algae cultivation does not compete with food crops for land and water resources, addressing food security concerns associated with other biofuel feedstocks (Renzaho et al., 2017). Microalgae-derived biofertilizers improve soil fertility, enhance plant growth, and reduce dependence on chemical fertilizers, which contribute to soil degradation and greenhouse gas emissions. Integrating algal biorefineries with other industrial processes, such as carbon capture and utilization, creates synergistic systems that maximize resource efficiency and environmental benefits. This approach supports soil nutrient replenishment, enhances agricultural productivity, and ensures food security for the growing global population (Goncalves et al., 2023).

In summary, algal biorefineries offer a sustainable and innovative solution to pressing environmental challenges. By converting algal biomass into valuable products, they align with circular economy principles and contribute to climate change mitigation. As we advance toward a net-zero future, the role of algal biorefineries in promoting sustainability and resilience will become increasingly significant. The following sections of this chapter will delve deeper into the specific advantages of different algae species, innovative cultivation techniques, and comprehensive assessments of their environmental and economic impacts, providing a detailed roadmap for the successful implementation of algal biorefineries.

6.4 ALGAL BIOMASS: A RENEWABLE RESOURCE

Algal biomass stands as a versatile and renewable resource, essential for the development of sustainable biorefineries. With the growing need for alternative energy sources and environmentally friendly materials, algae offer significant potential due to their rapid growth rates, high productivity, and ability to thrive in diverse environments. This section delves into the various types of algae utilized in biorefineries, their cultivation and harvesting methods, and the processing techniques essential for optimizing yield and efficiency.

6.4.1 Types of Algae Used in Biorefineries

6.4.1.1 Microalgae vs. Macroalgae

Algal biorefineries utilize both microalgae and macroalgae to convert biomass into valuable bioproducts. Microalgae, characterized by their microscopic size (0.2–200 μm) and unicellular organization (Caroppo & Pagliara, 2022), contrast with macroalgae, which are larger, multicellular, and visible to the naked eye, ranging from 30 cm to 60 m in length (Biris-Dorhoi et al., 2020; Sudhakar & Viswanaathan, 2019). Microalgae, such as *Chlorella* and *Spirulina* species, are known for their high lipid content and rapid growth rates, making them ideal for biofuel production (Chhandama et al., 2023; Patle et al., 2021). They thrive with sufficient sunlight, nutrients, and suitable environmental conditions (Abdur Razzak et al., 2024), and their lipid accumulation can be further enhanced under stress conditions (Khoo et al., 2023; Morales et al., 2021). Macroalgae, commonly known as seaweed, contain polysaccharides like cellulose or alginate, which can be converted into mono-sugars for biofuel production (Kim et al., 2019; Sudhakar & Viswanaathan, 2019). The absence of lignin in macroalgae makes them easier to digest by microbes during biorefinery processes compared to terrestrial plants (S et al., 2023; V et al., 2021). Additionally, macroalgae are relatively easier to harvest, although they may be susceptible to natural disasters and anthropogenic activities (Hanley et al., 2024; Monagail et al., 2017). Despite their slower growth rates compared to microalgae, macroalgae can produce substantial biomass yields, making them a valuable resource for biorefineries (Arias et al., 2023b).

6.4.1.2 Common Species and Their Characteristics

Several species of microalgae and macroalgae are prominent in biorefinery applications due to their distinct characteristics. For instance, *Chlorella* species are noted for their high lipid accumulation and biomass yield (Je & Yamaoka, 2022), while *Spirulina* is favored for its rapid growth rate (Palanisamy et al., 2023). Among macroalgae, *Laminaria japonica* is rich in polysaccharides, making it suitable for biodiesel and bioethanol production (Kim et al., 2019). Another important macroalgae species is *Saccharina latissima*, which is also high in polysaccharides and has demonstrated potential for bioethanol production (Schultze-Jena et al., 2022). These characteristics underline the suitability of these algae species for various biofuel production processes, including biodiesel, bioethanol, biohydrogen, and bio-oil. Each species offers unique advantages that make them integral to the biorefinery industry, driving the development of sustainable energy solutions and materials. A detailed list of common algae species and their characteristics is provided in Table 6.1.

6.4.2 Cultivation Techniques

6.4.2.1 Open Ponds vs. Closed Photobioreactors

Algae cultivation systems, applicable to both macroalgae and microalgae, are divided into open and closed systems, each offering unique benefits and challenges. Open cultivation systems, such as raceway ponds, fiberglass tanks, and artificial ponds, expose algae cultures to the external environment (Liu et al., 2022a; Wimmerova et al., 2023). These systems are cost-effective but offer minimal control over environmental factors like light, temperature, and pH, leading to potential contamination and generally supporting low-density cultures (Khan et al., 2018; Mahmood et al., 2023). To mitigate contamination risks, batch harvesting is often employed. This method is particularly suitable for large-scale macroalgae cultivation, as macroalgae, being more resilient to environmental

TABLE 6.1
Common Algae Species and Their Characteristics in Biofuel Production

Common Species	Types of Biofuels	Characteristics	References
***Chlorella vulgaris, Chlorella* sp.**	Biodiesel, biohydrogen	High accumulation of lipids under nitrogen starvation, high biomass yield, good H_2 production	Guccione et al. (2014); Je and Yamaoka (2022)
***Spirulina* sp.**	biodiesel, bioethanol	High growth rate, moderate lipid content	Palanisamy et al. (2023); Rempel et al. (2019)
Dunaliella salina, Dunaliella tertiolecta,	biodiesel, bioethanol	High protein, high carbohydrate (some strains), good lipid accumulation	Ahmed et al. (2017); Hopkins et al. (2019)
***Scenedesmus* sp.**	Bioethanol, biohydrogen	High protein content, high carbohydrate content, thin cell wall	Mercado et al. (2020); Sánchez-García et al. (2020)
***Chlamydomonas* sp.**	Biohydrogen, biodiesel	Good lipid accumulation under nitrogen starvation, good H_2 production under sulfur starvation	Choi et al. (2022); Scranton et al. (2015)
***Nannochloropsis* sp.**	Biodiesel, bio-oil	Excellent lipid accumulation under nitrogen starvation	Li et al. (2020); Roncaglia et al. (2021)
Laminaria japonica	Biodiesel, bioethanol	Rich in polysaccharides	Kim et al. (2019); Yin et al. (2019)
Saccharina latissimi	Biohydrogen, bioethanol	High in polysaccharides, potential for high bioethanol production	Lin et al. (2019); Schultze-Jena et al. (2022)

fluctuations, can thrive in these open systems, although their growth can still be affected by weather conditions and water quality (Khan et al., 2024; Tullberg et al., 2022). While microalgae can also be cultivated in open systems, they face greater risks of contamination and require careful management to maintain productivity (Laezza et al., 2022).

In contrast, closed cultivation systems, such as photobioreactors, provide a controlled environment with either artificial or natural light (Chanquia et al., 2022; Obando et al., 2022). Although initially more expensive, photobioreactors are highly recommended for producing high-quality, uncontaminated biomass. These systems support continuous operation and high-density cultures, with low evaporation and water loss, making them highly efficient (Egloff et al., 2018). Photobioreactors are particularly advantageous for microalgae cultivation due to their fast growth rates and high lipid content, which are best maintained under controlled conditions, enhancing the efficiency of biofuel production and other valuable products (Abdur Razzak et al., 2024). Similarly, macroalgae benefit from the optimized environment, resulting in higher yields and consistent quality (Obando et al., 2022; Yong et al., 2014). Despite the complexity and maintenance requirements of photobioreactors, their ability to optimize growth conditions and maximize biomass yield makes them a valuable tool in the cultivation of algae for biofuel production and other biorefinery applications.

6.4.2.2 Innovations in Algae Cultivation

Innovations in algae cultivation are essential for advancing the production and utilization of both macroalgae and microalgae in biotechnological applications, particularly in achieving net-zero emissions. Scaling from laboratory to industrial scale is necessary to meet commercial demands. One innovative application is the development of a novel sequential flow baffled microalgal-bacterial photobioreactor, which combines microalgae and microbial communities to treat wastewater while producing biomass and lipids (Leong et al., 2021). This system not only enhances wastewater treatment efficiency but also generates valuable by-products, contributing to a circular economy and reducing overall carbon footprint. Additionally, advancements in genetic engineering have led to the development of genetically modified microalgae strains with higher lipid and biomass yields, further optimizing biofuel production and supporting carbon-neutral energy solutions (Khoo et al., 2023; Sproles et al., 2021).

For macroalgae, innovations such as integrated multi-trophic aquaculture (IMTA) systems have shown promise. IMTA systems cultivate macroalgae alongside other aquaculture species like fish and shellfish, creating a balanced ecosystem where the waste products of one species serve as nutrients for another (Checa et al., 2024; Nissar et al., 2023). This not only improves the overall efficiency of the aquaculture system but also reduces environmental impact. Furthermore, advancements in seaweed farming techniques, including the use of floating rafts, submerged longlines, and automation, have increased the productivity and scalability of macroalgae cultivation (Khan et al., 2024; Solvang et al., 2021). These innovations support the sustainable production of biofuels, bioplastics, and other high-value products, thereby enhancing the economic viability of algae-based industries and aiding in the achievement of net-zero goals. For instance, algae-based bioplastics offer biodegradable and environmentally friendly alternatives to petroleum-based plastics, aligning with global efforts to achieve net-zero emissions by reducing reliance on fossil fuels and decreasing overall carbon emissions (Arora et al., 2023; Elkaliny et al., 2024).

6.4.3 Harvesting and Processing

6.4.3.1 Methods of Algae Harvesting

The methods of harvesting vary significantly between macroalgae and microalgae. Macroalgae are generally easier to harvest due to their larger size. Commercial harvesting of wild seaweeds is typically done through hand cutting, hand gathering, and mechanical harvesting (Monagail et al., 2017). Hand cutting and gathering can include baling the seaweed to be towed by boat or using a boat to rake seaweed (Vance et al., 2023). In contrast, microalgae often require more sophisticated methods to achieve efficient biomass recovery (Mathimani & Mallick, 2018; Najjar & Abu-Shamleh, 2020). Optimizing these harvesting methods is crucial for minimizing environmental impact and maximizing biomass yield in the effort to transition to net zero. Efficient harvesting techniques like flotation, flocculation/coagulation, and bio-flocculation/auto-flocculation are particularly suitable for large-scale microalgae operations due to their high biomass recovery rates and low operational costs (Matter et al., 2019; Tan et al., 2020). These methods enhance the economic viability of algae-based biofuels and bioproducts by reducing energy consumption and waste (de Morais et al., 2023; Taghavijeloudar et al., 2024). Additionally, innovative approaches such as magnetic separation offer high recovery rates, though they come with higher costs (Wang et al., 2015; Zhu et al., 2024). Table 6.2 summarizes the various harvesting methods and their suitability for large-scale operations.

6.4.3.2 Processing Techniques for Optimal Yield

Processing harvested algae biomass to extract valuable components is a critical step in optimizing yield and advancing toward net-zero emissions. One of the most important of these components is lipids, which are essential for biofuel production. For both macroalgae and microalgae, pretreatment techniques are necessary due to the presence of thick cell walls that hinder solvent penetration and lipid release (Gallego et al., 2021; Ramola et al., 2019). Methods such as microwave treatment, homogenization, ultrasound, autoclaving, and lyophilization effectively disrupt cell walls, facilitating the extraction process (AlMohamadi et al., 2023; Krishnamoorthy et al., 2022; Rokicka et al., 2021). Additionally, chemical treatments involving acids or alkalis, as well as biochemical reactions using enzymes, can further enhance cell wall disruption and improve lipid recovery rates (Elnajjar et al., 2023; Maneein et al., 2018; Zuorro et al., 2019).

TABLE 6.2
Summary of Microalgae Harvesting Methods and Their Suitability for Large-Scale Operations

Harvesting Methods	Solid Concentration (%)	Biomass Recovery Rate (%)	Suitability for Large Scale	Operational Costs
Centrifugation	10–25	> 90	No	High
Sedimentation	0.5–3	10–50	No	Low
Filtration	5–27	20–87	No	High
Flotation	< 7	50–90	Yes	Low
Flocculation/coagulation	< 6	50–98	Yes	Low
Bio-flocculation/auto-flocculation	–	40–99	Yes	Low
Magnetic separation	–	> 95	Yes	High

Sources: de Morais et al. (2023), Mathimani and Mallick (2018), and Najjar and Abu-Shamleh (2020).

Once the biomass is pretreated, various extraction methods are employed to obtain lipids. Techniques such as chloroform extraction, the Bligh and Dyer method, enzymatic extraction, and supercritical fluid extraction are commonly used (Zhou et al., 2022a). After lipid extraction, transesterification is performed to produce biodiesel (Ghedini et al., 2021; Pandey et al., 2024). This process involves converting lipids into biodiesel using catalysts (either chemical or biological) under specific conditions, including optimal temperature, reaction time, and solvent-to-lipid ratios (Farouk et al., 2024). For macroalgae, processing often focuses on polysaccharides and other valuable compounds, using similar extraction and processing methods to enhance yield (Dobrinčić et al., 2020; Perez-Vazquez et al., 2023). The advancements in processing techniques not only improve the efficiency and quality of biofuel production but also support the sustainable and eco-friendly utilization of algae, aligning with global efforts to achieve net-zero emissions (Krishnan et al., 2024).

6.5 BIOREFINERY PROCESSES AND TECHNOLOGIES

Biorefinery processes and technologies play a pivotal role in transforming algae biomass into a wide range of valuable products, contributing significantly to sustainable development and net-zero emissions goals. By integrating various conversion pathways and advanced technologies, biorefineries optimize the utilization of algal biomass, enhancing economic viability and environmental benefits. This section explores the key methods involved in converting algal biomass into biofuels, biochemicals, and other high-value products, emphasizing both biochemical and thermochemical conversion pathways, as well as the extraction of valuable compounds.

6.5.1 Algal Biomass Conversion Pathways

6.5.1.1 Biochemical Conversion (Fermentation, Anaerobic Digestion)

Biochemical conversion of algal biomass utilizes microorganisms to transform organic materials into valuable products. This process involves harvesting and pretreating the algal biomass, followed by microbial fermentation to produce biofuels, biochemicals, and bio-based materials (Babich et al., 2024). For example, the fermentation of microalgae like *Chlorella* and macroalgae such as brown seaweed (*Laminaria*) can yield bioethanol, which serves as a renewable biofuel (Hou et al., 2017; Shenbagamuthuraman & Kasianantham, 2023). The use of bioreactors optimizes growth conditions for microorganisms, significantly enhancing production efficiency (Granata et al., 2024; Lian et al., 2018). Advanced technologies, such as enzymatic hydrolysis and metabolic engineering, further improve yields by breaking down complex molecules and optimizing metabolic pathways (Ezeorba et al., 2024; Nath, 2024). These innovations contribute to sustainability, versatility, and environmental benefits by utilizing renewable biomass and reducing emissions.

Anaerobic digestion decomposes organic matter in the absence of oxygen, producing biogas, a mix of methane and carbon dioxide. This process includes stages such as hydrolysis, acidogenesis, acetogenesis, and methanogenesis, each supported by specific microbial communities (Faisal et al., 2022; Harirchi et al., 2022). Both macroalgae and microalgae are viable for anaerobic digestion, showing promising energy yields (Milledge et al., 2019; Sheng et al., 2022). For instance, the anaerobic co-digestion of *Scenedesmus* sp. and *Chlorella* sp. can generate significant amounts of methane (Zhen et al., 2016), while the digestion of *Laminaria digitata* can produce substantial quantities of biogas (Vanegas & Bartlett, 2015). These methods offer sustainable energy solutions but face challenges such as contamination control and scalability (Gaurav et al., 2024). Ongoing research aims to improve efficiency and address these limitations, contributing to the overall goal of net-zero emissions by providing cleaner energy alternatives.

6.5.1.2 Thermochemical Conversion (Pyrolysis, Gasification)

Thermochemical conversion transforms biomass into energy and chemicals through high-temperature processes, contributing significantly to net-zero goals by providing alternative energy sources and reducing reliance on fossil fuels. Pyrolysis decomposes organic materials at 400–600°C in the absence of oxygen, resulting in bio-oil, syngas, and biochar (Osman et al., 2023a; Sekar et al., 2021). Fast pyrolysis heats biomass rapidly to produce bio-oil, which can be refined into fuels (Aboelela et al., 2023; Li et al., 2021). Gasification partially oxidizes biomass at 800–1200°C to produce syngas, a mixture of carbon monoxide, hydrogen, and other gases (Milledge et al., 2014; Raheem et al., 2021). This syngas can be used for power generation or as a feedstock for chemical synthesis. Both macroalgae and microalgae can be utilized in thermochemical conversion processes, although precise control of operating conditions is required to manage challenges like feedstock variability and reactor design complexity (Samoraj et al., 2024). For example, the high moisture content of kelp requires drying before pyrolysis to prevent inefficient combustion and excessive tar formation (Osman et al., 2023b), while the high lipid content of microalgae like *Spirulina* necessitates careful reactor design to handle the increased production of bio-oil and manage heat distribution effectively during conversion (Deepika et al., 2024).

Innovations in reactor technology and process integration, including the development of fluidized bed reactors, continue to enhance the viability of these methods (Das et al., 2021). Fluidized bed reactors offer improved heat and mass transfer rates, leading to higher conversion efficiencies and more stable operation (Al-Rabiah et al., 2022; Zapater et al., 2024). Advancements such as catalytic pyrolysis, which uses catalysts to improve the quality of bio-oil and reduce unwanted by-products, are being explored (Lee et al., 2020a, 2020b). Similarly, the integration of gasification with Fischer-Tropsch synthesis can convert syngas into liquid fuels, offering a pathway to produce drop-in replacements for conventional fossil fuels (Blay-Roger et al., 2024; Jenčík et al., 2021). These technological improvements not only increase the efficiency and yield of bioenergy production but also enhance the flexibility of feedstock utilization, allowing a wider range of algal biomass to be processed effectively (Malode et al., 2021; Yahyazadeh et al., 2024). By optimizing these thermochemical processes, we can better harness the potential of algae as a renewable energy source, thereby supporting the transition to a more sustainable and carbon-neutral energy system.

6.5.1.3 Extraction of Valuable Compounds (Lipids, Proteins, Carbohydrates)

Extracting valuable compounds such as lipids, proteins, and carbohydrates from algal biomass maximizes its economic potential and supports net-zero goals by providing sustainable alternatives to conventional resources. Lipid extraction, crucial for biodiesel production, often employs solvent extraction methods such as the Bligh and Dyer technique, which utilizes a chloroform-methanol mixture to effectively separate lipids from the biomass (Saini et al., 2021b). Mechanical disruption methods like bead milling and ultrasonic extraction enhance lipid release by breaking down the tough cell walls of algae (Liu et al. 2022b). Additionally, supercritical fluid extraction using carbon dioxide offers an advanced method that ensures high purity and environmental benefits by avoiding the use of harmful solvents (Tzima et al, 2023).

Protein extraction involves the use of alkaline solutions or enzymes to break down cell walls, followed by membrane filtration to purify the proteins (de Souza Celente et al., 2023; Soto-Sierra et al., 2018). This process is essential for producing high-quality protein supplements and feed. Carbohydrate extraction typically involves acid or enzymatic hydrolysis to convert complex carbohydrates into fermentable sugars, which can then be used for bioethanol production (Tse et al., 2021; Vasić et al., 2021). Both macroalgae and microalgae are rich sources of these compounds. For instance, microalgae like *Dunaliella* are processed for their high-value pigments and proteins (de Souza Celente et al., 2022), while macroalgae such as *Laminaria* and *Gracilaria* are harvested for their polysaccharides and bioactive compounds (Xie et al., 2024). Continuous advancements in extraction technologies aim to improve yield, purity, and sustainability. Techniques such as supercritical fluid extraction and enzymatic hydrolysis are being optimized to enhance the efficiency and environmental friendliness of these processes (Arias et al., 2023b; Asanka Sanjeewa et al., 2023).

Integrating these advanced extraction methods into algal biorefineries not only enhances their economic viability but also contributes to achieving net-zero emissions. By providing sustainable and renewable sources of biofuels, biochemicals, and bio-based materials, these extraction processes reduce reliance on fossil fuels and lower greenhouse gas emissions (Mahmood et al., 2023; Nath, 2024). This aligns with the broader goal of creating a sustainable energy landscape and advancing toward net-zero objectives.

6.5.2 Technological Advancements

6.5.2.1 Recent Innovations in Biorefinery Technologies

Recent innovations have significantly advanced the efficiency and sustainability of biorefinery technologies, aligning with net-zero objectives by enhancing resource utilization and reducing emissions. In biochemical processes, genetic engineering tools like CRISPR have optimized microbial strains, resulting in higher yields and enhanced tolerance to by-products (Garg et al., 2023; Javed et al., 2019). This technology has notably improved lipid production in microalgae like *Chlorella*, which is essential for biodiesel production. Continuous fermentation systems have also emerged as superior alternatives to traditional batch processes, offering improved productivity and lower operational costs (Figueroa-Torres & Theodoropoulos, 2023; Gaur et al., 2024). These systems maintain a constant environment, ensuring optimal growth conditions for microorganisms and reducing downtime between production cycles.

In the realm of thermochemical processes, new reactor designs, including fluidized bed reactors, have significantly improved heat and mass transfer, resulting in better conversion rates and product quality (Azizi & Mowla, 2016; Das et al., 2021). The application of fluidized bed reactors in the pyrolysis of macroalgae, such as *Laminaria*, has increased both the yield and quality of bio-oil, demonstrating a more effective approach to biomass processing (Choi et al., 2015). Moreover, hybrid systems that integrate biochemical and thermochemical pathways maximize biomass utilization by combining the strengths of both approaches (Choudhary et al., 2020; Osman et al., 2021). After the biochemical extraction of lipids from microalgae, the residual biomass can be subjected to pyrolysis or gasification to produce additional biofuels and biochemicals, ensuring no part of the biomass is wasted (Sambasivam et al., 2023; Shahid et al., 2019).

Extraction technologies have also seen significant advancements. Novel solvents like ionic liquids and supercritical CO_2 provide high selectivity and environmental benefits, offering greener alternatives to traditional solvent extraction methods (Irianto et al., 2024; Tzima et al., 2023). Case studies from commercial operations have demonstrated the successful implementation of these innovations. Supercritical CO_2 extraction, for instance, has been effectively used to isolate high-value compounds from *Spirulina*, highlighting its economic and environmental benefits (Crampon et al., 2017; Pinto et al., 2022). These technological advancements contribute to the overall goal of net zero by improving the efficiency and sustainability of algal biorefineries, making them more competitive and viable in the global market. By integrating advanced biochemical, thermochemical, and extraction technologies, biorefineries can more effectively convert algal biomass into a wide range of valuable products, reducing reliance on fossil fuels and lowering greenhouse gas emissions (Allen et al., 2018; Fan et al., 2020).

6.5.2.2 Integration of Multi-Product Biorefineries

Multi-product biorefineries are designed to maximize resource utilization by producing multiple valuable products from a single biomass feedstock, thereby supporting net-zero objectives through enhanced efficiency and reduced waste. The integration of biochemical and thermochemical processes ensures complete biomass utilization, boosting efficiency and yield (Choudhary et al., 2020; Koyande et al., 2019). Residual biomass from fermentation, such as microalgae like *Chlorella* or macroalgae like *Laminaria*, can be repurposed as feedstock for pyrolysis or gasification, effectively closing the loop in biomass conversion (Faisal et al., 2022; Raheem et al., 2021). This approach not only maximizes output but also minimizes waste, aligning with sustainability goals.

The incorporation of advanced process control technologies, including sensors, the Internet of Things (IoT), and artificial intelligence (AI), significantly enhances the monitoring, data analysis, and decision-making processes in biorefineries (Abu Sofian et al., 2024; Oruganti et al., 2023). These technologies ensure optimal conditions for biomass conversion, leading to higher efficiency and reduced operational costs. For example, IoT-enabled sensors can monitor the moisture content in kelp before pyrolysis, ensuring proper drying to prevent inefficient combustion and excessive tar formation (Ibidoja et al., 2023). Similarly, AI algorithms can optimize fermentation conditions for microalgae to maximize lipid extraction for biofuel production (Coşgun et al., 2021; S et al., 2024).

Economically, multi-product biorefineries generate multiple revenue streams, reducing dependency on single products and enhancing financial viability (Dutta et al., 2023; Murthy Konda et al., 2015; Nishshanka et al., 2022). Environmentally, they contribute to waste reduction, lower greenhouse gas emissions, and improved resource utilization (Gurreri et al., 2023; Samoraj et al., 2024). Future trends in biorefinery technologies include the discovery of new biomass feedstocks, advancements in biotechnological tools, and supportive policies to drive adoption and growth. Exploring novel algae species with high lipid or carbohydrate content could further enhance biofuel production (Ahmed et al., 2024; Huang et al., 2022). Research and development (R&D) efforts are crucial to addressing current challenges and optimizing processes for cost-effective and scalable solutions (Ezhumalai et al., 2024; Johnston et al., 2023). These integrated approaches ensure that biorefineries contribute effectively to the transition toward net-zero emissions by providing sustainable and economically viable alternatives to fossil fuels.

6.6 ENVIRONMENTAL AND ECONOMIC BENEFITS

Algal biorefineries present a promising pathway toward achieving net-zero emissions, offering substantial environmental and economic benefits that align with the overarching goals of sustainability and climate change mitigation (Figure 6.2). By harnessing the natural ability of algae to efficiently capture CO_2, these systems convert atmospheric or industrial CO_2 into valuable biomass for various industrial applications. Integrating algal biorefineries within existing infrastructures can significantly reduce greenhouse gas emissions, contributing to climate change mitigation and promoting sustainable development. Economically, algal biorefineries open new

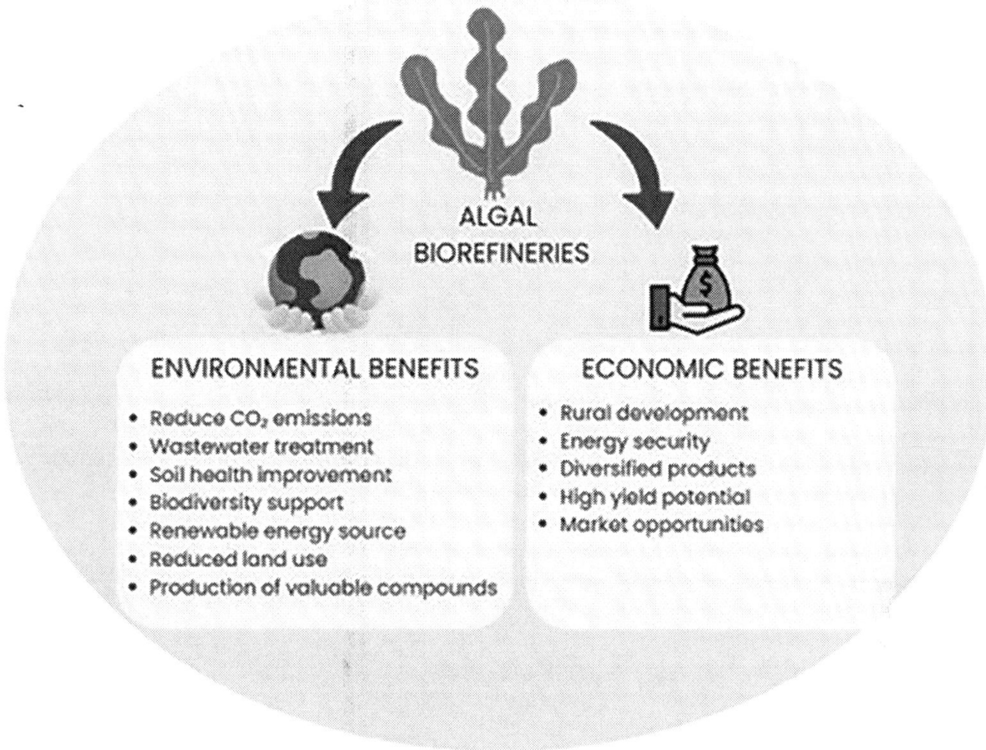

FIGURE 6.2 Overview of the environmental and economic benefits associated with algal biorefineries.

market opportunities and create jobs, thereby fuelling the growth of the bioeconomy (Katiyar et al., 2021; Ullmann & Grimm, 2021). This section explores these environmental and economic benefits, highlighting the potential of algal biorefineries to drive both ecological sustainability and economic prosperity.

6.6.1 Reduction of Greenhouse Gas Emissions

6.6.1.1 Carbon Sequestration Potential of Algae

Algal biorefineries provide a viable solution to environmental challenges through effective carbon sequestration. Algae have the remarkable ability to absorb and convert atmospheric CO_2 into organic compounds during photosynthesis (Figure 6.3), a process further enhanced by their rapid growth and high photosynthetic efficiency compared to land plants (Onyeaka et al., 2021; Singh & Ahluwalia, 2013). Using carbonic anhydrase, algae convert CO_2 into bicarbonate (HCO_3^-), facilitating the photosynthesis process within chloroplasts (Davoodbasha et al., 2024). This process begins with light-dependent reactions, where chlorophyll captures light energy, leading to the splitting of water molecules into oxygen, hydrogen ions, and electrons. These electrons travel through the electron transport chain, producing ATP and reducing $NADP^+$ to NADPH. Subsequently, the Calvin cycle reduces CO_2 to ribulose-1,5-bisphosphate (RuBP) with the aid of the enzyme ribulose-1,5-bisphosphate carboxylase/oxygenase (RuBisCO), generating 3-phosphoglycerate (3-PGA), which is then converted into glucose and other carbohydrates (Prasad et al., 2021). The harvested algal biomass can be used for biofuel production, animal feed, and bioplastics, contributing to a carbon-neutral cycle. Additionally, carbon derived from algae can be

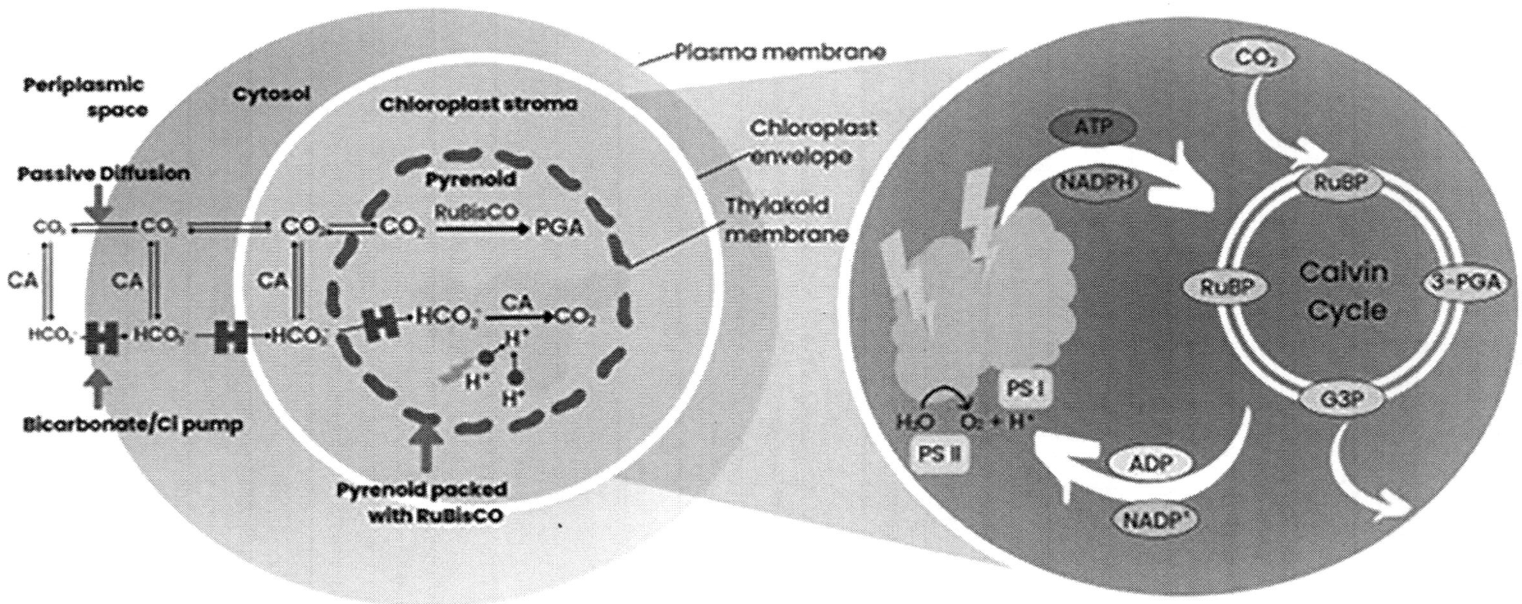

FIGURE 6.3 Mechanism of carbon sequestration in algae, illustrating the conversion of CO_2 into organic compounds during photosynthesis. (Adapted from Prasad et al., 2021.)

converted into biochar for long-term soil storage or utilized in CCS technologies to ensure stable carbon retention (Tripathi et al., 2023).

Recent advancements focus on enhancing algal strains and optimizing cultivation conditions to improve carbon sequestration. Genetically engineered microalgae have shown increased CO_2 fixation rates by targeting pathways that boost carbonic anhydrase activity, thus enhancing CO_2 uptake and biomass productivity, which can be further utilized in producing biofuels and other valuable bioproducts (Koh et al., 2023; Priyadharsini et al., 2022). Additionally, integrating algae cultivation with industrial processes not only reduces industrial carbon emissions but also produces useful biomass. Certain algae species can utilize industrial flue gases as a CO_2 source, capturing emissions that would otherwise contribute to atmospheric CO_2 levels (Iglina et al., 2022; Yang et al., 2024). This approach has been successfully tested in pilot-scale projects, where algae are cultivated in photobioreactors directly connected to industrial sources, allowing the CO_2 from flue gases to be converted into biomass (Cutshaw et al., 2020; Li et al., 2023).

6.6.1.2 Life Cycle Assessment of Algal Biorefinery Processes

Beyond carbon sequestration, algae offer numerous environmental benefits, such as nutrient absorption. Algae can absorb excess nutrients like nitrogen and phosphorus from wastewater, thereby mitigating eutrophication in aquatic ecosystems (Abdelfattah et al., 2023; Cakmak et al., 2022). This nutrient assimilation not only promotes algal growth and biomass accumulation but also creates a synergistic interaction between wastewater treatment and carbon capture (Farghali et al., 2023; Ospina and Darwich-Cedeño, 2024). Algae also contribute to water quality enhancement and biodiversity by providing habitats for aquatic organisms and helping maintain ecological balance (Cotas et al., 2023; Mulenga et al., 2024). This is particularly relevant in integrated aquaculture systems, where algae serve as natural feed and habitat, advancing sustainable aquaculture practices (Ma & Hu, 2024; Roleda & Hurd, 2019). Additionally, using algal biomass as biofertilizer improves soil health by enhancing soil structure, moisture retention, and nutrient levels, while also promoting the growth of beneficial soil microorganisms (Ammar et al., 2022).

Algae can also be converted into various biofuels, such as biodiesel, bioethanol, and biogas, providing a renewable alternative to fossil fuels (Devi et al., 2024). With a high energy yield per unit area, algal biofuels enhance efficiency and sustainability as an energy source (Prabakaran & Karthikeyan, 2024). Moreover, algae can be cultivated in non-arable regions, including deserts and saline areas, thereby minimizing competition with food crops for arable land (Diaz et al., 2023; Tahir et al., 2024). This approach optimizes land use through techniques like vertical farming (Abdur Razzak et al., 2024; Saad et al., 2019). Algae also produce valuable compounds such as omega-3 fatty acids, pigments, and bioplastics, generating multiple revenue streams and adding value to algal biorefineries (Hoang et al., 2023; Uma et al., 2023; Yadav et al., 2022). These diverse applications and benefits position algal biorefineries as a crucial technology for addressing environmental challenges and advancing toward net-zero emissions.

6.6.2 Economic Viability

6.6.2.1 Cost Analysis and Market Potential

The economic viability of algal biorefineries hinges on a comprehensive assessment of factors, including cost, market potential, and supportive policies (Devi et al., 2024; Kashyap et al., 2023). Ensuring that these biorefineries are both environmentally sustainable and economically competitive with conventional and emerging technologies requires addressing these key areas. Initial investments in algal biorefineries are substantial, particularly for infrastructure such as photobioreactors and open ponds, along with corresponding downstream processing facilities (Qin et al., 2019; Sirohi et al., 2022). However, advancements in technology and the realization of economies of scale are expected to reduce these costs over time. Recent developments in cultivation systems, harvesting techniques, and extraction methods are already contributing to a reduction in operational expenses. For instance, research has demonstrated a 30% reduction in production costs through the development of low-cost, large-scale cultivation systems like open raceway ponds and hybrid systems (Kumar et al., 2024; Malik et al., 2022; Shahid et al., 2020).

Algal biorefineries are poised to become a vital component of the growing bioeconomy, offering significant economic benefits. One of the major advantages is job creation across various sectors, including research, cultivation, processing, and distribution (Olabi et al., 2023; V et al., 2021). This is particularly advantageous for rural areas, where the establishment of algal biorefineries can stimulate local economies and create employment opportunities, thus promoting rural development and reducing migration pressures toward urban centers (Bhatia et al., 2022a; Tahir et al., 2024). Additionally, algal biorefineries contribute to energy security by producing biofuels that decrease dependence on imported fossil fuels. Diversifying energy sources in this way mitigates the risks associated with energy supply fluctuations and provides long-term price stability (Barragán-Ocaña et al., 2023; Gaurav et al., 2024).

The versatility of algae as a raw material significantly enhances the economic viability of algal biorefineries. Algae can be used to produce a wide range of valuable products, including pharmaceuticals, cosmetics, animal feed, and biopolymers, thereby generating multiple revenue streams (Kammler et al., 2024; Nishshanka et al., 2022). Additionally, the rapid growth rates of algae contribute to high productivity, which can reduce production costs and increase profitability, positioning algal biorefineries as competitive players in the bio-based industry (Bošnjaković & Sinaga, 2020; Wiatrowski et al., 2022). The global shift toward sustainability and eco-friendliness is also creating new market opportunities for algal-derived products (Samoraj et al., 2024). As consumers and industries increasingly seek greener alternatives, the demand for these products is expected to rise, driving economic growth and solidifying the role of algal biorefineries as key contributors to the emerging bioeconomy (Shahid et al., 2020; Barragán-Ocaña et al., 2023).

6.6.2.2 Economic Incentives and Policies Supporting Algal Biorefineries

Government policies and economic incentives are critical to advancing the research and commercialization of algal biorefineries. In many countries, policies promoting renewable energy and carbon emission reductions provide significant opportunities for algal biorefinery projects (Saini et al., 2021a; Sandaka & Kumar, 2023). These policies often include tax credits, grants, and subsidies for R&D, as well as mandates for renewable or low-carbon fuel blending with transportation fuels. For instance, the U.S. has introduced various support mechanisms for bioenergy research, such as the Advanced Research Projects Agency-Energy (ARPA-E), which funds innovative energy technologies, including algal biofuels (Blaustein, 2012; Chen et al., 2020; V et al., 2021). In the European Union, the Horizon Europe program provides substantial funding for research and innovation in bioeconomy sectors, including algae-based technologies, under its Green Deal and Circular Economy initiatives (Gatto & Re, 2021; Johnson et al., 2021; Lange et al., 2021). These policy frameworks help accelerate the development and commercialization of algal biorefineries by providing the necessary financial support and creating a favorable environment for technological advancements.

Carbon pricing mechanisms, such as carbon taxes or cap-and-trade systems, also play a crucial role in enhancing the competitiveness of algal biorefineries (Daneshvar et al., 2022). By assigning a financial cost to carbon emissions, these mechanisms incentivize low-carbon technologies, making algal biorefineries more economically viable (Ali, 2023; Le & Azhgaliyeva, 2023). Additionally, public–private partnerships and collaborations among academia, industry, and government are essential for accelerating the development of algal technologies (Makepa & Chihobo, 2024; Samoraj et al., 2024). These partnerships provide the necessary funding, expertise, and infrastructure to scale up operations from pilot projects to commercial levels. A prominent example is the collaboration between ExxonMobil and Synthetic Genomics, which has made significant strides in producing biofuels from algae on a commercial scale (Ajjawi et al., 2017). Their joint efforts have led to notable achievements, including a 40% increase in lipid production in genetically modified algae strains.

6.7 CHALLENGES AND SOLUTIONS

Algal biorefineries offer promising opportunities for sustainable energy production and value-added products. However, the path to their widespread adoption is fraught with challenges that must be addressed to unlock their full potential. These challenges span technical, economic, and regulatory domains, each requiring tailored solutions to ensure the viability and scalability of algal biorefineries. This section explores the primary obstacles encountered in scaling up production, improving biomass conversion efficiency, securing investment, and navigating regulatory frameworks. It also proposes potential solutions through R&D strategies, policy recommendations, and international collaboration.

6.7.1 Technical Challenges

6.7.1.1 Scaling Up Production

Scaling up algal biorefineries from laboratory to commercial levels presents a significant technical challenge. Replicating controlled lab conditions on an industrial scale is complex and often leads to difficulties in maintaining optimal growth environments, such as light intensity, nutrient availability, and CO_2 concentration, across large cultivation systems (Benner et al., 2022). Additionally, the risk of contamination increases with scale, posing threats to biomass yield and quality (Gaurav et al., 2024). The energy-intensive nature of large-scale algae cultivation and harvesting processes also raises concerns about the overall energy balance and economic feasibility of algal biorefineries (Koyande et al., 2019). Addressing these challenges requires innovative cultivation systems that can maintain the delicate balance of growth conditions at scale, along with advanced monitoring and control technologies (Abdur Razzak et al., 2024; Samoraj et al., 2024). Moreover, integrating renewable energy sources and optimizing resource use within the biorefinery process can help mitigate energy consumption issues, making large-scale algae production more sustainable and economically viable (Allen et al., 2018; Dutta et al., 2023).

6.7.1.2 Efficiency of Biomass Conversion Processes

The efficiency of converting algal biomass into biofuels and other valuable products is another critical technical challenge. Algal cells have complex structures that make the extraction of lipids, proteins, and carbohydrates energy-intensive and costly (Jothibasu et al., 2022; Kostas et al., 2021). Conventional methods such as solvent extraction, mechanical disruption, and chemical hydrolysis often lead to incomplete recovery of these components, reducing the overall yield and profitability of the biorefinery process (Matos et al., 2021; Tse et al., 2021). Furthermore, the co-production of multiple products from a single biomass source introduces additional complexities, as each product stream may require different processing conditions (Premaratne et al., 2022; Zafar et al., 2023). To overcome these challenges, advancements in biomass pretreatment and fractionation techniques are essential (Bhatia et al., 2022b; Sharma et al., 2024). The development of integrated biorefinery processes that can efficiently separate and process multiple components in a single step can improve conversion efficiency (Dong et al., 2016; Narayanan, 2024). Innovations in enzymatic and microbial conversion processes also hold promise for enhancing the yield and reducing the cost of biomass conversion, thereby improving the overall viability of algal biorefineries (Nath, 2024; Parakh et al., 2023; Zuliani et al., 2021).

6.7.2 Economic and Regulatory Hurdles

6.7.2.1 Investment and Funding Challenges

Securing adequate investment and funding remains a significant hurdle for the development and commercialization of algal biorefineries. The high capital costs associated with building and maintaining large-scale algae cultivation and processing facilities often deter private investors, especially given the long timelines required for returns on investment (Biloria & Thakkar, 2020; Johnston et al., 2023). Additionally, the perceived risks related to the scalability, market demand, and regulatory uncertainties further exacerbate funding challenges (Makepa & Chihobo, 2024; Samoraj et al., 2024). Public funding for algal biorefineries is also limited and highly competitive, with many projects competing for a limited pool of resources. To address these challenges, innovative funding mechanisms such as public–private partnerships, venture capital, and green bonds can be explored (Hegeman & Sørheim, 2021; Jian, 2023; Lange et al., 2021). Additionally, demonstrating the economic viability and environmental benefits of algal biorefineries through pilot projects can help attract more investors (Kingsley et al., 2023). Establishing clear and consistent communication about the potential of algal technologies can also play a crucial role in building investor confidence and securing the necessary financial support (Gatto & Re, 2021; van Dijk et al., 2024).

6.7.2.2 Regulatory Frameworks and Standards

Navigating the complex and often inconsistent regulatory frameworks governing algal biorefineries poses a significant challenge to their commercialization. Regulations related to environmental impacts, product safety, and market approval vary widely across regions, creating barriers to entry and scaling (Lo et al., 2020; Smith et al., 2024). For instance, biofuel production from algae may require compliance with both renewable energy standards and environmental regulations, which can be cumbersome and costly (Khoo et al., 2023; Malode et al., 2021). Additionally, the lack of standardized methods for assessing the sustainability and life cycle impacts of algal biorefineries further complicates regulatory approval processes (Braud et al., 2023; Talwar & Holden, 2022). To overcome these challenges, harmonizing regulatory frameworks at regional and international levels is crucial (Camarena-Gómez et al., 2022; Rumin et al., 2021). Developing clear guidelines and standards specific to algal biorefineries can streamline approval processes and reduce compliance costs. Moreover, fostering dialog between industry stakeholders and regulatory bodies can lead to more pragmatic and supportive policies that encourage innovation and commercialization in the algal bioeconomy (Makepa & Chihobo, 2024).

6.7.3 Potential Solutions

6.7.3.1 Research and Development Strategies

R&D plays a pivotal role in overcoming the challenges faced by algal biorefineries. Focused R&D efforts on optimizing cultivation techniques, improving biomass conversion processes, and developing cost-effective harvesting methods are essential for enhancing the overall efficiency and economic viability of algal biorefineries (Hasnain et al., 2024; Kombe, 2023). Additionally, exploring genetic engineering and synthetic biology approaches can lead to the development of algae strains with enhanced growth rates, higher lipid content, and greater resilience to environmental stressors, thereby improving productivity and reducing costs (Gupta et al., 2024; Naduthodi et al., 2021). Collaboration between academia, industry, and government in R&D initiatives can accelerate the pace of innovation, leading to breakthroughs that can be rapidly translated into commercial applications (Makepa & Chihobo, 2024; Samoraj et al., 2024). Furthermore, expanding research into the co-production of high-value bioproducts alongside biofuels can diversify revenue streams and make algal biorefineries more economically sustainable (Kim et al., 2019; Zafar et al., 2023; Zhou et al., 2022b).

6.7.3.2 Policy Recommendations and International Collaboration

Policy recommendations and international collaboration are critical to creating an enabling environment for the growth of algal biorefineries. Governments can support the sector by implementing policies that incentivize research, development, and commercialization of algae-based technologies, such as tax credits, grants, and subsidies (Moshood et al., 2021; Osman et al., 2024). Additionally, establishing clear and consistent regulations that facilitate market access for algal products can reduce uncertainties and encourage investment (Lähteenmäki-Uutela et al., 2021; Rumin et al., 2021). International collaboration is also essential, as it allows for the sharing of knowledge, resources, and best practices, which can accelerate the development of algal biorefineries globally (Samoraj et al., 2024; V et al., 2021). Collaborative efforts, such as international research consortia and cross-border pilot projects, can help overcome technical and regulatory barriers by pooling expertise and resources (McLaughlin et al., 2023; Zhang & Thomsen, 2021). By aligning policy frameworks and fostering global cooperation, the potential of algal biorefineries can be fully realized, contributing to a more sustainable and resilient bioeconomy.

6.8 CASE STUDIES

The development and implementation of algal biorefineries have seen varying degrees of success across the globe, providing valuable insights into the potential and challenges of these systems. This section highlights successful algal biorefinery projects from different regions, examining the lessons learned and best practices that can be adopted. Additionally, it explores emerging projects and innovations, showcasing the latest advancements in technology and offering a glimpse into the future prospects of algal biorefineries. These case studies serve as practical examples of how algal biorefineries can be effectively integrated into the bioeconomy and contribute to sustainability goals.

6.8.1 Successful Algal Biorefinery Projects

In the U.S., Algenol Biotech has pioneered the use of algae for bioethanol production, demonstrating a significant reduction in greenhouse gas emissions compared to traditional fossil fuels (Algenol Biotech LLC., 2024). Their patented technology uses algae to convert CO_2 into ethanol, achieving high productivity with minimal environmental impact (Mody et al., 2015). In Europe, the Allmicroalgae project in Portugal, supported by the European Union, focuses on the large-scale cultivation of microalgae for producing biofuels, animal feed, and high-value products such as omega-3 fatty acids (Allmicroalgae Natural Products, 2024). This initiative has successfully demonstrated the scalability of microalgae cultivation and its integration into existing agricultural systems, underscoring its economic and environmental viability (Conde et al., 2021). In Asia, Euglena Co., Ltd. in Japan has developed advanced microalgae cultivation systems for producing biofuels and high-value bioproducts (Euglena Co., Ltd., 2024). Their approach integrates large-scale cultivation with advanced processing to produce biodiesel and nutritional supplements, reflecting significant advancements in sustainable technology (Suzuki, 2017). These examples illustrate how regional approaches, driven by local resources and needs, have successfully adapted algal biorefinery technologies to meet both environmental and economic objectives.

The successful implementation of these algal biorefinery projects offers several key lessons and best practices that can guide future initiatives. First, the importance of leveraging patented technologies, as demonstrated by Algenol Biotech in the U.S., cannot be overstated. By tailoring innovations to local resources, these technologies can achieve high productivity while minimizing environmental impacts, particularly in reducing greenhouse gas emissions (Yadav et al., 2022). The integration of algal biorefineries with existing agricultural or industrial systems, as seen in the Allmicroalgae project in Portugal, is another critical lesson. This approach not only reduces the initial capital investment but also ensures the effective utilization of existing resources, such as CO_2 emissions from nearby industries, which can be repurposed to enhance the efficiency of algal cultivation (Guerra et al., 2021; Miranda et al., 2024). Continuous R&D efforts are equally vital, particularly in optimizing algal strains, improving cultivation techniques, and scaling operations. Euglena Co., Ltd. in Japan exemplifies how advanced processing systems and ongoing innovation can lead to significant advancements in sustainable biofuel and bioproduct production (Qin et al., 2023). Moreover, the collaboration between industry, academia, and government entities has been a cornerstone of success, facilitating the acceleration of technological advancements and their broader adoption (Moon et al., 2022; Vickram et al., 2023). Finally, maintaining a diversified product portfolio, including biofuels, animal feed, and high-value bioproducts, is essential for ensuring economic resilience and adapting to fluctuating market demands (Gaurav et al., 2024; Samoraj et al., 2024).

6.8.2 Emerging Projects and Innovations

Emerging technologies and pilot projects are redefining the capabilities of algal biorefineries, setting the stage for a more sustainable and efficient future. A notable innovation is the development of floating photobioreactors, specifically designed to optimize light exposure and nutrient delivery in aquatic environments (Huang et al., 2016; Zhu et al., 2019). These reactors are being tested on a large scale and have the potential to significantly enhance algal cultivation by fully utilizing available space and natural resources, such as sunlight and water currents (Francke et al., 2022; Khor et al., 2022). This approach not only maximizes biomass production but also reduces land usage, making it an attractive option for regions with limited terrestrial space. In parallel, advancements in metabolic engineering are enabling the development of microalgae strains with enhanced lipid production and greater tolerance to environmental stressors (Lu et al., 2021; Song et al., 2024; Sun et al., 2019). These genetically modified strains show great promise in commercial-scale operations, with the potential to revolutionize biofuel production by increasing yields and reducing costs. Additionally, the integration of algal biorefineries with carbon capture technologies is another groundbreaking development (Iglina et al., 2022; Singh & Dhar, 2019). This innovative approach allows algae to absorb CO_2 from industrial emissions, converting it into valuable bioproducts such as biofuels and bioplastics (Johnston et al., 2023; Yadav et al., 2022). This not only helps mitigate greenhouse gas emissions but also provides an economic incentive for the process. These emerging technologies are paving the way for more efficient, sustainable, and versatile algal biorefinery systems, promising significant environmental and economic benefits.

The future of algal biorefineries is promising, with several trends indicating their growing importance in the global bioeconomy. As technology advances, the scalability and efficiency of algal production are expected to improve, making these systems more economically viable. The increasing focus on sustainability and carbon neutrality is likely to drive demand for algal-based products, particularly in industries such as biofuels, bioplastics, and pharmaceuticals (Mohan et al., 2022; Sadvakasova et al., 2023). Moreover, international collaboration and policy support will play a critical role in overcoming current barriers, such as regulatory hurdles and high production costs (Makepa & Chihobo, 2024; Samoraj et al., 2024). The potential for algal biorefineries to contribute to the circular economy and mitigate climate change will likely spur further investment and innovation in this field (Sharmila et al., 2022; Zabochnicka et al., 2022). As these technologies mature, algal biorefineries could become a cornerstone of the sustainable development agenda, offering solutions to some of the most pressing environmental and economic challenges of our time.

6.9 FUTURE PERSPECTIVES

The future of algal biorefineries holds immense promise for driving global efforts toward net-zero carbon emissions and climate resilience. As the world urgently transitions to sustainable energy systems, the significance of algal biorefineries becomes increasingly apparent. Scaling up these biorefineries involves overcoming technical, economic, and policy challenges to establish efficient infrastructure and sustainable practices. Integrating algal biorefineries with other renewable technologies, such as solar, wind, and bioenergy, can create synergies that enhance overall system efficiency and sustainability. Long-term impacts on climate change mitigation are projected to be substantial, with algal biorefineries playing a pivotal role in achieving net-zero goals. By leveraging advanced cultivation techniques, innovative technologies, and comprehensive life cycle assessments, the potential of algal biorefineries can be fully realized. The collective efforts of the public and private sectors, supportive policies, and continuous innovation will be crucial in driving the global adoption of this transformative technology, ultimately contributing to a sustainable and resilient energy future.

To achieve large-scale implementation of algal biorefineries, a multifaceted approach is essential. This includes optimizing cultivation techniques by utilizing advanced systems like closed photobioreactors and hybrid models that combine open ponds with photobioreactors, which can significantly enhance algal biomass productivity (Penloglou et al., 2024; Yun et al., 2018). The development of genetically engineered algae strains with higher lipid and carbohydrate content is another critical strategy, as these can further increase yields (Khoo et al., 2023; Zhang et al., 2023). Additionally, the integration of automated monitoring and control systems for parameters such as nutrient delivery, light intensity, and CO_2 concentration is crucial for maximizing growth rates and ensuring high-quality biomass (Lim et al., 2022a; Reyes et al., 2024). On the conversion side, incorporating technologies like hydrothermal liquefaction and supercritical fluid extraction can improve the efficiency of converting algal biomass into biofuels, biochemicals, and other valuable products (Deepika et al., 2024; Duan et al., 2018). Addressing the logistical aspects, such as optimizing harvesting, dewatering, and transportation processes, can streamline the supply chain and reduce costs (Dębowski et al., 2023; Penloglou et al., 2024). Establishing decentralized biorefineries near cultivation sites can minimize transportation-related carbon footprints (López-Molina et al., 2020; Sarwer et al., 2022). Finally, economic and policy considerations must be addressed through public–private partnerships, financial incentives, and supportive regulations, which are vital for creating an enabling environment for scaling up algal biorefineries (Moshood et al., 2021).

The scaling up of algal biorefineries requires robust collaboration between the public and private sectors, each contributing uniquely to the process. The public sector plays a pivotal role in providing funding for research, infrastructure development, and pilot projects (Bošnjaković & Sinaga, 2020; Chanana et al., 2023). Governments can further stimulate industry growth by establishing clear regulatory frameworks and offering financial incentives that attract private investments. Public sector involvement in research consortia and public–private partnerships also facilitates knowledge sharing and collaborative innovation (Pessôa et al., 2021; Rumin et al., 2021). On the other hand, private companies are crucial for driving technological advancements and bringing algal biorefinery products to market (Gatto & Re, 2021; Su et al., 2017). Through investments in cutting-edge research, pilot projects, and the scaling up of production facilities, the private sector accelerates the commercialization of cost-effective and efficient biorefinery processes (Behera et al., 2022; Dębowski et al., 2023). Collaboration with academic institutions and government agencies enhances this innovation ecosystem, ensuring a smooth transition from research to commercial applications (Makepa & Chihobo, 2024). Large-scale demonstration projects, which often result from public–private collaboration, are essential for validating technologies and overcoming regulatory, technical, and economic barriers, thus paving the way for the widespread adoption of algal biorefineries (Mohammad et al., 2024; Venkatesh, 2022).

Integrating algal biorefineries with other renewable technologies, such as solar, wind, and bioenergy, can create powerful synergies that enhance overall system efficiency and sustainability. For example, solar energy can be harnessed to power algal cultivation systems, particularly photobioreactors, thereby reducing the carbon footprint associated with energy use (Villalba et al., 2023). Similarly, wind energy can complement solar power in providing a stable, renewable energy source for biorefinery operations

(Lim et al., 2022b). Bioenergy, derived from the organic residues of algal processing, can be used to generate additional power, further reducing reliance on external energy sources (Benedetti et al., 2018; Shahid et al., 2019). This holistic integration not only optimizes the use of renewable resources but also improves the economic viability of algal biorefineries by reducing operational costs and enhancing energy self-sufficiency (Saran et al., 2024; Sarwer et al., 2022). Such synergies are critical for creating a more resilient and sustainable energy system, where multiple renewable technologies work in harmony to achieve greater environmental and economic benefits.

A holistic approach to sustainable energy systems involves the seamless integration of algal biorefineries into broader renewable energy networks. This approach emphasizes the interconnectedness of different energy technologies and the need for a cohesive strategy that maximizes resource efficiency and minimizes environmental impact. For instance, integrating algal biorefineries with wastewater treatment plants not only provides a nutrient-rich environment for algae cultivation but also purifies water, contributing to a circular economy (Fal et al., 2023; Hasan et al., 2023). Additionally, coupling algal biorefineries with CCS technologies can address both energy production and climate change mitigation simultaneously (Onyeaka et al., 2021; Roy et al., 2023). By incorporating life cycle assessments and sustainability metrics into the planning and operation of these systems, stakeholders can ensure that algal biorefineries contribute positively to global sustainability goals (Arias et al., 2023a; Kumar & Verma, 2021). This holistic perspective is essential for transitioning to a low-carbon future, where diverse renewable energy sources are integrated to create a balanced, resilient, and sustainable energy ecosystem.

Algal biorefineries are emerging as a pivotal technology in the global pursuit of net-zero carbon emissions. Unlike traditional energy sources, algae-based biofuels offer the potential to achieve carbon neutrality by capturing and utilizing CO_2 during cultivation, which offsets emissions during fuel combustion (Gao & Beardall, 2022; Sadvakasova et al., 2023). This carbon cycle not only mitigates the impact of fossil fuels but also provides a renewable source of energy that aligns with long-term climate goals. Furthermore, the adaptability of algal systems allows for their integration into existing industrial processes, enhancing their scalability and applicability across various sectors (Arora et al., 2021; Senthil Rathi et al., 2024). By contributing to the decarbonization of industries such as transportation and power generation, algal biorefineries offer a practical and sustainable pathway toward net-zero targets (Reddy et al., 2024; Sharmila et al., 2022). Additionally, the development of bio-based products from algal biomass, including biodegradable plastics and high-value chemicals, further reinforces the role of algae in reducing the carbon footprint (Cheah et al., 2023; Wu et al., 2023). As countries and industries intensify efforts to meet net-zero commitments, the expansion and optimization of algal biorefineries will be crucial in achieving these ambitious climate objectives.

The global outlook for algal biorefineries is increasingly optimistic, driven by the urgent need to meet net-zero carbon emission goals (DeAngelo et al., 2023; Paul et al., 2021). As the technology continues to advance, it is gaining recognition as a versatile solution that can be deployed in diverse geographic and industrial contexts. Regions with high levels of industrial CO_2 emissions and ample sunlight, such as those with established agricultural or manufacturing bases, are particularly well-positioned to adopt algal biorefinery systems (Iglina et al., 2022; Ou et al., 2021). International collaboration and investment are key to accelerating the deployment of these technologies, especially in areas where existing infrastructure can be leveraged to scale up operations rapidly (Khujamberdiev & Cho, 2024; Nath, 2024). Moreover, supportive policy frameworks and incentives, such as carbon credits and renewable energy mandates, are expected to play a significant role in driving the adoption of algal biorefineries on a global scale (De Bhowmick et al., 2019; McLaughlin et al., 2023). As nations strive to align their energy systems with net-zero goals, algal biorefineries are likely to become an integral part of the global energy landscape, offering a sustainable and economically viable alternative to fossil fuels (Hussain & Rittmann, 2023; Zafar et al., 2023). This widespread adoption will not only contribute to global climate targets but also foster new industries and job opportunities in the green economy.

6.10 CONCLUSION

The path to achieving net-zero emissions is intricate and demands innovative, sustainable solutions across multiple sectors. Algal biorefineries are poised to play a critical role in this global transition, offering a versatile and efficient means of converting algal biomass into valuable products such as biofuels, bioplastics, and other bioproducts. This chapter has highlighted the pivotal role of algal biorefineries within the broader framework of sustainable development and climate resilience, underscoring their potential to contribute significantly to the global effort toward achieving net-zero emissions.

DECLARATION

The authors have not used any generative AI tools during the writing of this manuscript.

REFERENCES

Abdelfattah, A., Ali, S. S., Ramadan, H., El-Aswar, E. I., Eltawab, R., Ho, S. H., Elsamahy, T., Li, S., El-Sheekh, M. M., Schagerl, M., Kornaros, M., & Sun, J. (2023). Microalgae-based wastewater treatment: Mechanisms, challenges, recent advances, and future prospects. *Environmental Science and Ecotechnology*, *13*, 100205. doi:10.1016/j.ese.2022.100205

Abdur Razzak, S., Bahar, K., Oajedul Islam, K. M., Haniffa, A. K., Faruque, M. O., Zakir Hossain, S. M., & Hossain, M. M. (2024). Microalgae cultivation in photobioreactors: Sustainable solutions for a greener future. *Green Chemical Engineering, 5*(4), 418–439. https://doi.org/10.1016/j.gce.2023.10.004

Aboelela, D., Saleh, H., Attia, A. M., Elhenawy, Y., Majozi, T., & Bassyouni, M. (2023). Recent advances in biomass pyrolysis processes for bioenergy production: Optimization of operating conditions. *Sustainability*, *15*(14), 11238. doi:10.3390/su151411238

Abu Sofian, A. D. A., Lim, H. R., Manickam, S., Ang, W. L., & Show, P. L. (2024). Towards a sustainable circular economy: Algae-based bioplastics and the role of internet-of-things and machine learning. *ChemBioEng Reviews*, *11*(1), 39–59. doi:10.1002/cben.202300028

AghaKouchak, A., Chiang, F., Huning, L. S., Love, C. A., Mallakpour, I., Mazdiyasni, O., Moftakhari, H., Papalexiou, S. M., Ragno, E., & Sadegh, M. (2020). Climate extremes and compound hazards in a warming world. *Annual Review of Earth and Planetary Sciences*, *48*, 519–548. doi:10.1146/annurev-earth-071719-055228

Ahmed, N., Sheikh, M. A., Ubaid, M., Chauhan, P., Kumar, K., & Choudhary, S. (2024). Comprehensive exploration of marine algae diversity, bioactive compounds, health benefits, regulatory issues, and food and drug applications. *Measurement: Food*, *14*, 100163. doi:10.1016/j.meafoo.2024.100163

Ahmed, R. A., He, M., Aftab, R. A., Zheng, S., Nagi, M., Bakri, R., & Wang, C. (2017). Bioenergy application of *Dunaliella salina* SA 134 grown at various salinity levels for lipid production. *Scientific Reports*, *7*, 8118. doi:10.1038/s41598-017-07540-x

Ajjawi, I., Verruto, J., Aqui, M., Soriaga, L. B., Coppersmith, J., Kwok, K., Peach, L., Orchard, E., Kalb, R., Xu, W., Carlson, T. J., Francis, K., Konigsfeld, K., Bartalis, J., Schultz, A., Lambert, W., Schwartz, A. S., Brown, R., & Moellering, E. R. (2017). Lipid production in *Nannochloropsis gaditana* is doubled by decreasing expression of a single transcriptional regulator. *Nature Biotechnology*, *35*, 647–652. doi:10.1038/nbt.3865

Algenol Biotech LLC. (2024). The Algenol Advantage. www.algenol.com/. Accessed 10 August 2024.

Ali, J. (2023). Road map for sustainable and effective carbon pricing: Bridging the gap of realities and ambitions. *Environmental Science and Pollution Research*, *30*, 94070–94080. doi:10.1007/s11356-023-29034-2

Allen, J., Unlu, S., Demirel, Y., Black, P., & Riekhof, W. (2018). Integration of biology, ecology and engineering for sustainable algal-based biofuel and bioproduct biorefinery. *Bioresources and Bioprocessing*, *5*, 47. doi:10.1186/s40643-018-0233-5

Allen, M. R., Friedlingstein, P., Girardin, C. A., Jenkins, S., Malhi, Y., Mitchell-Larson, E., Peters, G. P., & Rajamani, L. (2022). Net zero: Science, origins, and implications. *Annual Review of Environment and Resources*, *47*, 849–887. doi:10.1146/annurev-environ-112320-105050

Allmicroalgae Natural Products. (2024). Cultivating Sustainable Microalgae Solutions. www.allmicroalgae.com/en/. Accessed 12 August 2024.

AlMohamadi, H., Alamoudi, M., Yameen, M. Z., & Naqvi, S. R. (2023). An integrated approach for the extraction of lipids from marine macroalgae consortium using RSM optimization and thermo-kinetic analysis. *Chemosphere*, *338*, 139623. doi:10.1016/j.chemosphere.2023.139623

Al-Rabiah, A. A., Al-Dawsari, J. N., Ajbar, A. M., Al Darwish, R. K., & Abdelaziz, O. Y. (2022). Development of a biomass gasification process for the coproduction of methanol and power from red sea microalgae. *Energies*, *15*(21), 7890. doi:10.3390/en15217890

Ammar, E. E., Aioub, A. A. A., Elesawy, A. E., Karkour, A. M., Mouhamed, M. S., Amer, A. A., & EL-Shershaby, N. A. (2022). Algae as bio-fertilizers: Between current situation and future prospective. *Saudi Journal of Biological Sciences*, *29*(5), 3083–3096. doi:10.1016/j.sjbs.2022.03.020

Arent, D. J., Green, P., Abdullah, Z., Barnes, T., Bauer, S., Bernstein, A., Berry, D., Berry, J., Burrell, T., Carpenter, B., Cochran, J., Cortright, R., Curry-Nkansah, M., Denholm, P., Gevorian, V., Himmel, M., Livingood, B., Keyser, M., King, J., Kroposki, B., Mai, T., Mehos, M., Muratori, M., Narumanchi, S., Pivovar, B., Romero-Lankao, P., Ruth, M., Stark, G., & Turchi, C. (2022). Challenges and opportunities in decarbonizing the U.S. energy system. *Renewable and Sustainable Energy Reviews*, *169*, 112939. doi:10.1016/j.rser.2022.112939

Arias, A., Feijoo, G., & Moreira, M. T. (2023a). Biorefineries as a driver for sustainability: Key aspects, actual development and future prospects. *Journal of Cleaner Production*, *418*, 137925. doi:10.1016/j.jclepro.2023.137925

Arias, A., Feijoo, G., & Moreira, M. T. (2023b). Macroalgae biorefineries as a sustainable resource in the extraction of value-added compounds. *Algal Research*, *69*, 102954. doi:10.1016/j.algal.2022.102954

Arora, K., Kumar, P., Bose, D., Li, X., & Kulshrestha, S. (2021). Potential applications of algae in biochemical and bioenergy sector. *3 Biotech*, *11*, 296. doi:10.1007/s13205-021-02825-5

Arora, P., Chance, R., Hendrix, H., Realff, M. J., Thomas, V. M., & Yuan, Y. (2020). Life cycle greenhouse gas emissions of different CO_2 supply options for an algal biorefinery. *Journal of CO2 Utilization*, *40*, 101213. doi:10.1016/j.jcou.2020.101213

Arora, Y., Sharma, S., & Sharma, V. (2023). Microalgae in bioplastic production: A comprehensive review. *Arabian Journal for Science and Engineering*, *48*, 7225–7241. doi:10.1007/s13369-023-07871-0

Asanka Sanjeewa, K. K., Herath, K. H. I. N. M., Kim, Y. S., Jeon, Y. J., & Kim, S. K. (2023). Enzyme-assisted extraction of bioactive compounds from seaweeds and microalgae. *TrAC Trends in Analytical Chemistry*, *167*, 117266. doi:10.1016/j.trac.2023.117266

Azizi, S., & Mowla, D. (2016). CFD modeling of algae flash pyrolysis in the batch fluidized bed reactor including heat carrier particles. *International Journal of Chemical Reactor Engineering*, *14*(1), 463–480. doi:10.1515/ijcre-2014-0185

Babich, O., Ivanova, S., Michaud, P., Budenkova, E., Kashirskikh, E., Anokhova, V., & Sukhikh, S. (2024). Fermentation of micro- and macroalgae as a way to produce value-added products. *Biotechnology Reports*, *41*, e00827. doi:10.1016/j.btre.2023.e00827

Barbhuiya, S., Das, B. B., & Adak, D. (2024). Roadmap to a net-zero carbon cement sector: Strategies, innovations and policy imperatives. *Journal of Environmental Management*, *359*, 121052. doi:10.1016/j.jenvman.2024.121052

Barragán-Ocaña, A., Merritt, H., Sánchez-Estrada, O. E., Méndez-Becerril, J. L., & Longar-Blanco, M. P. (2023). Biorefinery and sustainability for the production of biofuels and value-added products: A trends analysis based on network and patent analysis. *PLoS One*, *18*(1), e0279659. doi:10.1371/journal.pone.0279659

Behera, B., Mari Selvam. S., & Paramasivan, B. (2022). Research trends and market opportunities of microalgal biorefinery technologies from circular bioeconomy perspectives. *Bioresource Technology*, *351*, 127038. doi:10.1016/j.biortech.2022.127038

Benedetti, M., Vecchi, V., Barera, S., & Dall'Osto, L. (2018). Biomass from microalgae: The potential of domestication towards sustainable biofactories. *Microbial Cell Factories*, *17*, 173. doi:10.1186/s12934-018-1019-3

Benner, P., Meier, L., Pfeffer, A., Krüger, K., Vargas, J. E. O., & Weuster-Botz, D. (2022). Lab-scale photobioreactor systems: Principles, applications, and scalability. *Bioprocess and Biosystems Engineering*, *45*, 791–813. doi:10.1007/s00449-022-02711-1

Bhatia, L., Bachheti, R. K., Garlapati, V. K., & Chandel, A. K. (2022a). Third-generation biorefineries: A sustainable platform for food, clean energy, and nutraceuticals production. *Biomass Conversion and Biorefinery*, *12*, 4215–4230. doi:10.1007/s13399-020-00843-6

Bhatia, S. K., Ahuja, V., Chandel, N., Gurav, R., Bhatia, R. K., Govarthanan, M., Tyagi, V. K., Kumar, V., Pugazendhi, A., Banu, J. R., & Yang, Y. H. (2022b). Advances in algal biomass pretreatment and its valorisation into biochemical and bioenergy by the microbial processes. *Bioresource Technology*, *358*, 127437. doi:10.1016/j.biortech.2022.127437

Biloria, N., & Thakkar, Y. (2020). Integrating algae building technology in the built environment: A cost and benefit perspective. *Frontiers of Architectural Research*, *9*(2), 370–384. doi:10.1016/j.foar.2019.12.004

Biris-Dorhoi, E. S., Michiu, D., Pop, C. R., Rotar, A. M., Tofana, M., Pop, O. L., Socaci, S. A., & Farcas, A. C. (2020). Macroalgae – A sustainable source of chemical compounds with biological activities. *Nutrients*, *12*(10), 3085. doi:10.3390/nu12103085

Blaustein, R. (2012). Can biology transform our energy future? ARPA-E infuses innovative research ventures with fresh funds. *BioScience*, *62*(2), 115–119. doi:10.1525/bio.2012.62.2.4

Blay-Roger, R., Nawaz, M. A., Baena-Moreno, F. M., Bobadilla, L. F., Reina, T. R., & Odriozola, J. A. (2024). Tandem catalytic approaches for CO_2 enriched Fischer-Tropsch synthesis. *Progress in Energy and Combustion Science*, *103*, 101159. doi:10.1016/j.pecs.2024.101159

Bošnjaković, M., & Sinaga, N. (2020). The perspective of large-scale production of algae biodiesel. *Applied Sciences*, *10*(22), 8181. doi:10.3390/app10228181

Braud, L., McDonnell, K., & Murphy, F. (2023). Environmental life cycle assessment of algae systems: Critical review of modelling approaches. *Renewable and Sustainable Energy Reviews*, *179*, 113218. doi:10.1016/j.rser.2023.113218

Cakmak, E. K., Hartl, M., Kisser, J., & Cetecioglu, Z. (2022). Phosphorus mining from eutrophic marine environment towards a blue economy: The role of bio-based applications. *Water Research*, *219*, 118505. doi:10.1016/j.watres.2022.118505

Camarena-Gómez, M. T., Lähteenmäki-Uutela, A., & Spilling, K. (2022). Macroalgae production in Northern Europe: Business and government perspectives on how to regulate a novel blue bioeconomy. *Aquaculture*, *560*, 738434. doi:10.1016/j.aquaculture.2022.738434

Caroppo, C., & Pagliara, P. (2022). Microalgae: A promising future. *Microorganisms*, *10*(8), 1488. doi:10.3390/microorganisms10081488

Chanana, I., Kaur, P., Kumar, L., Kumar, P., & Kulshreshtha, S. (2023). Advancements in microalgal biorefinery technologies and their economic analysis and positioning in energy resource market. *Fermentation*, *9*(3), 202. doi:10.3390/fermentation9030202

Chanquia, S. N., Vernet, G., & Kara, S. (2022). Photobioreactors for cultivation and synthesis: Specifications, challenges, and perspectives. *Engineering in Life Sciences*, *22*(12), 712–724. doi:10.1002/elsc.202100070

Cheah, W. Y., Er, A. C., Aiyub, K., Mohd Yasin, N. H., Ngan, S. L., Chew, K. W., Khoo, K. S., Ling, T. C., Juan, J. C., Ma, Z., & Show, P. L. (2023). Current status and perspectives of algae-based bioplastics: A reviewed potential for sustainability. *Algal Research*, *71*, 103078. doi:10.1016/j.algal.2023.103078

Checa, D., Macey, B. M., Bolton, J. J., Brink-Hull, M., O'Donohoe, P., Cardozo, A., Poersch, L. H., & Sánchez, I. (2024). Circularity assessment in aquaculture: The case of integrated multi-trophic aquaculture (IMTA) systems. *Fishes*, *9*(5), 165. doi:10.3390/fishes9050165

Chen, H., Wang, X., & Wang, Q. (2020). Microalgal biofuels in China: The past, progress and prospects. *GCB Bioenergy*, *12*(12), 1044–1065. doi:10.1111/gcbb.12741

Chhandama, M. V. L., Ruatpuia, J. V. L., Ao, S., Chetia, A. C., Satyan, K. B., & Rokhum, S. L. (2023). Microalgae as a sustainable feedstock for biodiesel and other production industries: Prospects and challenges. *Energy Nexus*, *12*, 100255. doi:10.1016/j.nexus.2023.100255

Chin, G. J. W. L., Andrew, A. R., Abdul-Sani, E. R., Yong, W. T. L., Misson, M., & Anton, A. (2023). The effects of light intensity and nitrogen concentration to enhance lipid production in four tropical microalgae. *Biocatalysis and Agricultural Biotechnology*, *48*, 102660. doi:10.1016/j.bcab.2023.102660

Choi, B. Y., Shim, D., Kong, F., Auroy, P., Lee, Y., Li-Beisson, Y., Lee, Y., & Yamaoka, Y. (2022). The *Chlamydomonas* transcription factor MYB1 mediates lipid accumulation under nitrogen depletion. *New Phytologist*, *235*(2), 595–610. doi:10.1111/nph.18141

Choi, J. W., Choi, J. H., Suh, D. J., & Kim, H. (2015). Feasibility of *Laminaria japonica* as a feedstock for fast pyrolysis in a bubbling fluidized-bed reactor. *Journal of Analytical and Applied Pyrolysis*, *112*, 141–149. doi:10.1016/j.jaap.2015.02.004

Choudhary, P., Assemany, P. P., Naaz, F., Bhattacharya, A., Castro, J. S., Couto, E. A. C., Calijuri, M. L., Pant, K. K., & Malik, A. (2020). A review of biochemical and thermochemical energy conversion routes of wastewater grown algal biomass. *Science of the Total Environment*, *726*, 137961. doi:10.1016/j.scitotenv.2020.137961

Conde, T. A., Neves, B. F., Couto, D., Melo, T., Neves, B., Costa, M., Silva, J., Domingues, P., & Domingues, M. R. (2021). Microalgae as sustainable bio-factories of healthy lipids: Evaluating fatty acid content and antioxidant activity. *Marine Drugs*, *19*(7), 357. https://doi.org/10.3390/md19070357

Coşgun, A., Günay, M. E., & Yıldırım, R. (2021). Exploring the critical factors of algal biomass and lipid production for renewable fuel production by machine learning. *Renewable Energy*, *163*, 1299–1317. doi:10.1016/j.renene.2020.09.034

Cotas, J., Gomes, L., Pacheco, D., & Pereira, L. (2023). Ecosystem services provided by seaweeds. *Hydrobiology*, *2*(1), 75–96. doi:10.3390/hydrobiology2010006

Crampon, C., Nikitine, C., Zaier, M., Lépine, O., Tanzi, C. D., Vian, M. A., Chemat, F., & Badens, E. (2017). Oil extraction from enriched *Spirulina platensis* microalgae using supercritical carbon dioxide. *The Journal of Supercritical Fluids*, *119*, 289–296. doi:10.1016/j.supflu.2016.10.006

Cutshaw, A., Daiek, C., Zheng, Y., Frost, H., Marks, A., Clements, D., Uludag-Demirer, S., Verhanovitz, N., Pavlik, D., Clary, W., Liu, Y., & Liao, W. (2020). A long-term pilot-scale algal cultivation on power plant flue gas – Cultivation stability and biomass accumulation. *Algal Research*, *52*, 102115. doi:10.1016/j.algal.2020.102115

Daneshvar, E., Wicker, R. J., Show, P. L., & Bhatnagar, A. (2022). Biologically-mediated carbon capture and utilization by microalgae towards sustainable CO_2 biofixation and biomass valorization – A review. *Chemical Engineering Journal*, *427*, 130884. doi:10.1016/j.cej.2021.130884

Das, P., V. P., Chandramohan., Mathimani, T., & Pugazhendhi, A. (2021). Recent advances in thermochemical methods for the conversion of algal biomass to energy. *Science of the Total Environment*, *766*, 144608. doi:10.1016/j.scitotenv.2020.144608

Davoodbasha, M., Kathiravan, N., Jayakannan, A., Raghunathan, S., Kim, J. W., & Nooruddin, T. (2024). An evidence of carbonic anhydrase activity in native microalgae for CO_2 capture application. *Applied Biochemistry and Biotechnology*, 196, 7064–7073. doi:10.1007/s12010-024-04908-4

De Bhowmick, G., Sarmah, A. K., & Sen, R. (2019). Zero-waste algal biorefinery for bioenergy and biochar: A green leap towards achieving energy and environmental sustainability. *Science of the Total Environment*, *650*(Part 2), 2467–2482. doi:10.1016/j.scitotenv.2018.10.002

de Morais, E. G., Sampaio, I. C. F., Gonzalez-Flo, E., Ferrer, I., Uggetti, E., & García, J. (2023). Microalgae harvesting for wastewater treatment and resources recovery: A review. *New Biotechnology*, *78*, 84–94. doi:10.1016/j.nbt.2023.10.002

de Souza Celente, G., Rizzetti, T. M., Sui, Y., & de Souza Schneider, R. C. (2022). Potential use of microalga *Dunaliella salina* for bioproducts with industrial relevance. *Biomass and Bioenergy*, *167*, 106647. doi:10.1016/j.biombioe.2022.106647

de Souza Celente, G., Sui, Y., & Acharya, P. (2023). Seaweed as an alternative protein source: Prospective protein extraction technologies. *Innovative Food Science and Emerging Technologies*, *86*, 103374. doi:10.1016/j.ifset.2023.103374

DeAngelo, J., Saenz, B. T., Arzeno-Soltero, I. B., Frieder, C. A., Long, M. C., Hamman, J., Davis, K. A., & Davis, S. J. (2023). Economic and biophysical limits to seaweed farming for climate change mitigation. *Nature Plants*, *9*, 45–57. doi:10.1038/s41477-022-01305-9

Dębowski, M., Świca, I., Kazimierowicz, J., & Zieliński, M. (2023). Large scale microalgae biofuel technology—Development perspectives in light of the barriers and limitations. *Energies*, *16*(1), 81. doi:10.3390/en16010081

Deepika, C., Mrinal, Pohrmen, C. B., Jaiswal, K. S., Sangmesh, B., Jaiswal, K. K., Ramasamy, A. P., & Jaiswal, A. K. (2024). Hydrothermal liquefaction of wet microalgal biomass for biofuels and platform chemicals: Advances and future prospects. *Discover Applied Sciences*, *6*, 245. doi:10.1007/s42452-024-05911-4

Devi, A., Saran, C., Ferreira, L. F. R., Mulla, S. I., & Bharagava, R. N. (2024). Sustainable approaches to algal biofuels: Opportunities, key challenges and current status. In: Arya, S. K., Khatri, M., & Singh, G., editors. *Value Added Products from Bio-algae Based Biorefineries: Opportunities and Challenges*. Singapore: Springer, pp. 163–188. doi:10.1007/978-981-97-1662-3_8

Diaz, C. J., Douglas, K. J., Kang, K., Kolarik, A. L., Malinovski, R., Torres-Tiji, Y., Molino, J. V., Badary, A., & Mayfield, S. P. (2023). Developing algae as a sustainable food source. *Frontiers in Nutrition*, *9*, 1029841. doi:10.3389/fnut.2022.1029841

Dobrinčić, A., Balbino, S., Zorić, Z., Pedisić, S., Kovačević, D. B., Garofulić, I. E., & Dragović-Uzelac, V. (2020). Advanced technologies for the extraction of marine brown algal polysaccharides. *Marine Drugs*, *18*(3), 168. doi:10.3390/md18030168

Dong, T., Knoshaug, E. P., Davis, R., Laurens, L. M. L., Wychen, S. V., Pienkos, P. T., & Nagle, N. (2016). Combined algal processing: A novel integrated biorefinery process to produce algal biofuels and bioproducts. *Algal Research, 19*, 316–323. doi:10.1016/j.algal.2015.12.021

Duan, P. G., Yang, S. K., Xu, Y. P., Wang, F., Zhao, D., Weng, Y. J., & Shi, X. L. (2018). Integration of hydrothermal liquefaction and supercritical water gasification for improvement of energy recovery from algal biomass. *Energy, 155*, 734–745. doi:10.1016/j.energy.2018.05.044

Dutta, N., Kundu, P., Lee, J. T. E., & Bhattacharya, S. (2023). Implementation and optimization of algal biomass in value-added products recovery: A step towards algae-based green economy. *Hydrobiology, 2*(2), 326–346. doi:10.3390/hydrobiology2020021

Egloff, S., Tschudi, F., Schmautz, Z., & Refardt, D. (2018). High-density cultivation of microalgae continuously fed with unfiltered water from a recirculating aquaculture system. *Algal Research, 34*, 68–74. doi:10.1016/j.algal.2018.07.004

Elkaliny, N. E., Alzamel, N. M., Moussa, S. H., Elodamy, N. I., Madkor, E. A., Ibrahim, E. M., Elshobary, M. E., & Ismail, G. A. (2024). Macroalgae bioplastics: A sustainable shift to mitigate the ecological impact of petroleum-based plastics. *Polymers, 16*(9), 1246. doi:10.3390/polym16091246

Elnajjar, E., Purayil, S. T. P., Alnuaimi, F., Al Khawaja, H., Shaikhoun, L., Arnaoud, N., & Almutawa, S. (2023). Mechanistic investigation of efficient cell disruption methods for lipid extraction from various macro and micro species of algae. *Bioresource Technology Reports, 22*, 101482. doi:10.1016/j.biteb.2023.101482

Euglena Co., Ltd. (2024). Sustainability First. www.euglena.jp/en/. Accessed 15 August 2024.

Ezeorba, T. P. C., Okeke, E. S., Mayel, M. H., Nwuche, C. O., & Ezike, T. C. (2024). Recent advances in biotechnological valorization of agro-food wastes (AFW): Optimizing integrated approaches for sustainable biorefinery and circular bioeconomy. *Bioresource Technology Reports, 26*, 101823. doi:10.1016/j.biteb.2024.101823

Ezhumalai, G., Arun, M., Manavalan, A., Rajkumar, R., & Heese, K. (2024). A holistic approach to circular bioeconomy through the sustainable utilization of microalgal biomass for biofuel and other value-added products. *Microbial Ecology, 87*, 61. doi:10.1007/s00248-024-02376-1

Faisal, S., Zaky, A., Wang, Q., Huang, J., & Abomohra, A. (2022). Integrated marine biogas: A promising approach towards sustainability. *Fermentation, 8*(10), 520. doi:10.3390/fermentation8100520

Fal, S., Smouni, A., & El Arroussi, H. (2023). Integrated microalgae-based biorefinery for wastewater treatment, industrial CO_2 sequestration and microalgal biomass valorization: A circular bioeconomy approach. *Environmental Advances, 12*, 100365. doi:10.1016/j.envadv.2023.100365

Fan, L., Zhang, H., Li, J., Wang, Y., Leng, L., Li, J., Yao, Y., Lu, Q., Yuan, W., & Zhou, W. (2020). Algal biorefinery to value-added products by using combined processes based on thermochemical conversion: A review. *Algal Research, 47*, 101819. doi:10.1016/j.algal.2020.101819

Fankhauser, S., Smith, S. M., Allen, M., Axelsson, K., Hale, T., Hepburn, C., Kendall, J. M., Khosla, R., Lezaun, J., Mitchell-Larson, E., Obersteiner, M., Rajamani, L., Rickaby, R., Seddon, N., & Wetzer, T. (2022). The meaning of net zero and how to get it right. *Nature Climate Change, 12*, 15–21. doi:10.1038/s41558-021-01245-w

Farghali, M., Mohamed, I. M. A., Osman, A. I., & Rooney, D. W. (2023). Seaweed for climate mitigation, wastewater treatment, bioenergy, bioplastic, biochar, food, pharmaceuticals, and cosmetics: A review. *Environmental Chemistry Letters, 21*, 97–152. doi:10.1007/s10311-022-01520-y

Farouk, S. M., Tayeb, A. M., Abdel-Hamid, S. M. S., & Osman, R. M. (2024). Recent advances in transesterification for sustainable biodiesel production, challenges, and prospects: A comprehensive review. *Environmental Science and Pollution Research, 31*, 12722–12747. doi:10.1007/s11356-024-32027-4

Figueroa-Torres, G. M., & Theodoropoulos, C. (2023). Techno-economic analysis of a microalgae-based biorefinery network for biofuels and value-added products. *Bioresource Technology Reports*, 23, 101524. doi:10.1016/j.biteb.2023.101524

Francke, L., Löhn, S., Weiderer, P., Kosheleva, A., Wieczorek, N., & Kuchta, K. (2022). A novel tubular photobioreactor immersed in open waters for passive temperature control and operated with the microalga *Tetradesmus obliquus*. *Algal Research, 67*, 102832. doi:10.1016/j.algal.2022.102832

Gallego, R., Bueno, M., Chourio, A. M., Ibáñez, E., Saldaña, M. D. A., & Herrero, M. (2021). Use of high and ultra-high pressure based-processes for the effective recovery of bioactive compounds from *Nannochloropsis oceanica* microalgae. *The Journal of Supercritical Fluids, 167*, 105039. doi:10.1016/j.supflu.2020.105039

Gao, K., & Beardall, J. (2022). Using macroalgae to address UN Sustainable Development goals through CO_2 remediation and improvement of the aquaculture environment. *Applied Phycology, 3*(1), 360–367. doi:10.1080/26388081.2022.2025617

Garg, D., Samota, M. K., Kontis, N., Patel, N., Bala, S., & Rosado, A. S. (2023). Revolutionizing biofuel generation: Unleashing the power of CRISPR-Cas mediated gene editing of extremophiles. *Microbiological Research*, *274*, 127443. doi:10.1016/j.micres.2023.127443

Gatto, F., & Re, I. (2021). Circular bioeconomy business models to overcome the Valley of Death. A systematic statistical analysis of studies and projects in emerging bio-based technologies and trends linked to the SME instrument support. *Sustainability*, *13*(4), 1899. doi:10.3390/su13041899

Gaur, S., Kaur, M., Kalra, R., Rene, E. R., & Goel, M. (2024). Application of microbial resources in biorefineries: Current trend and future prospects. *Heliyon*, *10*(8), e28615. doi:10.1016/j.heliyon.2024.e28615

Gaurav, K., Neeti, K., & Singh, R. (2024). Microalgae-based biodiesel production and its challenges and future opportunities: A review. *Green Technologies and Sustainability*, *2*(1), 100060. doi:10.1016/j.grets.2023.100060

Ghedini, E., Taghavi, S., Menegazzo, F., & Signoretto, M. (2021). A review on the efficient catalysts for algae transesterification to biodiesel. *Sustainability*, *13*(18), 10479. doi:10.3390/su131810479

Gonçalves, J., Freitas, J., Fernandes, I., & Silva, P. (2023). Microalgae as biofertilizers: A sustainable way to improve soil fertility and plant growth. *Sustainability*, *15*(16), 12413. doi:10.3390/su151612413

Granata, T., Rattenbacher, B., Kehl, F., & Egli, M. (2024). Microbial factories and exploiting synergies of bioreactor technologies to produce bioproducts. *Fermentation*, *10*(3), 135. doi:10.3390/fermentation10030135

Guccione, A., Biondi, N., Sampietro, G., Rodolfi, L., Bassi, N., & Tredici, M. R. (2014). *Chlorella* for protein and biofuels: From strain selection to outdoor cultivation in a Green Wall Panel photobioreactor. *Biotechnology for Biofuels*, *7*, 84. doi:10.1186/1754-6834-7-84

Guerra, I., Pereira, H., Costa, M., Silva, J. T., Santos, T., Varela, J., Mateus, M., & Silva, J. (2021). Operation regimes: A comparison based on *Nannochloropsis oceanica* biomass and lipid productivity. *Energies*, *14*(6), 1542. doi:10.3390/en14061542

Gupta, A., Kang, K., Pathania, R., Saxton, L., Saucedo, B., Malik, A., Torres-Tiji, Y., Diaz, C. J., Dutra Molino, J. V., & Mayfield, S. P. (2024). Harnessing genetic engineering to drive economic bioproduct production in algae. *Frontiers in Bioengineering and Biotechnology*, *12*, 1350722. doi:10.3389/fbioe.2024.1350722

Gür, T. M. (2022). Carbon dioxide emissions, capture, storage and utilization: Review of materials, processes and technologies. *Progress in Energy and Combustion Science*, *89*, 100965. doi:10.1016/j.pecs.2021.100965

Gurreri, L., Rindina, M. C., Luciano, A., Lima, S., Scargiali, F., Fino, D., & Mancini, G. (2023). Environmental sustainability of microalgae-based production systems: Roadmap and challenges towards the industrial implementation. *Sustainable Chemistry and Pharmacy*, *35*, 101191. doi:10.1016/j.scp.2023.101191

Hanley, M. E., Firth, L. B., & Foggo, A. (2024). Victim of changes? Marine macroalgae in a changing world. *Annals of Botany*, *133*(1), 1–16. doi:10.1093/aob/mcad185

Harirchi, S., Wainaina, S., Sar, T., Nojoumi, S. A., Parchami, M., Parchami, M., Varjani, S., Khanal, S. K., Wong, J., Awasthi, M. K., & Taherzadeh, M. J. (2022). Microbiological insights into anaerobic digestion for biogas, hydrogen or volatile fatty acids (VFAs): A review. *Bioengineered*, *13*(3), 6521–6557. doi:10.1080/21655979.2022.2035986

Hariz, H. B., & Takriff, M. S. (2017). Growth and biomass production of native microalgae *Chlorella* sp., *Chlamydomonas* sp. and *Scenedesmus* sp. cultivated in palm oil mill effluent (POME) at different cultivation conditions. *Transactions on Science and Technology*, *4*(298–311.

Hasan, H. A., Muhamad, M. H., Ji, B., Nazairi, N. A., Jiat, K. W., Sim, S. I. S. W. A., & Poh, A. F. M. S. (2023). Revolutionizing wastewater treatment with microalgae: Unveiling resource recovery, mechanisms, challenges, and future possibilities. *Ecological Engineering*, *197*, 107117. doi:10.1016/j.ecoleng.2023.107117

Hasnain, M., Munir, N., Sharif, N., Dias, D. A., Abideen, Z., Rizwana, H., Abid, I., Jamil, F., Hussain, M., & Choi, Y. J. (2024). Potential of algal biomass and their cultivation for biofuels production as plausible bio-resource for economic sustainability. *Korean Journal of Chemical Engineering*, *41*, 2567–2580. doi:10.1007/s11814-024-00157-7

Hegeman, P. D., & Sørheim, R. (2021). Why do they do it? Corporate venture capital investments in cleantech startups. *Journal of Cleaner Production*, *294*, 126315. doi:10.1016/j.jclepro.2021.126315

Hoang, A. T., Sirohi, R., Pandey, A., Nižetić, S., Lam, S. S., Chen, W. H., Luque, R., Thomas, S., Arıcı, M., & Pham, V. V. (2023). Biofuel production from microalgae: Challenges and chances. *Phytochemistry Reviews*, *22*, 1089–1126. doi:10.1007/s11101-022-09819-y

Hopkins, T. C., Sullivan Graham, E. J., & Schuler, A. J. (2019). Biomass and lipid productivity of *Dunaliella tertiolecta* in a produced water-based medium over a range of salinities. *Journal of Applied Phycology, 31*, 3349–3358. doi:10.1007/s10811-019-01836-3

Horton, B. P., Khan, N. S., Cahill, N., Lee, J. S., Shaw, T. A., Garner, A. J., Kemp, A. C., Engelhart, S. E., & Rahmstorf, S. (2020). Estimating global mean sea-level rise and its uncertainties by 2100 and 2300 from an expert survey. *npj Climate and Atmospheric Science, 3*, 18. doi:10.1038/s41612-020-0121-5

Hou, X., From, N., Angelidaki, I., Huijgen, W. J. J., & Bjerre, A. B. (2017). Butanol fermentation of the brown seaweed *Laminaria digitata* by *Clostridium beijerinckii* DSM-6422. *Bioresource Technology, 238*, 16–21. doi:10.1016/j.biortech.2017.04.035

Huang, J. J., Bunjamin, G., Teo, E. S., Ng, D. B., & Lee, Y. K. (2016). An enclosed rotating floating photobioreactor (RFP) powered by flowing water for mass cultivation of photosynthetic microalgae. *Biotechnology for Biofuels and Bioproducts, 9*, 218. doi:10.1186/s13068-016-0633-8

Huang, M. M., Xin, X., Shi, H. L., Lin, Y. M., Ling, H. Z., & Ge, J. (2022). Identification of six novel microalgal strains and characterization of their potential for development of high-value compounds. *South African Journal of Botany, 147*, 1–7. doi:10.1016/j.sajb.2021.12.007

Hussain, J., & Rittmann, B. E. (2023). Algae as a source of renewable energy: Opportunities, challenges, and recent developments. *Sustainable Energy & Fuels, 7*(11), 2515–2544. doi:10.1039/D2SE01599D

Iacovidou, E., Hahladakis, J. N., & Purnell, P. (2021). A systems thinking approach to understanding the challenges of achieving the circular economy. *Environmental Science and Pollution Research, 28*, 24785–24806. doi:10.1007/s11356-020-11725-9

Ibidoja, O. J., Shan, F. P., Suheri, M. E., Sulaiman, J., & Ali, M. K. M. (2023). Intelligence system via machine learning algorithms in detecting the moisture content removal parameters of seaweed big data. *Pertanika Journal of Science & Technology, 31*(6), 2783–2803. doi:10.47836/pjst.31.6.09

Iglina, T., Iglin, P., & Pashchenko, D. (2022). Industrial CO_2 capture by algae: A review and recent advances. *Sustainability, 14*(7), 3801. doi:10.3390/su14073801

Irianto, I., Naryaningsih, A., Trisnawati, N. W., Astuti, A., Komariyah, K., Qomariyah, L., Chaidir, C., Saputri, A., Wulandari, R., Rizkiyah, D. N., & Putra, N. R. (2024). From sea to solution: A review of green extraction approaches for unlocking the potential of brown algae. *South African Journal of Chemical Engineering, 48*, 1–21. doi:10.1016/j.sajce.2024.01.001

Javed, M. R., Noman, M., Shahid, M., Ahmed, T., Khurshid, M., Rashid, M. H., Ismail, M., Sadaf, M., & Khan, F. (2019). Current situation of biofuel production and its enhancement by CRISPR/Cas9-mediated genome engineering of microbial cells. *Microbiological Research, 219*, 1–11. doi:10.1016/j.micres.2018.10.010

Je, S., & Yamaoka, Y. (2022). Biotechnological approaches for biomass and lipid production using microalgae *Chlorella* and its future perspectives. *Journal of Microbiology and Biotechnology, 32*(11), 1357–1372. doi:10.4014/jmb.2209.09012

Jenčík, J., Hönig, V., Obergruber, M., Hájek, J., Vráblík, A., Černý, R., Schlehöfer, D., & Herink, T. (2021). Advanced biofuels based on Fischer–Tropsch synthesis for applications in diesel engines. *Materials, 14*(11), 3077. doi:10.3390/ma14113077

Jian, Y. (2023). Green bonds and green environment: Exploring innovative financing mechanisms for environmental project sustainability. *Environmental Science and Pollution Research, 30*, 122293–122303. doi:10.1007/s11356-023-30580-y

Johnson, C., Sierra, A. R., Dettmer, J., Sidiropoulou, K., Zicmane, E., Canalis, A., Llorente, P., Paiano, P., Mengal, P., & Puzzolo, V. (2021). The Bio-Based Industries Joint Undertaking as a catalyst for a green transition in Europe under the European Green Deal. *EFB Bioeconomy Journal, 1*, 100014. doi:10.1016/j.bioeco.2021.100014

Johnston, K. G., Abomohra, A., French, C. E., & Zaky, A. S. (2023). Recent advances in seaweed biorefineries and assessment of their potential for carbon capture and storage. *Sustainability, 15*(17), 13193. doi:10.3390/su151713193

Jothibasu, K., Muniraj, I., Jayakumar, T., Ray, B., Dhar, D. W., Karthikeyan, S., & Rakesh, S. (2022). Impact of microalgal cell wall biology on downstream processing and nutrient removal for fuels and value-added products. *Biochemical Engineering Journal, 187*, 108642. doi:10.1016/j.bej.2022.108642

Justine, I., Chin, G. J. W. L., Yong, W. T. L., & Misson, M. (2023). Characterization and optimization of *Rhodotorula toruloides* and *Ankistrodesmus falcatus* co-culture in palm oil mill effluent for efficient COD removal and lipid production. *Biocatalysis and Agricultural Biotechnology, 51*, 102782. doi:10.1016/j.bcab.2023.102782

Kammler, S., Romero, A. M., Burkhardt, C., Baruth, L., Antranikian, G., Liese, A., & Kaltschmitt, M. (2024). Macroalgae valorization for the production of polymers, chemicals, and energy. *Biomass and Bioenergy, 183*, 107105. doi:10.1016/j.biombioe.2024.107105

Kashyap, M., Chakraborty, S., Kumari, A., Rai, A., Varjani, S., & Vinayak, V. (2023). Strategies and challenges to enhance commercial viability of algal biorefineries for biofuel production. *Bioresource Technology, 387*, 129551. doi:10.1016/j.biortech.2023.129551

Katiyar, R., Banerjee, S., & Arora, A. (2021). Recent advances in the integrated biorefinery concept for the valorization of algal biomass through sustainable routes. *Biofuels, Bioproducts and Biorefining, 15*(3), 879–898. doi:10.1002/bbb.2187

Khan, M. I., Shin, J. H., & Kim, J. D. (2018). The promising future of microalgae: Current status, challenges, and optimization of a sustainable and renewable industry for biofuels, feed, and other products. *Microbial Cell Factories, 17*, 36. doi:10.1186/s12934-018-0879-x

Khan, N., Sudhakar, K., & Mamat, R. (2024). Macroalgae farming for sustainable future: Navigating opportunities and driving innovation. *Heliyon, 10*(7), e28208. doi:10.1016/j.heliyon.2024.e28208

Khoo, K. S., Ahmad, I., Chew, K. W., Iwamoto, K., Bhatnagar, A., & Show, P. L. (2023). Enhanced microalgal lipid production for biofuel using different strategies including genetic modification of microalgae: A review. *Progress in Energy and Combustion Science, 96*, 101071. doi:10.1016/j.pecs.2023.101071

Khor, W. H., Kang, H. S., Lim, J. W., Iwamoto, K., Tang, C. H. H., Goh, P. S., Quen, L. K., Shaharuddin, N. M. R., & Lai, N. Y. G. (2022). Microalgae cultivation in offshore floating photobioreactor: State-of-the-art, opportunities and challenges. *Aquacultural Engineering, 98*, 102269. doi:10.1016/j.aquaeng.2022.102269

Khujamberdiev, R., & Cho, H. M. (2024). Biofuels in aviation: Exploring the impact of sustainable aviation fuels in aircraft engines. *Energies, 17*(11), 2650. doi:10.3390/en17112650

Kim, G. Y., Seo, Y. H., Kim, I., & Han, J. I. (2019). Co-production of biodiesel and alginate from *Laminaria japonica. Science of the Total Environment, 673*, 750–755. doi:10.1016/j.scitotenv.2019.04.049

Kingsley, P. R., Braud, L., Mediboyina, M. K., McDonnell, K., & Murphy, F. (2023). Prospects for commercial microalgal biorefineries: Integrated pilot demonstrations and process simulations based techno-economic assessment of single and multi-product value chains. *Algal Research, 74*, 103190. doi:10.1016/j.algal.2023.103190

Koh, H. G., Cho, J. M., Jeon, S., Chang, Y. K., Lee, B., & Kang, N. K. (2023). Transcriptional insights into *Chlorella* sp. ABC-001: A comparative study of carbon fixation and lipid synthesis under different CO_2 conditions. *Biotechnology for Biofuels and Bioproducts, 16*, 113. doi:10.1186/s13068-023-02358-4

Kombe, G. G. (2023). An overview of algae for biodiesel production using bibliometric indicators, *International Journal of Energy Research, 2023*, 9596398. doi:10.1155/2023/9596398

Kostas, E. T., Adams, J. M. M., Ruiz, H. A., Durán-Jiménez, G., & Lye, G. J. (2021). Macroalgal biorefinery concepts for the circular bioeconomy: A review on biotechnological developments and future perspectives. *Renewable and Sustainable Energy Reviews, 151*, 111553. doi:10.1016/j.rser.2021.111553

Koyande, A. K., Show, P. L., Guo, R., Tang, B., Ogino, C., & Chang, J. S. (2019). Bio-processing of algal bio-refinery: A review on current advances and future perspectives. *Bioengineered, 10*(1), 574–592. doi:10.1080/21655979.2019.1679697

Krishnamoorthy, A., Rodriguez, C., & Durrant, A. (2022). Sustainable approaches to microalgal pre-treatment techniques for biodiesel production: A review. *Sustainability, 14*(16), 9953. doi:10.3390/su14169953

Krishnan, L., Ravi, N., Mondal, A. K., Akter, F., Kumar, M., Ralph, P., & Kuzhiumparambil, U. (2024). Seaweed-based polysaccharides – Review of extraction, characterization, and bioplastic application. *Green Chemistry, 26*(10), 5790–5823. doi:10.1039/D3GC04009G

Kumar, B., & Verma, P. (2021). Life cycle assessment: Blazing a trail for bioresources management. *Energy Conversion and Management: X, 10*, 100063. doi:10.1016/j.ecmx.2020.100063

Kumar, C., Sharma, M., Kaur, M., Arya, S. K., & Khatri, M. (2024). Cultivation of algae: Techniques and challenges. In: Arya, S. K., Khatri, M., & Singh, G., editors. *Value Added Products from Bio-algae Based Biorefineries: Opportunities and Challenges*. Singapore: Springer, pp. 43–65. doi:10.1007/978-981-97-1662-3_3

Laezza, C., Salbitani, G., & Carfagna, S. (2022). Fungal contamination in microalgal cultivation: Biological and biotechnological aspects of fungi-microalgae interaction. *Journal of Fungi, 8*(10), 1099. doi:10.3390/jof8101099

Lähteenmäki-Uutela, A., Rahikainen, M., Camarena-Gómez, M. T., Piiparinen, J., Spilling, K., & Yang, B. (2021). European Union legislation on macroalgae products. *Aquaculture International, 29*, 487–509. doi:10.1007/s10499-020-00633-x

Lange, L., Connor, K. O., Arason, S., Bundgård-Jørgensen, U., Canalis, A., Carrez, D., Gallagher, J., Gøtke, N., Huyghe, C., Jarry, B., Llorente, P., Marinova, M., Martins, L. O., Mengal, P., Paiano, P., Panoutsou, C., Rodrigues, L., Stengel, D. B., van der Meer, Y., & Vieira, H. (2021). Developing a sustainable and circular bio-based economy in EU: By partnering across sectors, upscaling and using new knowledge faster, and for the benefit of climate, environment & biodiversity, and people & business. *Frontiers in Bioengineering and Biotechnology*, *8*, 619066. doi:10.3389/fbioe.2020.619066

Le, H., & Azhgaliyeva, D. (2023). Carbon pricing and firms' GHG emissions: Firm-level empirical evidence from East Asia. *Journal of Cleaner Production*, *429*, 139504. doi:10.1016/j.jclepro.2023.139504

Lee, J., Kwon, E. E., & Park, Y. K. (2020a). Recent advances in the catalytic pyrolysis of microalgae. *Catalysis Today*, 355, 263–271. doi:10.1016/j.cattod.2019.03.010

Lee, X. J., Ong, H. C., Gan, Y. Y., Chen, W. H., & Mahlia, T. M. I. (2020b). State of art review on conventional and advanced pyrolysis of macroalgae and microalgae for biochar, bio-oil and bio-syngas production. *Energy Conversion and Management*, *210*, 112707. doi:10.1016/j.enconman.2020.112707

Leong, W. H., Lim, J. W., Lam, M. K., Lam, S. M., Sin, J. C., & Samson, A. (2021). Novel sequential flow baffled microalgal-bacterial photobioreactor for enhancing nitrogen assimilation into microalgal biomass whilst bioremediating nutrient-rich wastewater simultaneously. *Journal of Hazardous Materials*, *409*, 124455. doi:10.1016/j.jhazmat.2020.124455

Li, G., Xiao, W., Yang, T., & Lyu, T. (2023). Optimization and process effect for microalgae carbon dioxide fixation technology applications based on carbon capture: A comprehensive review. *C*, *9*(1), 35. doi:10.3390/c9010035

Li, P., Shi, X., Wang, X., Song, J., Fang, S., Bai, J., Zhang, G., Chang, C., & Pang, S. (2021). Bio-oil from biomass fast pyrolysis: Yields, related properties and energy consumption analysis of the pyrolysis system. *Journal of Cleaner Production*, *328*, 129613. doi:10.1016/j.jclepro.2021.129613

Li, T., Wang, W., Yuan, C., Zhang, Y., Xu, J., Zheng, H., Xiang, W., & Li, A. (2020). Linking lipid accumulation and photosynthetic efficiency in *Nannochloropsis* sp. under nutrient limitation and replenishment. *Journal of Applied Phycology*, *32*, 1619–1630. doi:10.1007/s10811-020-02056-w

Lian, J., Wijffels, R. H., Smidt, H., & Sipkema, D. (2018). The effect of the algal microbiome on industrial production of microalgae. *Microbial Biotechnology*, *11*(5), 806–818. doi:10.1111/1751-7915.13296

Lim, H. R., Khoo, K. S., Chia, W. Y., Chew, K. W., Ho, S. H., & Show, P. L. (2022a). Smart microalgae farming with internet-of-things for sustainable agriculture. *Biotechnology Advances*, *57*, 107931. doi:10.1016/j.biotechadv.2022.107931

Lim, J. Y., Teng, S. Y., How, B. S., Nam, K., Heo, S., Máša, V., Stehlík, P., & Yoo, C. K. (2022b). From microalgae to bioenergy: Identifying optimally integrated biorefinery pathways and harvest scheduling under uncertainties in predicted climate. *Renewable and Sustainable Energy Reviews*, *168*, 112865. doi:10.1016/j.rser.2022.112865

Lin, R., Deng, C., Ding, L., Bose, A., & Murphy, J. D. (2019). Improving gaseous biofuel production from seaweed *Saccharina latissima*: The effect of hydrothermal pretreatment on energy efficiency. *Energy Conversion and Management*, *196*, 1385–1394. doi:10.1016/j.enconman.2019.06.044

Liu, F., Wang, S., Huang, K., Yeager, K. M., Li, Y., Lv, L., Jia, G., & Ma, B. (2023). Great and fast increase in soil CH_4 uptake after reforestation in karst cropland area is linked to the environmental and microbial factors. *Agriculture, Ecosystems & Environment*, *347*, 108367. doi:10.1016/j.agee.2023.108367

Liu, J., Tong, Y., Xia, J., Sun, Y., Zhao, X., Sun, J., Zhao, S., Zhuang, M., Zhang, J., & He, P. (2022a). *Ulva* macroalgae within local aquaculture ponds along the estuary of Dagu River, Jiaozhou Bay, Qingdao. *Marine Pollution Bulletin*, 174, 113243. doi:10.1016/j.marpolbul.2021.113243

Liu, Y., Liu, X., Cui, Y., & Yuan, W. (2022b). Ultrasound for microalgal cell disruption and product extraction: A review. *Ultrasonics Sonochemistry*, *87*, 106054. doi:10.1016/j.ultsonch.2022.106054

Lo, C. W. H., Liu, N., Pang, X., & Li, P. H. Y. (2020). Unpacking the complexity of environmental regulatory governance in a globalizing world: A critical review for research agenda setting. *Journal of Environmental Policy & Planning*, *22*(5), 594–607. doi:10.1080/1523908X.2020.1767550

López-Molina, A., Sengupta, D., Shi, C., Aldamigh, E., Alandejani, M., & El-Halwagi, M. M. (2020). An integrated approach to the design of centralized and decentralized biorefineries with environmental, safety, and economic objectives. *Processes*, *8*(12), 1682. doi:10.3390/pr8121682

Lu, Y., Zhang, X., Gu, X., Lin, H., & Melis, A. (2021). Engineering microalgae: Transition from empirical design to programmable cells. *Critical Reviews in Biotechnology*, *41*(8), 1233–1256. doi:10.1080/07388551.2021.1917507

Ma, M., & Hu, Q. (2024). Microalgae as feed sources and feed additives for sustainable aquaculture: Prospects and challenges. *Reviews in Aquaculture*, *16*(2): 818–835. doi:10.1111/raq.12869

Mahmood, T., Hussain, N., Shahbaz, A., Mulla, S. I., Iqbal, H. M. N., & Bilal, M. (2023). Sustainable production of biofuels from the algae-derived biomass. *Bioprocess and Biosystems Engineering*, *46*, 1077–1097. doi:10.1007/s00449-022-02796-8

Makepa, D. C., & Chihobo, C. H. (2024). Barriers to commercial deployment of biorefineries: A multi-faceted review of obstacles across the innovation chain. *Heliyon*, *10*(12), e32649. doi:10.1016/j.heliyon.2024.e32649

Malik, S., Shahid, A., Haider, M. N., Amin, M., Betenbaugh, M. J., Mehmood, M. A., Ul Haq, M. A., Syafiuddin, A., & Boopathy, R. (2022). Prospects of multiproduct algal biorefineries involving cascading processing of the biomass employing a zero-waste approach. *Current Pollution Reports*, *8*, 147–158. doi:10.1007/s40726-022-00213-y

Malode, S. J., Prabhu, K. K., Mascarenhas, R. J., Shetti, N. P., & Aminabhavi, T. M. (2021). Recent advances and viability in biofuel production. *Energy Conversion and Management: X*, *10*, 100070. doi:10.1016/j.ecmx.2020.100070

Maneein, S., Milledge, J. J., Nielsen, B. V., & Harvey, P. J. (2018). A review of seaweed pre-treatment methods for enhanced biofuel production by anaerobic digestion or fermentation. *Fermentation*, *4*(4), 100. doi:10.3390/fermentation4040100

Mathimani, T., & Mallick, N. (2018). A comprehensive review on harvesting of microalgae for biodiesel – Key challenges and future directions. *Renewable and Sustainable Energy Reviews*, *91*, 1103–1120. doi:10.1016/j.rser.2018.04.083

Matos, G. S., Pereira, S. G., Genisheva, Z. A., Gomes, A. M., Teixeira, J. A., & Rocha, C. M. R. (2021). Advances in extraction methods to recover added-value compounds from seaweeds: Sustainability and functionality. *Foods*, *10*(3), 516. doi:10.3390/foods10030516

Matter, I. A., Bui, V. K. H., Jung, M., Seo, J. Y., Kim, Y. E., Lee, Y. C., & Oh, Y. K. (2019). Flocculation harvesting techniques for microalgae: A review. *Applied Sciences*, *9*(15), 3069. doi:10.3390/app9153069

McLaughlin, H., Littlefield, A. A., Menefee, M., Kinzer, A., Hull, T., Sovacool, B. K., Bazilian, M. D., Kim, J., & Griffiths, S. (2023). Carbon capture utilization and storage in review: Sociotechnical implications for a carbon reliant world. *Renewable and Sustainable Energy Reviews*, *177*, 113215. doi:10.1016/j.rser.2023.113215

Mercado, I., Álvarez, X., Verduga, M. E., & Cruz, A. (2020). Enhancement of biomass and lipid productivities of *Scenedesmus* sp. cultivated in the wastewater of the dairy industry. *Processes*, *8*(11), 1458. doi:10.3390/pr8111458

Milledge, J. J., Nielsen, B. V., Maneein, S., & Harvey, P. J. (2019). A brief review of anaerobic digestion of algae for bioenergy. *Energies*, *12*(6), 1166. doi:10.3390/en12061166

Milledge, J. J., Smith, B., Dyer, P. W., & Harvey, P. (2014). Macroalgae-derived biofuel: A review of methods of energy extraction from seaweed biomass. *Energies*, *7*(11), 7194–7222. doi:10.3390/en7117194

Miranda, A. M., Hernandez-Tenorio, F., Villalta, F., Vargas, G. J., & Sáez, A. A. (2024). Advances in the development of biofertilizers and biostimulants from microalgae. *Biology*, *13*(3), 199. doi:10.3390/biology13030199

Mody, K. H., Ghosh, P. K., Sana, B., Gnanasekaran, G., Shukla, A. D., Eswaran, K., Brahmbhatt, H. R., Shah, B. G., Thampy, S., & Jha, B. (2015-03-03). Process for integrated production of ethanol and seaweed sap from *Kappaphycus alvarezi*i (Patent no. US20130005009A1). https://patents.google.com/patent/US20130005009A1/en

Mohammad, R. E. A., Abdullahi, S. S., Muhammed, H. A., Musa, H., Habibu, S., Jagaba, A. H., & Birniwa, A. H. (2024). Recent technical and non-technical biorefinery development barriers and potential solutions for a sustainable environment: A mini review. *Case Studies in Chemical and Environmental Engineering*, *9*, 100586. doi:10.1016/j.cscee.2023.100586

Mohan, A. A., Antony, A. R., Greeshma, K., Yun, J. H., Ramanan, R., & Kim, H. S. (2022). Algal biopolymers as sustainable resources for a net-zero carbon bioeconomy. *Bioresource Technology*, *344*(Part B), 126397. doi:10.1016/j.biortech.2021.126397

Monagail, M. M., Cornish, L., Morrison, L., Araújo, R., & Critchley, A. T. (2017). Sustainable harvesting of wild seaweed resources. *European Journal of Phycology*, *52*(4), 371–390. doi:10.1080/09670262.2017.1365273

Moon, M., Park, W. K., Lee, S. Y., Hwang, K. R., Lee, S., Kim, M. S., Kim, B., Oh, Y. K., & Lee, J. S. (2022). Utilization of whole microalgal biomass for advanced biofuel and biorefinery applications. *Renewable and Sustainable Energy Reviews*, *160*, 112269. doi:10.1016/j.rser.2022.112269

Mora, C., Spirandelli, D., Franklin, E. C., Lynham, J., Kantar, M. B., Miles, W., Smith, C. Z., Freel, K., Moy, J., Louis, L. V., Barba, E. W., Bettinger, K., Frazier, A. G., Colburn IX, J. F., Hanasaki, N., Hawkins, E., Hirabayashi, Y., Knorr, W., Little, C. M., Emanuel, K., Sheffield, J., Patz, J. A., & Hunter, C. L. (2018). Broad threat to humanity from cumulative climate hazards intensified by greenhouse gas emissions. *Nature Climate Change*, *8*, 1062–1071. doi:10.1038/s41558-018-0315-6

Morales, M., Aflalo, C., & Bernard, O. (2021). Microalgal lipids: A review of lipids potential and quantification for 95 phytoplankton species. *Biomass and Bioenergy*, *150*, 106108. doi:10.1016/j.biombioe.2021.106108

Moshood, T. D., Nawanir, G., & Mahmud, F. (2021). Microalgae biofuels production: A systematic review on socioeconomic prospects of microalgae biofuels and policy implications. *Environmental Challenges*, *5*, 100207. doi:10.1016/j.envc.2021.100207

Mulenga, M., Monde, C., Johnson, T., Ouma, K. O., & Syampungani, S. (2024). Advances in the integration of microalgal communities for biomonitoring of metal pollution in aquatic ecosystems of sub-Saharan Africa. *Environmental Science and Pollution Research*, *31*, 40795–40817. doi:10.1007/s11356-024-33781-1

Murthy Konda, N. V. S. N., Singh, S., Simmons, B. A., & Klein-Marcuschamer, D. (2015). An investigation on the economic feasibility of macroalgae as a potential feedstock for biorefineries. *BioEnergy Research*, *8*, 1046–1056. doi:10.1007/s12155-015-9594-1

Mustafa, S., Bhatti, H. N., Maqbool, M., & Iqbal, M. (2021). Microalgae biosorption, bioaccumulation and biodegradation efficiency for the remediation of wastewater and carbon dioxide mitigation: Prospects, challenges and opportunities. *Journal of Water Process Engineering*, *41*, 102009. doi:10.1016/j.jwpe.2021.102009

Naduthodi, M. I. S., Claassens, N. J., D'Adamo, S., van der Oost, J., & Barbosa, M. J. (2021). Synthetic biology approaches to enhance microalgal productivity. *Trends in Biotechnology*, *39*(10), 1019–1036. doi:10.1016/j.tibtech.2020.12.010

Najjar, Y. S. H., & Abu-Shamleh, A. (2020). Harvesting of microalgae by centrifugation for biodiesel production: A review. *Algal Research*, *51*, 102046. doi:10.1016/j.algal.2020.102046

Narayanan, M. (2024). Biorefinery products from algal biomass by advanced biotechnological and hydrothermal liquefaction approaches. *Discover Applied Sciences*, *6*, 146. doi:10.1007/s42452-024-05777-6

Nath, S. (2024). Biotechnology and biofuels: Paving the way towards a sustainable and equitable energy for the future. *Discover Energy*, *4*, 8. doi:10.1007/s43937-024-00032-w

Nishshanka, G. K. S. H., Anthonio, R. A. D. P., Nimarshana, P. H. V., Ariyadasa, T. U., & Chang, J. S. (2022). Marine microalgae as sustainable feedstock for multi-product biorefineries. *Biochemical Engineering Journal*, *187*, 108593. doi:10.1016/j.bej.2022.108593

Nissar, S., Bakhtiyar, Y., Arafat, M. Y., Andrabi, S., Mir, Z. A., Khan, N. A., & Langer, S. (2023). The evolution of integrated multi-trophic aquaculture in context of its design and components paving way to valorization via optimization and diversification. *Aquaculture*, *565*, 739074. doi:10.1016/j.aquaculture.2022.739074

Obando, J. M. C., dos Santos, T. C., Martins, R. C. C., Teixeira, V. L., Barbarino, E., & Cavalcanti, D. N. (2022). Current and promising applications of seaweed culture in laboratory conditions. *Aquaculture*, *560*, 738596. doi:10.1016/j.aquaculture.2022.738596

Obobisa, E. S. (2022). Achieving 1.5°C and net-zero emissions target: The role of renewable energy and financial development. *Renewable Energy*, *188*, 967–985. doi:10.1016/j.renene.2022.02.056

Olabi, A. G., Shehata, N., Sayed, E. T., Rodriguez, C., Anyanwu, R. C., Russell, C., & Abdelkareem, M. A. (2023). Role of microalgae in achieving sustainable development goals and circular economy. *Science of the Total Environment*, *854*, 158689. doi:10.1016/j.scitotenv.2022.158689

Onyeaka, H., Miri, T., Obileke, K., Hart, A., Anumudu, C., & Al-Sharify, Z. T. (2021). Minimizing carbon footprint via microalgae as a biological capture. *Carbon Capture Science & Technology*, *1*, 100007. doi:10.1016/j.ccst.2021.100007

Oruganti, R. K., Biji, A. P., Lanuyanger, T., Show, P. L., Sriariyanun, M., Upadhyayula, V. K. K., Gadhamshetty, V., & Bhattacharyya, D. (2023). Artificial intelligence and machine learning tools for high-performance microalgal wastewater treatment and algal biorefinery: A critical review. *Science of the Total Environment*, *876*, 162797. doi:10.1016/j.scitotenv.2023.162797

Osman, A. I., Fang, B., Zhang, Y., Liu, Y., Yu, J., Farghali, M., Rashwan, A. K., Chen, Z., Chen, L., Ihara, I., Rooney, D. W., & Yap, P. S. (2024). Life cycle assessment and techno-economic analysis of sustainable bioenergy production: A review. *Environmental Chemistry Letters*, *22*, 1115–1154. doi:10.1007/s10311-023-01694-z

Osman, A. I., Farghali, M., Ihara, I., Elgarahy, A. M., Ayyad, A., Mehta, N., Ng, K. H., Abd El-Monaem, E. M., Eltaweil, A. S., Hosny, M., Hamed, S. M., Fawzy, S., Yap, P. S., & Rooney, D. W. (2023a). Materials, fuels, upgrading, economy, and life cycle assessment of the pyrolysis of algal and lignocellulosic biomass: A review. *Environmental Chemistry Letters*, *21*, 1419–1476. doi:10.1007/s10311-023-01573-7

Osman, A. I., Lai, Z. Y., Farghali, M. Yiin, C. L., Elgarahy, A. M., Hammad, A., Ihara, I., Al-Fatesh, A. S., Rooney, D. W., & Yap, P. S. (2023b). Optimizing biomass pathways to bioenergy and biochar application in electricity generation, biodiesel production, and biohydrogen production. *Environmental Chemistry Letters*, *21*, 2639–2705. doi:10.1007/s10311-023-01613-2

Osman, A. I., Mehta, N., Elgarahy, A. M., Al-Hinai, A., Al-Muhtaseb, A. H., & Rooney, D. W. (2021). Conversion of biomass to biofuels and life cycle assessment: A review. *Environmental Chemistry Letters*, *19*, 4075–4118. doi:10.1007/s10311-021-01273-0

Ospina, M. I., & Darwich-Cedeño, M. T. (2024). Removal of nitrogen and phosphorus from domestic wastewater by microalgal cultures. In: Martínez-Roldán, A. J., editor. *Biotechnological Processes for Green Energy, and High Value Bioproducts by Microalgae, and Cyanobacteria Cultures*. Developments in Applied Phycology, vol. 13. Cham: Springer, pp. 19–29. doi:10.1007/978-3-031-43969-8_3

Ou, L., Banerjee, S., Xu, H., Coleman, A. M., Cai, H., Lee, U., Wigmosta, M. S., & Hawkins, T. R. (2021). Utilizing high-purity carbon dioxide sources for algae cultivation and biofuel production in the United States: Opportunities and challenges. *Journal of Cleaner Production*, *321*, 128779. doi:10.1016/j.jclepro.2021.128779

Palanisamy, K. M., Bhuyar, P., Ab. Rahim, M. H., Govindan, N., & Maniam, G. P. (2023). Cultivation of microalgae *Spirulina platensis* biomass using palmoil mill effluent for phycocyanin productivity and future biomass refinery attributes. *International Journal of Energy Research*, *2023*, 2257271. doi:10.1155/2023/2257271

Pandey, S., Narayanan, I., Selvaraj, R., Varadavenkatesan, T., & Vinayagam, R. (2024). Biodiesel production from microalgae: A comprehensive review on influential factors, transesterification processes, and challenges. *Fuel*, *367*, 131547. doi:10.1016/j.fuel.2024.131547

Parakh, S. K., Tian, Z., Wong, J. Z. E., & Tong, Y. W. (2023). From microalgae to bioenergy: Recent advances in biochemical conversion processes. *Fermentation*, *9*(6), 529. doi:10.3390/fermentation9060529

Patle, D. S., Pandey, A., Srivastava, S., Sawarkar, A. N., & Kumar S. (2021). Ultrasound-intensified biodiesel production from algal biomass: A review. *Environmental Chemistry Letters*, *19*(1), 209–229. doi:10.1007/s10311-020-01080-z

Paul, S., Bera, S., Dasgupta, R., Mondal, S., & Roy, S. (2021). Review on the recent structural advances in open and closed systems for carbon capture through algae. *Energy Nexus*, *4*, 100032. doi:10.1016/j.nexus.2021.100032

Penloglou, G., Pavlou, A., & Kiparissides, C. (2024). Recent advancements in photo-bioreactors for microalgae cultivation: A brief overview. *Processes*, *12*(6), 1104. doi:10.3390/pr12061104

Perez-Vazquez, A., Carpena, M., Barciela, P., Cassani, L., Simal-Gandara, J., & Prieto, M. A. (2023). Pressurized liquid extraction for the recovery of bioactive compounds from seaweeds for food industry application: A review. *Antioxidants*, *12*(3), 612. doi:10.3390/antiox12030612

Pessôa, L. C., Deamici, K. M., Pontes, L. A. M., Druzian, J. I., & de Jesus Assis, D. (2021). Technological prospection of microalgae-based biorefinery approach for effluent treatment. *Algal Research*, *60*, 102504. doi:10.1016/j.algal.2021.102504

Pinto, L. F. R., Ferreira, G. F., Beatriz, F. P., Cabral, F. A., & Filho, R. M. (2022). Lipid and phycocyanin extractions from *Spirulina* and economic assessment. *The Journal of Supercritical Fluids*, *184*, 105567. doi:10.1016/j.supflu.2022.105567

Prabakaran, P., & Karthikeyan, S. (2024). Algae biofuel: A futuristic, sustainable, renewable and green fuel for I.C. engines. *Materials Today: Proceedings*, doi:10.1016/j.matpr.2023.03.579

Prasad, R., Gupta, S. K., Shabnam, N., Oliveira, C. Y. B., Nema, A. K., Ansari, F. A., & Bux, F. (2021). Role of microalgae in global CO_2 sequestration: Physiological mechanism, recent development, challenges, and future prospective. *Sustainability*, *13*(23), 13061. doi:10.3390/su132313061

Premaratne, M., Nishshanka, G. K. S. H., Anthonio, R. A. D. P., Liyanaarachchi, V. C., Thevarajah, B., Nimarshana, P. H. V., Malik, A., & Ariyadasa, T. U. (2022). Resource recovery from waste streams for production of microalgae biomass: A sustainable approach towards high-value biorefineries. *Bioresource Technology Reports*, *18*, 101070. doi:10.1016/j.biteb.2022.101070

Priyadharsini, P., Nirmala, N., Dawn, S. S., Baskaran, A., SundarRajan, P., Gopinath, K. P., & Arun, J. (2022). Genetic improvement of microalgae for enhanced carbon dioxide sequestration and enriched biomass productivity: Review on CO_2 bio-fixation pathways modifications. *Algal Research*, *66*, 102810. doi:10.1016/j.algal.2022.102810

Qin, L., Alam, M. A., & Wang, Z. (2019). Open pond culture systems and photobioreactors for microalgal biofuel production. In: Alam, M. A., & Wang, Z., editors. *Microalgae Biotechnology for Development of Biofuel and Wastewater Treatment*. Singapore: Springer, pp. 45–74. doi:10.1007/978-981-13-2264-8_3

Qin, S., Wang, K., Gao, F., Ge, B., Cui, H., & Li, W. (2023). Biotechnologies for bulk production of microalgal biomass: From mass cultivation to dried biomass acquisition. *Biotechnology for Biofuels and Bioproducts*, *16*, 131. doi:10.1186/s13068-023-02382-4

Raheem, A., Abbasi, S. A., Mangi, F. H., Ahmed, S., He, Q., Ding, L., Memon, A. A., Zhao, M., & Yu, G. (2021). Gasification of algal residue for synthesis gas production. *Algal Research*, *58*, 102411. doi:10.1016/j.algal.2021.102411

Ramola, B., Kumar, V., Nanda, M., Mishra, Y., Tyagi, T., Gupta, A., & Sharma, N. (2019). Evaluation, comparison of different solvent extraction, cell disruption methods and hydrothermal liquefaction of *Oedogonium* macroalgae for biofuel production. *Biotechnology Reports*, *22*, e00340. doi:10.1016/j.btre.2019.e00340

Rashid, S., & Malik, S. H. (2023). Transition from a linear to a circular economy. In: Bandh, S. A., Malla, F. A., & Hoang, A. T., editors. *Renewable Energy in Circular Economy, Circular Economy and Sustainability*. Cham: Springer, pp. 1–20. doi:10.1007/978-3-031-42220-1_1

Reddy, V. J., Hariram, N. P., Maity, R., Ghazali, M. F., & Kumarasamy, S. (2024). Sustainable vehicles for decarbonizing the transport sector: A comparison of biofuel, electric, fuel cell and solar-powered vehicles. *World Electric Vehicle Journal*, *15*(3), 93. doi:10.3390/wevj15030093

Rempel, A., Sossella, F. S. Margarites, A. C., Astolfi, A. L., Steinmetz, R. L. R., Kunz, A., Treichel, H., & Colla, L. M. (2019). Bioethanol from *Spirulina platensis* biomass and the use of residuals to produce biomethane: An energy efficient approach. *Bioresource Technology*, *288*, 121588. doi:10.1016/j.biortech.2019.121588

Renzaho, A. M. N., Kamara, J. K., & Toole, M. (2017). Biofuel production and its impact on food security in low and middle income countries: Implications for the post-2015 sustainable development goals. *Renewable and Sustainable Energy Reviews*, *78*, 503–516. doi:10.1016/j.rser.2017.04.072

Reyes, L. P., Havlik, I., & Beutel, S. (2024). Software sensors in the monitoring of microalgae cultivations. *Reviews in Environmental Science and Bio/Technology*, *23*, 67–92. doi:10.1007/s11157-023-09679-8

Rokicka, M., Zieliński, M., Dudek, M., & Dębowski M. (2021). Effects of ultrasonic and microwave pretreatment on lipid extraction of microalgae and methane production from the residual extracted biomass. *BioEnergy Research*, *14*, 752–760. doi:10.1007/s12155-020-10202-y

Roleda, M. Y., & Hurd, C. L. (2019). Seaweed nutrient physiology: Application of concepts to aquaculture and bioremediation. *Phycologia*, *58*(5), 552–562. doi:10.1080/00318884.2019.1622920

Roncaglia, B., Papini, A., Chini Zittelli, G., Rodolfi, L., & Tredici, M. R. (2021). Cell wall and organelle modifications during nitrogen starvation in *Nannochloropsis oceanica* F&M-M24. *Journal of Applied Phycology*, *33*, 2069–2080. doi:10.1007/s10811-021-02416-0

Roy, P., Mohanty, A. K., & Misra, M. (2023). Prospects of carbon capture, utilization and storage for mitigating climate change. *Environmental Science: Advances*, *2*(3), 409–423. doi:10.1039/D2VA00236A

Rumin, J., de Oliveira Junior, R. G., Bérard, J. B., & Picot, L. (2021). Improving microalgae research and marketing in the European Atlantic Area: Analysis of major gaps and barriers limiting sector development. *Marine Drugs*, *19*(6), 319. doi:10.3390/md19060319

S, Kavitha., Ravi, Y. K., Kumar, G., Nandabalan, Y. K., & J, R. B. (2024). Microalgal biorefineries: Advancement in machine learning tools for sustainable biofuel production and value-added products recovery. *Journal of Environmental Management*, *353*, 120135. doi:10.1016/j.jenvman.2024.120135

S, Shabarish., K, Tamilarasan., J, Rajesh Banu., & V, Godvin Sharmila. (2023). Biohydrogen production from macroalgae via sonic biosurfactant disintegration: An energy efficient approach. *Resources, Environment and Sustainability*, 11, 100093. doi:10.1016/j.resenv.2022.100093

Saad, M. G., Dosoky, N. S., Zoromba, M. S., & Shafik, H. M. (2019). Algal biofuels: Current status and key challenges. *Energies*, *12*(10), 1920. doi:10.3390/en12101920

Sadvakasova, A. K., Kossalbayev, B. D., Bauenova, M. O., Balouch, H., Leong, Y. K., Zayadan, B. K., Huang, Z., Alharby, H. F., Tomo, T., Chang, J. S., & Allakhverdiev, S. I. (2023). Microalgae as a key tool in achieving carbon neutrality for bioproduct production. *Algal Research*, *72*, 103096. doi:10.1016/j.algal.2023.103096

Saini, R. K., Prasad, P., Shang, X., & Keum, Y. S. (2021b). Advances in lipid extraction methods—A review. *International Journal of Molecular Sciences*, *22*(24), 13643. doi:10.3390/ijms222413643

Saini, R., Osorio-Gonzalez, C. S., Brar, S. K., & Kwong, R. (2021a). A critical insight into the development, regulation and future prospects of biofuels in Canada. *Bioengineered*, *12*(2), 9847–9859. doi:10.1080/21655979.2021.1996017

Sambasivam, K. M., Kuppan, P., Shashirekha, V., Tamilarasan, K., & Abinandan, S. (2023). Cascading utilization of residual microalgal biomass: Sustainable strategies for energy, environmental and value-added product applications. *Bioresource Technology Reports*, *23*, 101588. doi:10.1016/j.biteb.2023.101588

Samoraj, M., Çalış, D., Trzaska, K., Mironiuk, M., & Chojnacka, K. (2024). Advancements in algal biorefineries for sustainable agriculture: Biofuels, high-value products, and environmental solutions. *Biocatalysis and Agricultural Biotechnology*, *58*, 103224. doi:10.1016/j.bcab.2024.103224

Sánchez-García, L., Cabello, J., Jiménez-García, L. F., Revah, S., & Morales-Ibarría, M. (2020). Enhancing the lipid content of *Scenedesmus obtusiusculus* AT-UAM by controlled acidification under indoor and outdoor conditions. *Algal Research*, *51*, 102024. doi:10.1016/j.algal.2020.102024

Sandaka, B. P., & Kumar, J. (2023). Alternative vehicular fuels for environmental decarbonization: A critical review of challenges in using electricity, hydrogen, and biofuels as a sustainable vehicular fuel. *Chemical Engineering Journal Advances*, *14*, 100442. doi:10.1016/j.ceja.2022.100442

Saran, C., Devi, A., Saratale, G. D., Ferreira, L. F. R., & Bharagava, R. N. (2024). Bioenergy generation and cogeneration from algae-based biorefineries. In: Arya, S. K., Khatri, M., & Singh, G., editors. *Value Added Products from Bio-Algae Based Biorefineries: Opportunities and Challenges*. Singapore: Springer, pp. 189–211. doi:10.1007/978-981-97-1662-3_9

Sarwer, A., Hamed, S. M., Osman, A. I., Jamil, F., Al-Muhtaseb, A. H., Alhajeri, N. S., & Rooney, D. W. (2022). Algal biomass valorization for biofuel production and carbon sequestration: A review. *Environmental Chemistry Letters*, *20*, 2797–2851. doi:10.1007/s10311-022-01458-1

Sayed, E. T., Wilberforce, T., Elsaid, K., Rabaia, M. K. H., Abdelkareem, M. A., Chae, K. J., & Olabi, A. G. (2021). A critical review on environmental impacts of renewable energy systems and mitigation strategies: Wind, hydro, biomass and geothermal. *Science of the Total Environment*, *766*, 144505. doi:10.1016/j.scitotenv.2020.144505

Schultze-Jena, A., Vroon, R. C., Macleod, A. K. A., Hreggviðsson, G. Ó., Adalsteinsson, B. T., Engelen-Smit, N. P. E., de Vrije, T., Budde, M. A. W., van der Wal, H., López-Contreras, A. M., & Boon, M. A. (2022). Production of acetone, butanol, and ethanol by fermentation of *Saccharina latissima*: Cultivation, enzymatic hydrolysis, inhibitor removal, and fermentation. *Algal Research*, *62*, 102618. doi:10.1016/j.algal.2021.102618

Scranton, M. A., Ostrand, J. T., Fields, F. J., & Mayfield, S. P. (2015). *Chlamydomonas* as a model for biofuels and bio-products production. *The Plant Journal*, *82*(3), 523–531. doi:10.1111/tpj.12780

Sekar, M., Mathimani, T., Alagumalai, A., Chi, N. T. L., Duc, P. A., Bhatia, S. K., Brindhadevi, K., & Pugazhendhi, A. (2021). A review on the pyrolysis of algal biomass for biochar and bio-oil – Bottlenecks and scope. *Fuel*, *283*, 119190. doi:10.1016/j.fuel.2020.119190

Senthil Rathi, B., Dinesh Aravind, V., Ranjith, G., Kishore, V., Ewe, L. S., Yew, W. K., & Baskaran, R. (2024). Sustainability considerations in bio-hydrogen from bio-algae with the aid of bio-algae cultivation and harvesting: Critical review. *MRS Energy & Sustainability*, 11, 317–342. doi:10.1557/s43581-024-00096-0

Shahid, A., Ishfaq, M., Ahmad, M. S., Malik, S., Farooq, M., Hui, Z., Batawi, A. H., Shafi, M. E., Aloqbi, A. A., Gull, M., & Mehmood, M. A. (2019). Bioenergy potential of the residual microalgal biomass produced in city wastewater assessed through pyrolysis, kinetics and thermodynamics study to design algal biorefinery. *Bioresource Technology*, *289*, 121701. doi:10.1016/j.biortech.2019.121701

Shahid, A., Malik, S., Khan, A. Z., Liu, C. G., & Mehmood, M. A. (2020). Multiproduct algal biorefineries: Challenges and opportunities. In: Verma, P., editor. *Biorefineries: A Step towards Renewable and Clean Energy. Clean Energy Production Technologies*. Singapore: Springer, pp. 513–537. doi:10.1007/978-981-15-9593-6_20

Sharma, S., Sahu, S., & Arya, S. K. (2024). Pretreatment and fractionation of algae biomass for value-added extraction. In: Arya, S. K., Khatri, M., & Singh, G., editors. *Value Added Products from Bio-Algae Based Biorefineries: Opportunities and Challenges*. Singapore: Springer, pp. 95–115. doi:10.1007/978-981-97-1662-3_5

Sharmila, V. G., Banu, J. R., Kumar, M. D., Kumar, S. A., & Kumar, G. (2022). Algal biorefinery towards decarbonization: Economic and environmental consideration. *Bioresource Technology*, *364*, 128103. doi:10.1016/j.biortech.2022.128103

Shenbagamuthuraman, V., & Kasianantham, N. (2023). Microwave irradiation pretreated fermentation of bioethanol production from *Chlorella vulgaris* biomasses: Comparative analysis of response surface methodology and artificial neural network techniques. *Bioresource Technology*, *390*, 129867. doi:10.1016/j.biortech.2023.129867

Sheng, Y., Shanmugam, S., Chinnathambi, A., Salmen, S. H., Ge, S., Xia, C., & Brindhadevi, K. (2022). Feasibility of microalgal and macroalgal biomass co-digestion on biomethane production. *International Journal of Hydrogen Energy*, *47*(88), 37394–37400. doi:10.1016/j.ijhydene.2021.12.202

Singh, J., & Dhar, D. W. (2019) Overview of carbon capture technology: Microalgal biorefinery concept and state-of-the-art. *Frontiers in Marine Science*, *6*, 29. doi:10.3389/fmars.2019.00029

Singh, U. B., & Ahluwalia, A. S. (2013). Microalgae: A promising tool for carbon sequestration. *Mitigation and Adaptation Strategies for Global Change*, *18*, 73–95. doi:10.1007/s11027-012-9393-3

Sirohi, R., Pandey, A. K., Ranganathan, P., Singh, S., Udayan, A., Awasthi, M. K., Hoang, A. T., Chilakamarry, C. R., Kim, S. H., & Sim, S. J. (2022). Design and applications of photobioreactors – A review. *Bioresource Technology*, *349*, 126858. doi:10.1016/j.biortech.2022.126858

Smith, E. K., Kolcava, D., & Bernauer, T. (2024). Stringent sustainability regulations for global supply chains are supported across middle-income democracies. *Nature Communications*, *15*, 1049. doi:10.1038/s41467-024-45399-5

Solvang, T., Bale, E. S., Broch, O. J., Handå, A., & Alver, M. O. (2021). Automation concepts for industrial-scale production of seaweed. *Frontiers in Marine Science*, *8*, 613093. doi:10.3389/fmars.2021.613093

Song, Y., Wang, F., Chen, L., & Zhang, W. (2024). Engineering fatty acid biosynthesis in microalgae: Recent progress and perspectives. *Marine Drugs*, *22*(5), 216. doi:10.3390/md22050216

Soto-Sierra, L., Stoykova, P., & Nikolov, Z. L. (2018). Extraction and fractionation of microalgae-based protein products. *Algal Research*, *36*, 175–192. doi:10.1016/j.algal.2018.10.023

Sproles, A. E., Fields, F. J., Smalley, T. N., Le, C. H., Badary, A., & Mayfield, S. P. (2021). Recent advancements in the genetic engineering of microalgae. *Algal Research*, *53*, 102158. doi:10.1016/j.algal.2020.102158

Su, Y., Song, K., Zhang, P., Su, Y., Cheng, J., & Chen, X. (2017). Progress of microalgae biofuel's commercialization. *Renewable and Sustainable Energy Reviews*, *74*, 402–411. doi:10.1016/j.rser.2016.12.078

Sudhakar, M. P., & Viswanaathan, S. (2019). Chapter 6 – Algae as a sustainable and renewable bioresource for bio-fuel production. In: Singh, J. S., & Singh, D. P., editors. *New and Future Developments in Microbial Biotechnology and Bioengineering*. Elsevier, pp. 77–84. doi:10.1016/B978-0-444-64191-5.00006-7

Sun, X. M., Ren, L. J., Zhao, Q. Y., Ji, X. J., & Huang, H. (2019). Enhancement of lipid accumulation in microalgae by metabolic engineering. *Biochimica et Biophysica Acta (BBA) – Molecular and Cell Biology of Lipids*, *1864*(4), 552–566. doi:10.1016/j.bbalip.2018.10.004

Suzuki, K. (2017). Large-scale cultivation of *Euglena*. In: Schwartzbach, S. D., & Shigeoka, S., editors. *Euglena: Biochemistry, Cell and Molecular Biology*. Advances in Experimental Medicine and Biology, vol 979. Cham: Springer, pp. 285–293. doi:10.1007/978-3-319-54910-1_14

Taghavijeloudar, M., Shabanizadeh, H., Yaqoubnejad, P., & Safaei, M. (2024). Efficient microalgae harvesting toward biofuel production using a novel plant-based flocculant as a promising alternative to conventional flocculants. *Algal Research*, *81*, 103557. doi:10.1016/j.algal.2024.103557

Tahir, F., Ashfaq, H., Khan, A. Z. Amin, M., Akbar, I., Malik, H. A., Abdullah, M., Alessa, A. H., Alsaigh, A. A., Ralph, P. J., Mehmood, M. A., & Maliket, S. (2024). Emerging trends in algae farming on non-arable lands for resource reclamation, recycling, and mitigation of climate change-driven food security challenges. *Reviews in Environmental Science and Bio/Technology*, 23, 869–896. doi:10.1007/s11157-024-09697-0

Talwar, N., & Holden, N. M. (2022). The limitations of bioeconomy LCA studies for understanding the transition to sustainable bioeconomy. *The International Journal of Life Cycle Assessment*, *27*, 680–703. doi:10.1007/s11367-022-02053-w

Tan, J. S., Lee, S. Y., Chew, K. W., Lam, M. K., Lim, J. W., Ho, S. H., & Show, P. L. (2020). A review on microalgae cultivation and harvesting, and their biomass extraction processing using ionic liquids. *Bioengineered*, *11*(1), 116–129. doi:10.1080/21655979.2020.1711626

Tang, W., Mai, L., & Li, M. (2023). Green innovation and resource efficiency to meet net-zero emission. *Resources Policy*, *86*(Part B), 104231. doi:10.1016/j.resourpol.2023.104231

Taylor, E. N., Diele-Viegas, L. M., Gangloff, E. J., Hall, J. M., Halpern, B., Massey, M. D., Rödder, D., Rollinson, N., Spears, S., Sun, B., & Telemeco, R. S. (2021). The thermal ecology and physiology of reptiles and amphibians: A user's guide. *Journal of Experimental Zoology Part A: Ecological and Integrative Physiology*, *335*(1), 13–44. doi:10.1002/jez.2396

Thanigaivel, S., Priya, A. K., Kumar, P. S., Shiong, K. K., Hoang, T. K. A., Rajendran, S., & Soto-Moscoso, M. (2022). Exploration of effective biorefinery approach to obtain the commercial value-added products from algae. *Sustainable Energy Technologies and Assessments*, *53*(Part A), 102450. doi:10.1016/j.seta.2022.102450

Tripathi, S., Choudhary, S., Meena, A., & Poluri, K. M. (2023). Carbon capture, storage, and usage with microalgae: A review. *Environmental Chemistry Letters*, *21*, 2085–2128. doi:10.1007/s10311-023-01609-y

Tse, T. J., Wiens, D. J., & Reaney, M. J. T. (2021). Production of bioethanol—A review of factors affecting ethanol yield. *Fermentation*, *7*(4), 268. doi:10.3390/fermentation7040268

Tullberg, R. M., Nguyen, H. P., & Wang, C. M. (2022). Review of the status and developments in seaweed farming infrastructure. *Journal of Marine Science and Engineering*, *10*(10), 1447. doi:10.3390/jmse10101447

Tzima, S., Georgiopoulou, I., Louli, V., & Magoulas, K. (2023). Recent advances in supercritical CO_2 extraction of pigments, lipids and bioactive compounds from microalgae. *Molecules*, *28*(3), 1410. doi:10.3390/molecules28031410

Ullmann, J., & Grimm, D. (2021). Algae and their potential for a future bioeconomy, landless food production, and the socio-economic impact of an algae industry. *Organic Agriculture*, *11*, 261–267. doi:10.1007/s13165-020-00337-9

Uma, V. S., Usmani, Z., Sharma, M., Diwan, D., Sharma, M., Guo, M., Tuohy, M. G., Makatsoris, C., Zhao, X., Thakur, V. K., & Gupta, V. K. (2023). Valorisation of algal biomass to value-added metabolites: Emerging trends and opportunities. *Phytochemistry Reviews*, *22*, 1015–1040. doi:10.1007/s11101-022-09805-4

V, Godvin Sharmila., M, Dinesh Kumar., Pugazhendi, A., Bajhaiya, A. K., Gugulothu, P., & J, R. B. (2021). Biofuel production from macroalgae: Present scenario and future scope. *Bioengineered*, *12*(2), 9216–9238. doi:10.1080/21655979.2021.1996019

van Dijk, M., Goedegebure, R., & Nap, J. P. (2024). Public acceptance of biomass for bioenergy: The need for feedstock differentiation and communicating a waste utilization frame. *Renewable and Sustainable Energy Reviews*, *202*, 114670. doi:10.1016/j.rser.2024.114670

Vance, C., Pollard, P., Maguire, J., Sweeney, J., & Murphy, F. (2023). Sustainable scale-up of Irish seaweed production: Quantifying potential environmental, economic, and social impacts of wild harvesting and cultivation pathways. *Algal Research*, *75*, 103294. doi:10.1016/j.algal.2023.103294

Vanegas, C. H., & Bartlett, J. (2015). Biogas production from the anaerobic digestion of *Laminaria digitata* in a 10 L pilot-plant with digestate re-use as fertiliser. *International Journal of Ambient Energy*, *36*(4), 183–189. doi:10.1080/01430750.2013.842496

Vasić, K., Knez, Ž., & Leitgeb, M. (2021). Bioethanol production by enzymatic hydrolysis from different lignocellulosic sources. *Molecules*, *26*(3), 753. doi:10.3390/molecules26030753

Velenturf, A. P. M., & Purnell, P. (2021). Principles for a sustainable circular economy. *Sustainable Production and Consumption*, *27*, 1437–1457. doi:10.1016/j.spc.2021.02.018

Venkatesh, G. (2022). Circular bio-economy – Paradigm for the future: Systematic review of scientific journal publications from 2015 to 2021. *Circular Economy and Sustainability*, *2*, 231–279. doi:10.1007/s43615-021-00084-3

Vickram, S., Manikandan, S., Deena, S. R., Mundike, J., Subbaiya, R., Karmegam, N., Jones, S., Yadav, K. K., Chang, S. W., Ravindran, B., & Awasthi, M. K. (2023). Advanced biofuel production, policy and technological implementation of nano-additives for sustainable environmental management – A critical review. *Bioresource Technology*, *387*, 129660. doi:10.1016/j.biortech.2023.129660

Villalba, M. R., Cervera, R., & Sánchez, J. (2023). Green solutions for urban sustainability: Photobioreactors for algae cultivation on façades and artificial trees. *Buildings*, *13*(6), 1541. doi:10.3390/buildings13061541

Wang, S. K., Stiles, A. R., Guo, C., & Liu, C. Z. (2015). Harvesting microalgae by magnetic separation: A review. *Algal Research*, *9*, 178–185. doi:10.1016/j.algal.2015.03.005

Wiatrowski, M., Klein, B. C., Davis, R. W., Quiroz-Arita, C., Tan, E. C. D., Hunt, R. W., & Davis, R. E. (2022). Techno-economic assessment for the production of algal fuels and value-added products: Opportunities for high-protein microalgae conversion. *Biotechnology for Biofuels and Bioproducts*, *15*, 8. doi:10.1186/s13068-021-02098-3

Wimmerova, L., Keken, Z., Solcova, O., & Vavrova, K. (2023). A comparative analysis of environmental impacts of operational phases of three selected microalgal cultivation systems. *Sustainability*, *15*(1), 769. doi:10.3390/su15010769

Wu, J. Y., Tso, R., Teo, H. S., & Haldar, S. (2023). The utility of algae as sources of high value nutritional ingredients, particularly for alternative/complementary proteins to improve human health. *Frontiers in Nutrition*, *10*, 1277343. doi:10.3389/fnut.2023.1277343

Xie, C., Lee, Z. J., Ye, S., Barrow, C. J., Dunshea, F. R., & Suleria, H. A. R. (2024). A review on seaweeds and seaweed-derived polysaccharides: Nutrition, chemistry, bioactivities, and applications. *Food Reviews International*, *40*(5), 1312–1347. doi:10.1080/87559129.2023.2212055

Yadav, K., Vasistha, S., Nawkarkar, P., Kumar, S., & Rai, M. P. (2022). Algal biorefinery culminating multiple value-added products: Recent advances, emerging trends, opportunities, and challenges. *3 Biotech*, *12*, 244. doi:10.1007/s13205-022-03288-y

Yahyazadeh, A., Nanda, S., & Dalai, A. K. (2024). A critical review of the sustainable production and application of methanol as a biochemical and bioenergy carrier. *Reactions*, *5*(1), 1–19. doi:10.3390/reactions5010001

Yang, Y., Tang, S., & Chen, J. P. (2024). Carbon capture and utilization by algae with high concentration CO_2 or bicarbonate as carbon source. *Science of the Total Environment*, *918*, 170325. doi:10.1016/j.scitotenv.2024.170325

Yin, Y., Hu, J., & Wang, J. (2019). Fermentative hydrogen production from macroalgae *Laminaria japonica* pretreated by microwave irradiation. *International Journal of Hydrogen Energy*, *44*(21), 10398–10406. doi:10.1016/j.ijhydene.2019.03.034

Yong, W. T. L., Ting, S. H., Yong, Y. S., Thien, V. Y., Wong, S. H., Chin, W. L., Rodrigues, K. F., & Anton, A. (2014). Optimization of culture conditions for the direct regeneration of *Kappaphycus alvarezii* (*Rhodophyta, Solieriaceae*). *Journal of Applied Phycology*, *26*, 1597–1606. doi:10.1007/s10811-013-0191-4

Yun, J. H., Cho, D. H., Lee, S., Heo, J., Tran, Q. G., Chang, Y. K., & Kim, H. S. (2018). Hybrid operation of photobioreactor and wastewater-fed open raceway ponds enhances the dominance of target algal species and algal biomass production. *Algal Research*, *29*, 319–329. doi:10.1016/j.algal.2017.11.037

Zabochnicka, M., Krzywonos, M., Romanowska-Duda, Z., Szufa, S., Darkalt, A., & Mubashar, M. (2022). Algal biomass utilization toward circular economy. *Life*, *12*(10), 1480. doi:10.3390/life12101480

Zafar, S. U., Mehra, A., Nesamma, A. A., & Jutur, P. P. (2023). Innovations in algal biorefineries for production of sustainable value chain biochemicals from the photosynthetic cell factories. *Algal Research*, *69*, 102949. doi:10.1016/j.algal.2022.102949

Zapater, D., Kulkarni, S. R., Wery, F., Cui, M., Herguido, J., Menendez, M., Heynderickx, G. J., Van Geem, K. M., Gascon, J., & Castaño, P. (2024). Multifunctional fluidized bed reactors for process intensification. *Progress in Energy and Combustion Science*, *105*, 101176. doi:10.1016/j.pecs.2024.101176

Zhang, J., Xue, D., Wang, C., Fang, D., Cao, L., & Gong, C. (2023). Genetic engineering for biohydrogen production from microalgae. *iScience*, *26*(8), 107255. doi:10.1016/j.isci.2023.107255

Zhang, X., & Thomsen, M. (2021). Techno-economic and environmental assessment of novel biorefinery designs for sequential extraction of high-value biomolecules from brown macroalgae *Laminaria digitata*, *Fucus vesiculosus*, and *Saccharina latissima*. *Algal Research*, *60*, 102499. doi:10.1016/j.algal.2021.102499

Zhen, G., Lu, X., Kobayashi, T., Kumar, G., & Xu, K. (2016). Anaerobic co-digestion on improving methane production from mixed microalgae (*Scenedesmus* sp., *Chlorella* sp.) and food waste: Kinetic modeling and synergistic impact evaluation. *Chemical Engineering Journal*, *299*, 332–341. doi:10.1016/j.cej.2016.04.118

Zhou, J., Wang, M., Saraiva, J. A., Martins, A. P., Pinto, C. A., Prieto, M. A., Simal-Gandara, J., Cao, H., Xiao, J., & Barba, F. J. (2022a). Extraction of lipids from microalgae using classical and innovative approaches. *Food Chemistry*, *384*, 132236. doi:10.1016/j.foodchem.2022.132236

Zhou, Y., Liu, L., Li, M., & Hu, C. (2022b). Algal biomass valorisation to high-value chemicals and bioproducts: Recent advances, opportunities and challenges. *Bioresource Technology*, *344*(Part B), 126371. doi:10.1016/j.biortech.2021.126371

Zhu, C., Zhai, X., Xi, Y., Wang, J., Kong, F., Zhao, Y., & Chi, Z. (2019). Progress on the development of floating photobioreactor for microalgae cultivation and its application potential. *World Journal of Microbiology and Biotechnology*, *35*, 190. doi:10.1007/s11274-019-2767-x

Zhu, J., Wakisaka, M., Omura, T., Yang, Z., Yin, Y., & Fang, W. (2024). Advances in industrial harvesting techniques for edible microalgae: Recent insights into sustainable, efficient methods and future directions. *Journal of Cleaner Production*, 436, 140626. doi:10.1016/j.jclepro.2024.140626

Zuliani, L., Serpico, A., De Simone, M., Frison, N., & Fusco, S. (2021). Biorefinery gets hot: Thermophilic enzymes and microorganisms for second-generation bioethanol production. *Processes*, *9*(9), 1583. doi:10.3390/pr9091583

Zuorro, A., Malavasi, V., Cao, G., & Lavecchia, R. (2019). Use of cell wall degrading enzymes to improve the recovery of lipids from *Chlorella sorokiniana*. *Chemical Engineering Journal*, 377, 120325. doi:10.1016/j.cej.2018.11.023

7 Algae-Based Biofertilizer for Reducing Agricultural Carbon Footprints

Muhammad Uzair Javed, Hamid Mukhtar, and Tehreem Mahmood

7.1 INTRODUCTION

The fertility of the soil has a major impact on agricultural performance. Fertile soil provides vital nutrients that plants need, and it also harbors a diverse and dynamic biotic population that fortifies the soil against environmental deterioration. Although soil contains a variety of elements, the quantity of heavy metals in the environment has increased recently due to human activities such as mining, energy production, fuel making, electroplating, wastewater sludge treatment, and agriculture. Plant productivity is impacted by nitrogen deficiency in several ways, such as reduced yield, dwarf plant output, yellowing of the leaves, and stunted growth and development. Most cropping systems have recovery efficiencies of less than 50%, and soil erosion, volatilization, leaching, and denitrification all contribute significantly to the loss of applied nitrogen (Chapman, 2013). Thus, the necessity for appropriate and efficient environmentally friendly bio-based fertilizers that promote pollution-free farming practices and boost soil productivity has become critical worldwide.

Utilizing biofertilizers to increase the soil's nutrient content and boost productivity is known as biofertilization, a sustainable agricultural technique (Suleiman et al., 2020). One of the most unique organisms on Earth, algae, can be found in practically every terrestrial environment. They can be used in agriculture as biofertilizers and soil conditioners to increase soil fertility and plant productivity (Jaimes Duarte et al., 2018). Soil is a significant habitat for algal evolution. Soil algae are tiny photosynthetic microorganisms that grow their roots in the soil and survive just a few inches below the surface (Duarte et al., 2018). It can aid in the development of the soil's characteristics, including texture, aeration, carbon content, and nitrogen fixation. The presence of algae among other living things that can be found in a variety of soil types and conditions may be a sign of a healthy range for the soil environment. Through control of water flow, algal development also lessens soil erosion. Similarly, they perform a part in soil reclamation, soil fertility, creation of microbiological crust, bio-controlling of agricultural pests, and agricultural wastewater treatment (Abdel-Raouf et al., 2012). It has also been discovered that algal microflora functions as a true bio-based fertilizer for eco-friendly farming practices that are free from any kind of pollution. Despite being photosynthetic, these algae have evolved to withstand harsh conditions and have been shown to flourish in conditions with intense light, minimal water requirements, and limited nutrient levels, such as C and N. Because they are superb N- and C-fixing agents, they are very independent of alerts in terrestrial N pools that manufacture primary essential amino acids, growth promoters, and hormones required for successful crop growth in cropping systems (Ammar et al., 2022).

The current chapter has outlined the potential of algae-based biofertilizers in supporting plant growth, enriching the soil and making agriculture less damaging to the environment, making it ideal for today's farming. The effect of traditional chemical fertilizers on the environment has been

DOI: 10.1201/9781003685814-7

discussed, such as carbon emissions and degradation of the soil, that provides a reason to shift toward the adoption of algae fertilizers. The discussion has also included the production and use of algae biofertilizers and comparing it to the conventional fertilizers in terms of sustainability. Conclusively, the scientific justifications support the algae-based biofertilizers as a viable option in supporting sustainable agriculture.

7.2 CHEMICAL FERTILIZERS AND THEIR ENVIRONMENTAL AND HEALTH IMPLICATIONS

Fertility of the soil has a direct impact on plant growth and output. For farmers, productive fields are priceless resources. On the other hand, poor farming methods can cause soil deterioration and loss. Chemical fertilizers' administration is a typical technique to boost fertility of soil and deliver critical minerals, like nitrogen, potassium, and phosphorus, for the development of plants. Because these synthetic chemicals can swiftly provide plants with the nutrients they need for development and higher agricultural production, they are widely used in modern agriculture. They come in a variety of forms, including liquid or solid formulations (Kumar et al., 2019). But an over-reliance on chemical fertilizers can have a number of negative effects on the environment and the health of the soil, including changes in pH, the growth of pests, acidification, and the formation of soil crusts. It can also lead to a decrease in beneficial species and soil organic carbon, which can have an impact on plant growth and yield as well as possibly increase greenhouse gas (GHS) emissions (Ozlu & Kumar, 2018; Pahalvi et al., 2021; Park et al., 2023; Tayoh, 2020). Furthermore, a primary problem with traditional fertilizers is their propensity to encourage unbalanced plant uptake of nutrients (Bisht & Chauhan, 2020). Without adequate soil management, an ongoing dependence on chemical fertilizers may cause some nutrients to be depleted and others to accumulate, upsetting the processes involved in the natural cycle of nutrients in the soil (Li et al., 2022). Numerous studies have shown that improper application of chemical fertilizers can upset the balance of macronutrients including potassium, phosphorus, and nitrogen, which lowers crop yields (Li et al., 2019; Rutkowski & Łysiak, 2023; Sun et al., 2020). Sun et al. (Sun et al., 2020) assessed the effects of various nitrogen fertilization rates on soil qualities (soil organic matter, soil enzymes, and accessible nutrients), yield of banana crops, and physiological characteristics of banana leaves, such as chlorophyll, enzymes for plant metabolism, and soluble matter. In contrast to lower rates of nitrogen application, the findings showed that excessive application of nitrogen fertilizers (≥414 g of nitrogen per plant) significantly reduced the amounts of available potassium and phosphorus in the soil, as well as the amounts of sugar and soluble protein and glutamine synthetase activity. The scientists concluded that excessive nitrogen fertilization had a negative impact on crop productivity and soil fertility based on these results.

Nitrate buildup in the soil or even in plants can be caused by the continuous use of nitrogen fertilizers. Generally, accumulation of nitrates happens extensively in nonleguminous plants and leafy green vegetables, including lettuce (Ahmed et al., 2020; Liu et al., 2014). However, these nitrates have the potential to leak into streams if they are not taken up by plants immediately. Consuming plants high in nitrate content or groundwater tainted with nitrate might cause major health problems. Nitrate by itself is safe, but when nitric oxide, N-nitroso compounds, or nitrite are produced from oral bacteria as well as from salivary enzymes, there are some possible health hazards. These include the development of methemoglobinemia syndrome and stomach/bladder cancer (Ahmed et al., 2017).

Heavy metal buildup in agricultural soils is another prevalent issue brought on by the extensive use of chemical fertilizers. Chemical fertilizers, like superphosphate fertilizers, typically include significant concentrations of heavy metal pollutants, including lead, cadmium, chromium, nickel, mercury, cobalt, copper, zinc, and cobalt. Thomas et al. (Thomas et al., 2012) examined how phosphorus fertilizer affected the buildup of heavy metals in soil and in the plants of *Amaranthus caudatus*. As per their results, the zinc was reported to be increased both in the plant and soil used for its cultivation. Recently, it was examined by Wei et al. (2020) examined the concentration and availability of heavy metals in soil found in greenhouses. According to their findings, cadmium, copper, and zinc buildup was greatly enhanced by the frequent use of chemical fertilizers.

Cadmium is an extremely harmful element for plants because it prevents carbon from being fixed, lowers the amount of chlorophyll, and lowers photosynthetic activity. This element also interferes with the uptake and transport of other nutrients such as potassium, phosphorus, calcium, magnesium, and manganese. It also affects the uptake of iron and zinc, which causes chlorosis in leaves (Xu et al., 2017). Cadmium mostly causes toxicity to the kidney in humans. Exposure to cadmium has been closely associated with renal damage and malfunction, which can result in diseases including proteinuria and polyuria. Nevertheless, there have also been reports of further detrimental effects on the cardiovascular, musculoskeletal, and pulmonary systems. Moreover, cadmium is known to cause cancer in people (Yan & Allen, 2021).

Additional research has demonstrated that lead and arsenic, which are frequently present in traditional fertilizers, can build up in the soil and on plants, posing a health danger to people (Sun et al., 2021). Chronic exposure to arsenic has been connected with multiple detrimental health impacts, like arsenicosis, cardiovascular illnesses, neurological problems, respiratory disorders, diabetes, several forms of malignancies, and renal and reproductive ailments (Rahaman et al., 2022). Lead has a number of negative health impacts as well. Lead has an immediate negative effect on the hematological system. At extremely low blood lead levels (10 μgL^{-1} BPb—blood lead level), the production of hemoglobin becomes limited and the life span of erythrocytes gets shorter. The cardiovascular, renal, neurological, skeletal, and reproductive systems are also impacted by lead exposure (Collin et al., 2022).

7.3 CONVENTIONAL BIOFERTILIZERS TO ALGAE-BASED FERTILIZERS: DEFINITION AND SCOPE

A combination of living or dormant organisms possessing the unique capacity of fixing atmospheric nitrogen, solubilizing mineral phosphates, or breaking down organic matter is known as a biofertilizer. These fertilizers are applied to soil to boost the amount of an advantageous microflora that speeds up important mineralization and nitrogen uptake processes, thus increasing the amount of nutrients in the medium. The word "biofertilizer" refers to any concentrated organic resource for cultivation that is absorbed by plants via microbial and plant-associated processes (Singh, 2015). Composting and utilizing biodegradable materials (food scraps, cow dung, and agricultural leftovers) have historically signaled the beginning of the utilization of plant nutrients in the form of organic matter. This practice has been carried down as tradition to the contemporary community from the ancient agricultural communities. The current biofertilizers that quicken the movement of organic residue and agricultural byproduct breakdown are the embodiment of these applications' full potential (Singh, 2014). These advantageous bacteria have been produced industrially throughout Southeast Asia, the region's main agrarian belt. Numerous studies have examined the effects of algal communities, primarily cyanobacterial biofertilizers, on crop yield. The Rhizobium in legumes is one of the most extensively utilized biofertilizers that has been employed frequently for N fixation and assimilation into plants. However, there is also a lot of research done on the uses of *Azospirillum* and mycorrhiza. According to reports, mycorrhiza inoculums have actively been employed in the agricultural industry and have quickly adapted to the tropical Southeast Asian climate, including those of Malaysia and India. This promotes the commercial production of biofertilizers, which improve soil absorption of water and minerals, increase soil fertility, reduce soil toxicity, inhibit the growth of soil pests, and limit the spread of diseases. Furthermore, these microbial assemblages are environmentally benign and commercially feasible due to their ease of growth in substrates such as mine sands and agricultural wastes. However, because producing high-density inoculum and bioconverting waste substrates take a high level of ability, biofertilizers are said to be less cost-effective than chemical fertilizers. In addition, studies conducted on land applications have demonstrated that, in comparison to chemical fertilizers, the activity of biofertilizers is slower in terms of fertilization efficiency. To ensure the viability of the inoculums for prolonged use, great care and attention are given throughout their production during the homogenization, mixing, and storing of the dried inoculum forms as powders. Since the typical biofertilizers, as previously indicated, work with living bacteria, their growth activity and nutrient conversions can also be highly inconsistent due to their dependence on the surrounding environment. However, the currently known algal biofertilizers are not based on active algal biomass; rather, they are a stable biomass mix that has been highly loaded with nutrients and contains active growth stimulants along with nourishing vitamins, microelements, and hormones that have demonstrated encouraging results. Algal biomass is a superior option for a biofertilizer since it is less sensitive to temperature changes and is durable for storage and transportation. It can also function as a carrier material for nutrients in biofertilizers.

Because beneficial exudates and exopolysaccharides from algal growth serve as a source of carbon (C) for soil bacteria, they promote the growth of rhizosphere bacteria (Kapustka & DuBois, 1987). Following mineralization, algae in the dry form is added to crop soils, and their extracts play a role in stimulating the growth of the bacteria that creates siderophores (Amin et al., 2009). The process entails the photolysis of the iron siderophore, which in turn causes the assimilation of iron in damp environments. This ultimately fosters the relationship among bacteria and extracts algae and bacteria, since the organic materials released by algal members serve as food for bacterial members. This facilitates the bioavailability and assimilation of Fe in low Fe and N conditions, which are primarily found in aquatic environments or when there is waterlogging, through Fe chelators such as siderophores. In addition, Blue-Green Algae (BGA) contributes significantly to the chelating properties of organic acids, which increase the bioavailability of micronutrients for developing plants, and to the transformation of soil micronutrients (Fe and Mn). Strong copper-complexing agents (cK > 108) have been shown to be excreted by algal species such as ***Anabaena***. *cylindrica, Synechococcus leopoliensis*, and *Anacystis nidulans*, whereas eukaryotic algae have only been identified to secrete comparatively weak copper-complexing agents (cK < 107.5) (McKnight & Morel, 1980). These results offer a plethora of possibilities for additional research in an agricultural setting to comprehend micronutrient interactions including siderophore manufacturing as well as connections among the soil algal along with species of plants in the biological fertilizer activity for soil nutrient translocation, even though the majority of these findings were reported from studies in aquatic systems.

7.4 CARBON FOOTPRINTS IN AGRICULTURE

The assessment of GHG emissions associated with different economic processes and goods is done through the application of the carbon footprint (CF) approach (Fantozzi et al., 2016). It is described as the balance of GHG emissions during a product or process's whole life cycle. Global warming potential (GWP) is the parameter to characterize the environmental effect class (climate change). Numerous studies have demonstrated that fertilizer significantly affects the CF of plant production (Gasol et al., 2012; Iriarte et al., 2010). Gaseous N compounds are released with the application of nitrogen (N) fertilizers in plant cultivation. Because it is a chemically reactive substance, nitrogen contributes to global warming. After carbon dioxide and methane, nitrous oxide (N_2O) is the third most significant GHG in its chemical form and is hazardous to agriculture (Mancia et al., 2022). The high index of GWP of N_2O highlights the impact of this gas on the formation of the greenhouse effect.

The comparison of GHG emissions from Europe's different agricultural production systems has been the subject of extensive research due to the increased awareness of climate change. Sustainable agricultural development aligns with the principles of organic farming, which is regarded as an eco-friendly system. There are, however, differing views in the literature about the environmental elements of field crop production in conventional and organic systems, despite research employing the life cycle assessment technique. While some research produced inconsistent results, other studies, involving the production of plants, have demonstrated the ability of organic farming in lowering the CF (Biernat-Jarka & Trkebska, 2018; Moudry Jr & Moudry, 2014).

When compared to non-organic systems, organic farming may yield a higher soil organic carbon concentration. One of the greatest options for reducing GHG emissions is to employ algae in conjunction with other ecologically friendly farming methods.

7.5 PRODUCTION OF BIOFERTILIZERS: THE ALGAL INDUSTRY

A rapidly expanding sector of the agricultural economy is dedicated to producing biofertilizers based on algae, which aim to offer environmentally safe and sustainable substitutes for chemical fertilizers as depicted in Figure 7.1. The procedure entails multiple crucial phases, ranging from the identification and cultivation of suitable algae species to the creation, packing, and dissemination of the ultimate product. This section explores the nuances of each stage of manufacturing, emphasizing the difficulties and advancements that have shaped the algal biofertilizer market.

The careful selection of appropriate microbial assemblages and algal species is the first stage in the production of algae-based biofertilizers. Because different species of algae and microorganisms perform distinct agricultural tasks, like nitrogen fixation, phosphate solubilization, and potassium mobilization, this selection is essential (Gupta, 2004).

Finding the precise agricultural demands that the biofertilizer is intended to answer is the first step in the selection process. For instance, phosphorus-solubilizing bacteria are selected to increase phosphorus availability in the soil, while nitrogen-fixing algae, such as certain cyanobacteria, are perfect for raising soil nitrogen levels. Following their isolation from settings such as aquatic bodies, soil matrices, aquatic bodies, and marshes, the microbial assemblages are subsequently grown under carefully regulated laboratory conditions. The most efficient organisms for the specified biofertilizer function can be identified thanks to this isolation method (Kuraganti et al., 2020).

It is crucial to choose the correct organisms since the microbes' functional abilities determine how effective the biofertilizer will be. For example, because of their ability to fix nitrogen symbiotically, Rhizobium species are specifically chosen for leguminous crops, whereas phosphate-solubilizing bacteria are necessary for crops to grow in soils lacking in phosphorus. This meticulous selection procedure guarantees that the biofertilizer is customized to satisfy particular crop and soil needs, increasing its usefulness and efficacy (Khan et al., 2023).

7.5.1 Cultivation and Optimization of Algal Biomass

After the selection of suitable algae species and microbial assemblages, the emphasis switches to the cultivation and optimization of these organisms' growth. Small-scale cultures are cultivated in flasks or petri dishes in a controlled laboratory setting at the start of the cultivation process. Understanding the growth features of the chosen organisms, such as their food requirements, growth rates, and environmental tolerances, depends on these first cultures (Mahapatra et al., 2014).

Moving from flask-based cultures to larger bioreactors or open pond systems is a necessary step in scaling up from laboratory cultures to commercial output. The selection between these methods is contingent upon various considerations, including as cost, scalability, and the particular needs of the algae species. For example, although photobioreactors can optimize the provision of light, temperature, and nutrients, they are more expensive and require more complex maintenance than open pond systems. A number of variables, including temperature, pH levels, nutrient concentrations, and light intensity, must be adjusted for optimal algal biomass production. Achieving the highest possible algal biomass yield while maintaining essential functional characteristics like phosphorus solubilization and nitrogen fixation is the aim. Testing various growth media is another step in this

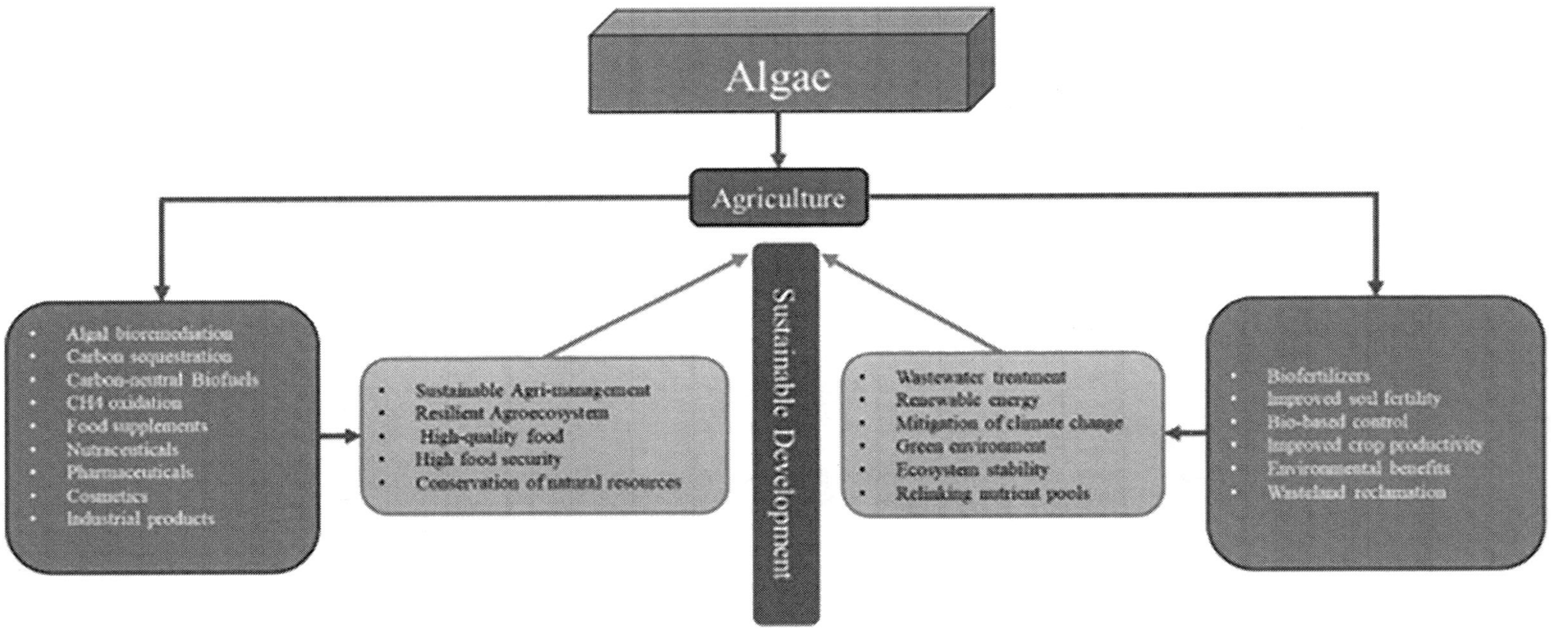

FIGURE 7.1 Selection of algal species and microbial assemblages.

process that aims to determine the best practical and economical choices for large-scale manufacturing (Javed et al., 2022).

The novel practice of treating wastewater and recovering nutrients through algae growth has drawn notice due to its dual advantages. The abundant nutrients found in wastewater can promote algae growth, as a result of which water gets cleaned by the action of algae to eliminate impurities and surplus nutrients. This symbiotic relationship makes algal biomass production more affordable and less harmful to the environment, which makes it a viable choice for sustainable agriculture (Mahapatra et al., 2018).

7.5.2 Carrier Material Selection and Formulation

The crucial next stage in creating the biofertilizer is choosing the right carrier materials once the growth of algal biomass has been optimized. The active algal or microbial biomass is combined with carrier materials, which offer stability, application simplicity, and extended shelf life. The planned form of the biofertilizer—whether it will be a powder, granule, or liquid—influences the choice of carrier material (Senthilnathan & Ganczarczyk, 1990).

Materials including peat, tapioca flour, and lignite are frequently utilized for powdered biofertilizers. These materials are chosen for their moisture retention qualities, ability to support microbial activity, and ease of processing. For example, peat has a great ability to retain moisture, which helps keep the microbes viable during application and storage. Conversely, tapioca flour offers a nutrient-rich media that promotes microbial development and is biodegradable (Saif et al., 2021). When it comes to liquid biofertilizers, stabilizing chemicals are usually added to the water-based carrier medium to keep the algae cells suspended. Because they are so simple to apply, liquid formulations are frequently chosen, especially in contemporary agricultural systems that use foliar spraying or drip irrigation. However, in order to avoid sedimentation, microbial contamination, and viability loss during storage, liquid biofertilizers must be carefully formulated (Mahapatra et al., 2018).

In order to ensure that the biofertilizer is easily applied to crops, the algal biomass and carrier material are combined in a way that maintains the microorganisms' functionality. This stage frequently calls for the addition of chemicals, like humic acid, to improve the nutrients' bioavailability or shield the microbes from external stresses (Chatterjee et al., 2017).

7.5.3 Inoculum Development and Quality Assurance

To guarantee that the finished biofertilizer product is dependable and efficient, great care must be taken in the development of the inoculum, which is a blend of active algal biomass and carrier material. This phase entails a number of procedures, including as combining the biomass and the carrier material, modifying the moisture content, and guaranteeing that the microbes are distributed uniformly throughout the formulation (Schmid, 2021).

An essential part of developing an inoculum is quality control. This entails evaluating the inoculum for a number of important factors, including as nutrient content, pH stability, microbial viability, and contaminant-free status. In order to stop the growth of undesirable microorganisms that could outcompete or limit the activities of the beneficial bacteria or algae, sterilization of the carrier material is frequently required. UV or gamma radiation are autoclaving, which uses high-pressure steam, are two common sterilizing techniques. After the inoculum is created, it is put through a thorough testing process to see how effective it is in different environmental settings. This includes biofertilizer trials in greenhouses, where test crops are given the fertilizer to see how it affects growth, nutrient uptake, and pest and disease resistance. These tests yield important information for improving the formulation and assist in identifying any potential drawbacks of the biofertilizer, such as decreased efficacy in specific soil types or climates (Herrmann & Lesueur, 2013).

Important considerations in the development of inoculums include also packaging and storage. The packaging of the biofertilizer needs to shield it from elements that can cause deterioration over time, like moisture, changes in temperature, and exposure to light. Prior to being applied to the fields, the biofertilizer's viability and effectiveness are maintained by meticulous regulation of storage conditions (Schmid, 2021).

7.5.4 Commercialization, Application, and Field Testing

Commercializing algae-based biofertilizers entails increasing production volume, reaching out to farmers, and making sure the product complies with laws and regulations. To get the product from the lab to the market, scientists, business leaders, and agricultural stakeholders must work together at this point.

The formulation of application guidelines that are simple to use and flexible enough to accommodate various farming techniques is a crucial component of commercialization. Whether it's soil integration, seed treatment, or foliar spraying, advice on when, how much, and how to apply it is included in this. These recommendations are frequently the result of in-depth field testing, which enhances the product's functionality and verifies its efficacy in actual situations.

The last stage before wide-scale dissemination is field testing. For evaluating the biofertilizer's effectiveness in a range of environmental circumstances, it must be applied to a variety of crops in a variety of geographical locations. Validating the biofertilizer's ability to increase soil fertility, encourage plant development, and lessen the need for chemical fertilizers requires this testing. Additionally, field trial data are utilized to bolster regulatory acceptance and foster trust among farmers and other end users (Mahapatra et al., 2022).

The algal biofertilizer business, while still in its fledgling phases, is poised for major expansion as breakthroughs in algae growing, biomass processing, and formulation technologies continue to emerge. Algal-based biofertilizers offer a viable substitute for conventional chemical fertilizers, making them a sustainable approach to addressing the issues facing modern agriculture, such as the need to improve soil health, lower CFs, and guarantee food security in the face of climate change.

7.6 EFFECTS OF ALGAL-BASED BIOFERTILIZER ON SOIL HEALTH

Certain microalgal species are widely recognized for their favorable influence on soil quality since they have the capacity to fix nitrogen in the atmosphere, enhance the accessibility of vital nutrients, or enhance the physiochemical characteristics of soil. Moreover, having the ability to perform photosynthesis, microalgae have the ability to recover nutrients from multiple sources, hence reducing the usage of conventional fertilizers. Different components of the algae contribute to the growth-promoting characteristics of plants whether directly or indirectly. Such constituents and their efficacy are discussed in Table 7.1.

7.6.1 Nitrogen Fixation

The majority of studies on the use of algae in agriculture have concentrated on cyanobacteria because of their capacity to fix atmospheric nitrogen and convert it into organic forms that higher plants can absorb (Bao et al., 2021; Shamim et al., 2020). All things considered, it has been shown that the utilization of cyanobacteria in soils can enhance crop micronutrient fraction enrichment while also promoting effective C-N sequestration and nutrient recycling with enhanced nitrogen availability. Cyanobacteria, either alone or in consortiums, can be added directly to the soil to increase nitrogen accessibility and create soil surface layer that prevent nitrogen runoff. This reduces the chance of water contamination, saves 25–30% on fertilizer, and greatly increases plant growth (Zhang et al., 2021). Furthermore, as shown for a number of crops, the cyanobacterial biofilms can further enhance

TABLE 7.1
The Biological Activities, Roles, and Applications in Agriculture of Bioactive Compounds Derived from Microalgal Species

Bioactive Compounds	Biological Activity	Role in Agriculture	Specific Agricultural Applications	Microalgae Sources	Additional Benefits	References
Phenolic compounds	• Antioxidant • Antibacterial • Antifungal	Protection of crops from biotic and abiotic stressors, including diseases.	• Fungicide • Pesticide • Plant stress alleviation	*Chlorella vulgaris*, *Isochrysis* sp., *Botryococcus braunii*, *Odontella sinensis*, *Chaetoceros calcitrans*, *Phaeodactylum tricornutum*, *Isochrysis galbana*, *Tetraselmis suecica*, *Saccharina japonica*, *Neochloris oleoabundans*, *Skeletonema costatum*, *Nannochloropsis gaditana*, *Tetraselmis chui*	Reduces oxidative stress in crops	Esquivel-Hernández et al. (2017), Pan et al. (2019), Singh et al. (2017)
Carotenoids	• Antioxidant • Anti-inflammatory • Anticancer	• Protection of crops from biotic and abiotic stress situations • Soil bioremediation and fertilization;	Enhancement of plant color, growth stimulation, soil enrichment	*Dunaliella salina*, *Chlorella pyrenoidosa*, *Haematococcus pluvialis*, *Chlorella protothecoides*, *Muriellopsis* sp., *Chlorella zofingiensis*, *P. tricornutum*, *Scenedesmus almeriensis*, *Nannochloropsis oculata*, *Porphyridium credendum*	• Protects against UV damage • Improves crop resilience	Cezare-Gomes et al. (2019), Pan et al. (2019)
Terpenoids	• Antioxidant • Antibacterial • Anticarcinogenic	• Protection of crops against bacteria, insects, and other creatures. • Drawing in pollinators. • Inducing growth and development in plants	Natural pesticides, attractants for bacterial insects, growth promotion	*Chondrococcus hornemanni*, *Hypnea pannosa*, *Plocamium cornutum*, *Sphaerococcus coronopifolius*, *Plocamium leptophyllum*, *Ulva lactuca*, *Codium fragile*, *Fucus vesiculosus*, *Halymenia durvillaei*, *Sargassum muticum*	• Enhances crop yield. • Reduces the need for chemical pesticides	Awasthi et al. (2018), Pan et al. (2019), Singh et al. (2017)
Polysaccharides	• Antioxidant • Anti-inflammatory • Antibacteria • Anticoagulant • Anticancer	• Protection of crops from biotic and abiotic stressors. • A rise in soil quality. • Growth stimulation for plants.	• Soil conditioner • Growth enhancer • Drought resistance	*Dunaliella* sp., *Chlorella* sp., *Navicula* sp., *Cylindrotheca* sp., *Rhodella* sp., *Phaeodactylum* sp., *Chlamydomonas* sp., *Porphyridium* sp., *Ulva* sp., *Gracilaria* sp., *Gelidium* sp.	• Enhances soil fertility. • Promotes root development.	El Arroussi et al. (2018), Gonçalves (2021), Pan et al. (2019), Singh et al. (2017)
Free fatty acids	• Antioxidant • Antifungal • Antiviral • Antibiotic • Anticancer	Protection of crops from several biotic and abiotic stress situations.	Fungicide Antiviral agent Soil amendment	*Dunaliella* sp., *Chlorella* sp., *Porphyridium* sp., *Nannochloropsis* sp., *Scenedesmus* sp., *Chaetoceros muelleri*, *T. suecica*, *Pavlova lutheri*, *P. tricornutum*	• Improves plant immunity. • Promotes healthy crop growth.	Gonçalves (2021), Pan et al. (2019), Singh et al. (2017)

plant development by manufacturing phytohormones (like cytokinins, auxins, and gibberellins), bioactive compounds (such as barbamide, carbamide, and aurilide), and an increase in the amount of glomalin-like protein in the soil (Prasanna et al., 2015).

Research has also been done on the application of microalgae as nitrogen-fixing biofertilizers. Different techniques of using microalgal monocultures of *Microcystis aeruginosa* and *Chlorella* sp. were investigated using willow and maize plants (Grzesik & Romanowska-Duda, 2014).

Research has also been done on the application of microalgae as nitrogen-fixing biofertilizers. Different techniques of using monocultures of *M. aeruginosa* and *Chlorella* sp. were investigated using willow and maize plants (Grzesik & Romanowska-Duda, 2014). Furthermore, Grzesik et al. (2017) investigated the foliar spraying of the mentioned monocultures, which demonstrated a notable enhancement in the physiological functions and plant development. Higher concentrations of activated chemicals from the microalgae strains may have contributed to the enhanced length, height, and quantity of shoots in both trials. Additionally, the plants' Nitrogen Phosphorus Potassium (NPK) content, activity of different enzymes (RNase, alkaline phosphatase, dehydrogenases, acid, or/and nitrate reductase), and the entire photosynthesis (stomatal conductance, cytomembrane stability, transpiration, chlorophyll quantity, and decreased intercellular carbon dioxide concentration) were all enhanced by the applied monocultures (Grzesik et al., 2017; Grzesik & Romanowska-Duda, 2014).

Lastly, research has also been done on the usage of consortia containing microalgae and cyanobacteria. Ramírez-López et al. (2019) demonstrated that the supplementary impacts of a photosynthetic and nitrogen-fixing microbial groups, comprising cyanobacteria and *Chlorophyta* species, which had positive impact on plant development and the soil, allowed for a 75% reduction in the chemical fertilization of wheat (*Triticum aestivum* L.). This was also demonstrated by Dineshkumar et al. (2018), as better outcomes were obtained by using microalgae (*Chlorella vulgaris* and *Arthrospira platensis*) in addition to lowering the chemical N fertilizer as much as 50 or 75% of the suggested amount. In conclusion, there is a lot of promise for sustainable agriculture in employing microalgae and cyanobacteria as biofertilizers, as they have been shown to improve nitrogen availability, encourage nutrient recycling, and boost plant growth thereby lowering the requirement for inorganic fertilizers.

7.6.2 Nutrient Availability

Secondary micronutrient deficits have occurred in the soil over the last 30 years as a result of crop residue recycling, uneven applications of macronutrients, reduced usage of organic manure, and bumper harvests. Photosynthetic microorganisms such as cyanobacteria and microalgae have the ability to absorb and retain vital nutrients (N and P) inside their biomass, even in locations where these resources are limited. They can also store other vital micronutrients that are engaged in redox processes in plant metabolism, such as potassium, magnesium, sulfur, and iron (Gonçalves, 2021). As a result, microalgae or cyanobacteria might be regarded as an alternate origin of macro- and micronutrients that are crucial for plant growth, as documented in the literature.

Cyanobacteria consortia or biofilms have the ability to fix nitrogen in soils and can also enhance the nutrients availability. According to reports, applying these biofilms increased the soil's composition by 40–60% in organic C and P, in addition to 50–90% in N. (Simranjit et al., 2019). In addition, these biofilms boosted the bioaccessibility of micronutrients like copper, manganese, iron, and zinc which exhibited a favorable effect on the pigments necessary for photosynthesis. When it comes to using microalgal species on various plants, *C. vulgaris* has been extensively researched (Dineshkumar, Subramanian, et al., 2020). By focusing on microalgae and cyanobacteria, Dineshkumar et al. (2019) evaluated using cow dung and a combination of *C. vulgaris* and *A. platensis* as a biofertilizer for corn and onions. Potassium levels along with the amounts of nitrogen, phosphorus, and manganese in maize were considerably raised by all of the treatments that were observed. Low amounts of Ca, Zn, Fe, CO_3, and Na contents, on the other hand, were noted. Early on, when the manure and microalgae were combined, maize grew most effectively, reaching 51.1% following 60 days of seeding (Dineshkumar et al., 2019). When onion plants were given the same conditions, the same pattern was seen. The optimal growth parameters for onion plants were obtained by combining the microalgal and cow dung to maximize the availability of both micro- and macronutrients. With enhanced biochemical (e.g., soluble sugars, phenols, free amino acids, and indoles) and mineral compositions (e.g., Na, K, P, Ca, Fe, Mg, and Mn), all treatments produced onions that were larger than the marketable size and had a higher market value. Mung bean plants were also tested by the same treatments under the application of a foliar spray (Dineshkumar, Duraimurugan, et al., 2020). Similarly, plants that were given the treatment of *C. vulgaris* and *A. platensis* produced higher levels of mung bean leaf and soil composition and enhanced the tangible properties of green gram, such as water and oil absorption and the index for water absorption as well as water solubility. Recently, Dineshkumar, Subramanian, et al. (2020) found that *C. vulgaris* extracts might be applied to black gram (*Vigna mungo* L.) as an environmentally friendly and possibly profitable foliar spray treatment. Improved production yields and soil qualities were accompanied by a fortification of the seed flour with consumer-friendly minerals like Na, K, Ca, Mg, and P. Due to constant cultivation, heavy fertilizer and pesticide use, and a lack of direct sunlight, nutrient depletion is a severe issue for crop productivity. As mentioned before, microalgae have the ability to retrieve nutrients from locations with restricted nutrient supply. In nutrient-poor soil substrates, Schreiber et al. (2018) examined the usage of mineral and fertilizers containing *C. vulgaris*. In comparison to the unfertilized controls (nutrient limitation), biofertilization increased the development of wheat plants under all experimental circumstances. There were no discernible variations between the fertilizers containing minerals and chlorella.

Wheat having comparable and lower amounts of nitrogen and phosphorus, respectively, grew more easily when fertilized with *C. vulgaris* as opposed to wheat plants cultivated without nutrient constraint. This implies that the soil nutrient shortfall can be compensated for by microalgae cells; however, whereas P may be released slowly or not at all, N can be produced in a configuration which is readily absorbable by wheat roots. However, the plants' root-hair characteristics were consistent among treatments, indicating that the microalgal P in the soil has been transformed into forms that are assimilated by the plants. These findings confirm that, when applied to marginal soils, microalgal biomass can promote crop development in a manner comparable to that of mineral fertilizers.

Other microalgal species have also been investigated; one such example includes the evaluation of the culture, dry mass and extracts of *Acutodesmus dimorphus* for use in tomato plant biofertilization, aerial spraying, and seed priming (Garcia-Gonzalez & Sommerfeld, 2016). With foliar application of the extract, the authors were able to obtain longer plants, more blooms per plant, and early sprouting with the seed treatment. Moreover, compared to applying at the time of transplant, the use of *A. dimorphus* dry biomass 22 days beforehand markedly improved plant growth. These findings imply that in order for the biomass to decompose and improve the nutrient accessibility for plants usage, an initial application is necessary. In another instance, Saadaoui et al. (2019) examined the results of using *Tetraselmis* sp. to fertilize date palms. Following treatment for three months, the addition of 0.5 g of *Tetraselmis* sp. to the soil enhanced its NPK content while keeping the level of heavy metals safe, according to the authors' observations. When compared to traditional fertilizer, this resulted in higher plant growth rates for date palms, with rates of survival up to 100%, more leaves (3.17 ± 0.14 vs. 3 ± 0.66), a higher ability to root the maximum stem thickness, longer shoots, and higher amounts of chlorophyll.

7.6.3 Soil Amendment

The qualities of soil (aggregation, porosity, permeability, ventilation, and humidity) can also be improved by microalgae and cyanobacteria. Certain algae species have the ability to release polysaccharides that combine to produce a gelatinous and sticky mesh that binds soil particles to their surface, enhancing soil aeration and water retention. This capacity is important for controlling temperature, pH, and protecting against erosion. These enhancements significantly impact the crop yield since they promote root development and soil microbial activity. Thus, cyanobacteria and microalgae can aid in restoring depleted soil crusts, preventing erosion, and storing harmful contaminants in the soil. Moreover, beneficial partnerships between microalgae and cyanobacteria and plant roots can be formed. By generating extracellular enzymes that decompose organic debris in the soil and liberate nutrients like nitrogen and phosphorus that are vital for plant growth, they help plants. Additionally, they are able to create hormones that promote plant development. Microorganisms utilize exudates like compounds secreted by plant roots, including simple sugars, amino acids, organic acids, and other materials as food sources (Ferreira et al., 2023). Lastly, by eliminating heavy metals (such as cadmium, lead, and chromium), trace elements (such as iron, zinc, copper, and manganese), hydrocarbons, and other substances, algae species can remediate soils and improve soil fertility and quality. Microalgae are capable of converting, detoxifying, and volatilizing heavy metals. Additionally, because they are not pathogenic, there is no chance of an unintentional release of heavy metals into the atmosphere. Because the polysaccharides, proteins, or lipids in their cell walls have a high metal binding ability, microalgae not only take pollutants onto their surfaces but also destroy them enzymatically (Kumar et al., 2015). This ability is essential for producing rice because there has been a documented rise in land contamination, which not only decreases the amount of arable land but also raises the chances of heavy metals to buildup in the rice grains and induce harmful health effects (Tripathi et al., 2008; Srivastava et al., 2018). Combining *C. vulgaris* and *Pseudomonas putida* reduced the availability of As and altered its absorption and detoxifying processes in rice plants, thereby mitigating the As stress under P-enriched environments (Srivastava et al., 2018).

7.7 APPLICATIONS IN PLANT GROWTH STIMULATION

By enhancing the rate of plant germination and its characteristics including shoot and root length, leaf area, and nutritional contents, the usage of microalgae and cyanobacteria (and/or their extracts) can directly promote the growth and development of plants. The metabolites obtained from microalgal/cyanobacterial are responsible for these enhancements because compounds like vitamins, polyamines, amino acids, and polysaccharides can initiate a variety of metabolic responses, including respiration, photosynthesis, nucleic acid synthesis, chlorophyll production, and ions uptake. Different types of microalgae and cyanobacteria (biomass and/or extracts) that have been investigated for their ability to directly stimulate the growth of higher plants are shown in Table 7.2.

Kholssi et al. (2019) investigated *Chlorella sorokiniana's* ability to stimulate wheat plants and its potential for usage as a chemical fertilizer substitute. In addition to resuspending the collected biomass in both fresh and spent medium, the researchers examined the filtered medium used for microalga cultivation and contrasted it with fresh media exclusively (control). The germination percentage of all treatments was higher than that of the control; however, the only treatments that differed statistically from the control were those that contained spent medium, whether it contained microalga or not. Although there was a general improvement in plant growth with the microalga treatments, the filtrate of *C. sorokiniana* (spent medium) produced the greatest increases in plant length (30%) and fresh weight (77%). This indicates that the compounds released from microalgae were beneficial for plant growth into the medium. These findings are highly intriguing from a biorefinery standpoint because the harvested biomass can be used for additional high-value uses like food and/or feed, while the supernatant may be utilized to biostimulate plants (Ferreira et al., 2022).

Uysal et al. (2015) examined the auto- and heterotrophic biostimulant capacity of *C. vulgaris* and found that, in comparison to the control, wheat and barley plants grew taller. In a co-cultivation

TABLE 7.2
Impacts of Algae/Algal Extracts on Direct Growth Stimulation of Plants

Algae/Algal Extracts	Source of Algal Extracts	Target Crop	Observed Improvements	References
Laurencia obtusa, *Corallina elongata*, and *Jania rubens*	Seaweed cultures	Maize	• Enhancement of plant length and fresh and dry weight. • Increased amount of leaves. • Increment in the nutritional value of plant (increase in potassium, phosphorus, and nitrogen contents)	Safinaz and Ragaa (2013)
Aqueous extracts from *Gracilaria corticata* and *Enteromorpha flexuosa*	Seaweed cultures	Maize and Sunflower	• Increase in shoot and root length and dry weight. • Improvement of plant nutritional value (increase in photosynthetic pigments, carbohydrate, proteins, and nutrients contents)	Omar et al. (2015)
Exopolysaccharide extracts from *Dunaliella salina*	Seaweed cultures	wheat	• Increase in germination rate and seedling growth. • Increase in root and coleoptiles height. • Increase in tolerance to salt stress.	El Arroussi et al. (2015)
Total polysaccharides extract from *Spirulina platensis*	Algal cultures	Tomato and Pepper	• Increase in plants' size. • Increase in roots' weight. • Increase in size and number of nodes	Elarroussia et al. (2016)
Chlorella pyrenoidosa	Algal cultures	Pepper	• Increase in fruit set and yield. • Improved nutrient uptake. • Enhanced plant vigor	Tian et al. (2022)
Codium fragile	Seaweed cultures	Lettuce	• Improvement in leaf growth. • Increase in nutrient content. • Enhanced resistance to pests.	Chaturvedi et al. (2022)
Liquid extracts from *Stoechospermum marginatum*	Seaweed cultures	Brinjal	• Increase in shoot and root length. • Increase in fresh and dry weight of leaves. • Increase in leaf area. • Improvement of leaves nutritional value (increase in moisture, photosynthetic pigments, protein, amino acids, reducing sugars, and ascorbic acid contents).	Ramya et al. (2015)
Chlorella vulgaris and *S. platensis*	Algal cultures	Maize	• Increase in germination rate and plant yield. • Increase in fresh and dry weights of shoot, root, and whole plant. • Increase in shoot length and in the number of leaves	Dineshkumar et al. (2019)
Ulva lactuca and *J. rubens*	Seaweed cultures	Spinach	• Increase in plant yield and height. • Improvement of plant nutritional value (increase in chlorophyll and nitrogen contents)	El-din and Hassan (2016)
Sargassum muticum	Seaweed cultures	Tomato	• Increase in plant growth and fruit quality. • Enhanced resistance to environmental stress.	Hernández-Herrera et al. (2022)
Arthrospira platensis	Algal cultures	Lettuce	• Increase in seedling growth. • Increase in spermine content in leaves.	Mógor et al. (2018)

system, Barone et al. (2019) observed a favorable impact on the biomass productivity of microalgae and tomato plants, particularly when employing biomass from *Scenedesmus quadricauda* in the availability of digestates made from the waste of an agrolivestock farm. The co-cultivation system containing solely *C. vulgaris* yielded the greatest values by weighing fresh plant shoot and dry whole plant, with a 2.5-fold increase over the control.

Barone et al. (2018) also discovered that the expression of root characteristics and genes linked to sugar beet nutrient acquisition (genes putatively implicated in sulfate deprivation) was biostimulated by *C. vulgaris* and *S. Quadricauda* extracts. Puglisi et al. (2020) evaluated *S. quadricauda* extracts in a related study and found that they had a positive effect on lettuce seedling growth, particularly at the shoot level, as well as higher dry matter, chlorophyll, carotenoids, and protein contents. Furthermore, it has been demonstrated that these extracts favorably affect the activity of a number of enzymes involved in the fundamental metabolisms of nitrogen and carbon.

The effects of applying *Scenedesmus* sp. and *A. platensis* hydrolysates topically to petunia plants were assessed by Puglisi et al. (2020). While *Arthrospira* increased the root dry matter, amount of flowers, and water quantity, *Scenedesmus* was used to improve the plant's nutrient status and hasten plant growth and blooming.

Godlewska et al. (2019) also investigated the application of *Arthrospira* in radish plants at different dosages. When seeds were treated homogenously and the filtrate was sprayed foliarly, an enhancement of length and fresh weight was noted. Chlorophyll increased most when the concentration was lower. It is also important to recognize that, in comparison to the commercial biostimulant, *Arthrospira*'s lowest dosages consistently produced superior outcomes. According to (Akgül, 2019) research, microalgal cell extract at concentrations of up to 75% increased the germination and seedling production of wheat and barley plants, whereas greater concentrations had the opposite impact. Michalak et al. (2016) conducted field tests to assess *A. platensis* supercritical extracts' impact on winter wheat seedlings. *A. platensis* extracts produced the best outcomes (longer shank and more grains in the ear). These extracts also demonstrated biostimulant qualities comparable to those of commercial products, with the added benefit of being a natural and eco-friendly source of biologically active substances.

7.8 CARBON SEQUESTRATION AND GREENHOUSE GAS MITIGATION

Algae-based fertilizers have now been found as a viable option for minimizing the levels of nitrous oxide (N_2O) emission, a long-lived GHG. These fertilities improve the fertility of the soils as they encourage nitrogen fixation external to the soil and better cycling of nutrients avoiding synthetic nitrogen-based fertilities. Because algae-based fertilizers reduce the amount of nitrogen more than what is required in the soil, they minimize the microbial reactions that give N_2O gas such as denitrification. Also, it has been found that the incorporated organic matter of algae can increase the population density of microbiota in the soil which will make the conditions for releases of N_2O less conducive. This approach encourages sustainable farming practices to be adopted and at the same time reduces the effects of agriculture on the environment. The application of algal biochar potentially reduces N_2O emissions. Several research have reported on the reduction of N_2O emissions with the addition of algae-based fertilizers (Wang et al., 2017; Zhang et al., 2019). The use of biochar in rice agriculture promotes carbon sequestration and reduces N_2O emissions because of its increased surface area. Applying biochar to agricultural areas is a more effective way to mitigate N_2O emissions (Liu et al., 2012; Zhang et al., 2010).

Since the use of fossil fuels, the primary GHG that should be worried about is carbon dioxide because it seriously jeopardizes the planet's environment. Diverse methods, including as filtration and chemical and mechanical processes, have been used worldwide to reduce CO_2 emissions (Rizwan et al., 2018). By taking up carbon from thermal power plants and turning it into a variety of useful products, such as biofertilizers, microalgae offer a fantastic way to sequester carbon and turn it into a carbon-negative strategy (Mona et al., 2013; Mondal et al., 2017). Compared to terrestrial plants, which absorb 0.03e0.06% of carbon dioxide, it demonstrates an efficient capture of the CO_2 content of 5e15% from the exhaust gases (Li et al., 2008). Compared to terrestrial plants, green algae have an efficiency of 10–15 times greater in absorbing solar energy. According to Morais and Costa (2007), *Spirulina* sp. fixed 53.3% of carbon dioxide for 6% CO_2 (v/v) and 45.6% for 12% CO_2 (v/v) from the flue, indicating a maximum average fixation rate of 37.9% for 6% CO_2 (v/v). According to Yang et al. (2015), *Gracilaria* and *Saccharina* sp. can fix 2.3×10^5 and 1.0×10^6 tdw (tons dry weight) $year^{-1}$, respectively. High CO_2 fixation rates are also shown by *Anabaena* sp. (1450 mg L^{-1} d^{-1}) and *C. vulgaris* (6240 mg L^{-1} d^{-1}) ((Selvan *et al*, 2024) The fact that 1.83 kg of CO_2 may be fixed by growing 1 kg of microalgae demonstrates the potential of algae for the mitigation of GHGs like carbon dioxide (Cheah et al., 2015). According to Kadam (2002), using algae for CO_2 mitigation has the advantage of a biorefinery approach as microalgae have the ability to lower CO_2, CH_2, NO_x, and SO_x. According to Moreira and Pires (2016), it can be used to create biofuels, biogas, and biofertilizers. Because algae can store double the amount of energy per carbon compared to other materials, they are regarded as a carbon-neutral method that emits no CO_2 (Moreira & Pires, 2016). Furthermore, the creation of algal biochar offers a model for sequestering carbon and a soil ameliorant that can enhance the microbial ecology, nutrient status, and water-holding capacity of various soils, in addition to its energy potential and value-added products as discussed earlier (Lehmann & Joseph, 2015).

7.9 CHALLENGES AND OPPORTUNITIES FOR ADOPTION

There are a number of issues that are experienced during the manufacturing and application of algae-based fertilizers. One of the challenges is the cost of production and investment, accompanied by high energy levels in the culture process as well as the extraction processes. For example, the culture of microorganisms such as *C. vulgaris*, in photobioreactors, is quite costly in terms of energy consumption and regular care (Li et al., 2011). Moreover, there is the problem of scalability in the production of the algae mass needed to support large-scale agricultural applications. The scarcity of suitable algal strains for the development of efficient nitrogen-fixing or multifunctional soil nutrients adds more challenges to the design of efficient, universal fertilizers (Ghosh et al., 2022).

Other constraints for the production of algae-based fertilizers are seasonal as well as geographical restrictions (Zrimec Berden et al., 2024). This poses a challenge in the cultivation of these species, especially in areas where the climate is not friendly, for instance, in areas of long winters or areas where there is little fresh water supply (Arts et al., 2015; Pankratz et al., 2019). In addition, the stability and shelf life of bio-formulated algae-based fertilizers are, at times, spoiled easily and cannot be used after some time, or when stored under improper conditions (Miranda et al., 2024). Another difficulty is the presence of toxins in algae-based fertilizers. For instance, *Anabaena* may release cyanotoxins if not sufficiently regulated (Chorus et al., 2021). This is very important especially when dealing with large quantities of the product and farmers need to be careful with using algae-based fertilizers because of the safety issues especially where they have no idea of the dangers or the benefits of using such products (Zou et al., 2021). The use of algae-based fertilizers is being muted by low farmer awareness as farmers prioritize using conventional fertilizer products, thus will likely hold back in switching to algae products if they are not informed enough about the benefits (Chatterjee et al., 2017). These problems need to be solved if algae-based fertilizers are ever to be considered even a marginally better solution to the agricultural world.

7.10 FUTURE DIRECTIONS AND RESEARCH NEEDS

Because of their abilities as a biostimulant and a biofertilizer, algae are a prospective source for bio-based organic fertilizers. Microalgae are the center of focus for agrochemical firms that aim to augment the sustainability of agriculture. The development of new, environmentally and economically viable methods has the potential to improve the commercial production of biofertilizers. Using wastewater treatment to create algae-based biofertilizers is an integrated technique that promotes the growth of a circular bioeconomy. To validate the benefits of biofertilizers, molecular methods including proteomics, metagenomics, and genomes should be used to study the mechanism of algae-based biofertilizers (Vishwakarma et al., 2024).

Moreover, artificial intelligence (AI) and machine learning are two promising approaches for predicting algal potential, optimizing production, and improving the quality and performance of biofertilizers (Mafat et al., 2024). Genetic algorithms have the potential to enhance biomass yields and productivity by designing ideal conditions in relation to light, heat, and nutrient input after sifting through huge amounts of data. In addition, through utilizing the AI systems in biofertilizers' processing, the quality, as well as the functionality of the biofertilizers, can be significantly increased through customization of processing methods as well as predicting their effect on various kinds of soil as well as crops (Raj et al., 2022). By these developments, AI and machine learning can greatly enhance the usage rates of algae-based fertilizers for agriculture thus enshrining its sustainable usage rates.

The combination of algae-based wastewater treatment and algal bio-fertilizer production is currently being focused on by scientists. As lignocellulosic wastes are numerous and renewable, biotechnological exploitation of them holds great promise for ecologically friendly and sustainable energy production. This approach can address the huge challenge of lowering the volume of waste available on Earth. It is possible for the green microalga *Scenedesmus obliquus* to use the nutrients in waste digestate for mass development and growth. Conversely, algal biomass has the potential to be utilized as a free or immobilized biofertilizer to promote the growth of some valued plants for food or other uses. Saving a significant amount of fresh water is associated with lowering the cost of microalgal production by the use of anaerobic digestate, a byproduct of bioenergy production processes, as a medium for algae growing (Hubenov et al., 2023).

Thus, it is not economically feasible to extract numerous products from microalgae using the downstream processing methods. Reducing nutrient costs can be achieved by combining microalgal cultivation with wastewater treatment. Additionally, there are several issues to consider when thinking about the possible uses of microalgae cultivated on wastewater, including public acceptability and consumer perception in addition to the safety of the biomass that is produced. Thus, extensive studies and additional studies are needed in this area (Su et al., 2023).

7.11 CONCLUSION

Conclusively, algae-based biofertilizers open a new perspective in the strategies of modern agriculture which combines innovation and responsibility. Compared to traditional chemical fertilizers, algae-based fertilizers induce a positive impact on soil health while reducing the CF of agriculture. These biofertilizers not only boost crop health but also the sustainability of the ecosystems. Given the future challenges and growing global call to make agriculture "greener" and sustainable, algae-based biofertilizers turn the lens to rethink agriculture for the climate change era. Although algae-based fertilizers offer multiple benefits including speedy germination by initiating a variety of metabolic responses, including respiration, photosynthesis, nucleic acid synthesis, chlorophyll production, and ions uptake, several challenges hinder the applicability of such fertilizers. Such limitations include the complex scalability of production, costs, and farmers' prioritization. The adoption of AI approaches and omics technology can further help to better understand the underlying mechanism of fertilizer application and will boost their effective production and beneficial effects. Future research should be fundamental to realizing the full potential of these crops and moving toward a more sustainable system of production.

DECLARATION

The authors have not used any generative AI tools during the writing of this manuscript.

REFERENCES

Abdel-Raouf, N., Al-Homaidan, A. A., & Ibraheem, I. B. M. (2012). Agricultural importance of algae. *African Journal of Biotechnology*, *11*(54), 11648–11658.

Ahmed, M., Rauf, M., Akhtar, M., Mukhtar, Z., & Saeed, N. A. (2020). Hazards of nitrogen fertilizers and ways to reduce nitrate accumulation in crop plants. *Environmental Science and Pollution Research*, *27*, 17661–17670.

Ahmed, M., Rauf, M., Mukhtar, Z., & Saeed, N. A. (2017). Excessive use of nitrogenous fertilizers: an unawareness causing serious threats to environment and human health. *Environmental Science and Pollution Research*, *24*, 26983–26987.

Akgül, F. (2019). Effect of *Spirulina platensis* (Gomont) geitler extract on seed germination of wheat and barley. *Alinteri Journal of Agriculture Science*, *34*(2), 148–153.

Amin, S. A., Green, D. H., Hart, M. C., Küpper, F. C., Sunda, W. G., & Carrano, C. J. (2009). Photolysis of iron–siderophore chelates promotes bacterial–algal mutualism. *Proceedings of the National Academy of Sciences*, *106*(40), 17071–17076.

Ammar, E. E., Aioub, A. A. A., Elesawy, A. E., Karkour, A. M., Mouhamed, M. S., Amer, A. A., & El-Shershaby, N. A. (2022). Algae as bio-fertilizers: between current situation and future prospective. *Saudi Journal of Biological Sciences*, *29*(5), 3083–3096.

Arts, M. T., Hixson, S. M., & Mercieca, S. (2015). *Climate warming reduces essential fatty acid production in algae. Journal of Environmental Indicators, 9,* 7–8. International Society for Environmental Indicators.

Awasthi, M., Upadhyay, A. K., Singh, S., Pandey, V. P., & Dwivedi, U. N. (2018). Terpenoids as promising therapeutic molecules against Alzheimer's disease: amyloid beta-and acetylcholinesterase-directed pharmacokinetic and molecular docking analyses. *Molecular Simulation*, *44*(1), 1–11.

Bao, J., Zhuo, C., Zhang, D., Li, Y., Hu, F., Li, H., Su, Z., Liang, Y., & He, H. (2021). Potential applicability of a cyanobacterium as a biofertilizer and biopesticide in rice fields. *Plant and Soil*, *463*, 97–112.

Barone, V., Baglieri, A., Stevanato, P., Broccanello, C., Bertoldo, G., Bertaggia, M., Cagnin, M., Pizzeghello, D., Moliterni, V. M. C., & Mandolino, G. (2018). Root morphological and molecular responses induced by microalgae extracts in sugar beet (*Beta vulgaris* L.). *Journal of Applied Phycology*, *30*, 1061–1071.

Barone, V., Puglisi, I., Fragalà, F., Lo Piero, A. R., Giuffrida, F., & Baglieri, A. (2019). Novel bioprocess for the cultivation of microalgae in hydroponic growing system of tomato plants. *Journal of Applied Phycology*, *31*, 465–470.

Biernat-Jarka, A., & Trkebska, P. (2018). The importance of organic farming in the context of sustainable development of rural areas in Poland. *Acta Scientiarum Polonorum. Oeconomia*, *17*(2), 39–47.

Bisht, N., & Chauhan, P. S. (2020). Excessive and disproportionate use of chemicals cause soil contamination and nutritional stress. *Soil Contamination-Threats and Sustainable Solutions*, *2020*, 1–10.

Cezare-Gomes, E. A., Mejia-da-Silva, L. del C., Pérez-Mora, L. S., Matsudo, M. C., Ferreira-Camargo, L. S., Singh, A. K., & de Carvalho, J. C. M. (2019). Potential of microalgae carotenoids for industrial application. *Applied Biochemistry and Biotechnology*, *188*, 602–634.

Chapman, R. L. (2013). Algae: the world's most important "plants"—An introduction. *Mitigation and Adaptation Strategies for Global Change*, *18*, 5–12.

Chatterjee, A., Singh, S., Agrawal, C., Yadav, S., Rai, R., & Rai, L. C. (2017). Role of algae as a biofertilizer. In *Algal Green Chemistry* (pp. 189–200). Elsevier.

Chaturvedi, S., Kulshrestha, S., & Bhardwaj, K. (2022). Role of seaweeds in plant growth promotion and disease management. In *New and Future Developments in Microbial Biotechnology and Bioengineering* (pp. 217–238). Elsevier.

Cheah, W. Y., Show, P. L., Chang, J.-S., Ling, T. C., & Juan, J. C. (2015). Biosequestration of atmospheric CO_2 and flue gas-containing CO_2 by microalgae. *Bioresource Technology*, *184*, 190–201.

Chorus, I., Fastner, J., & Welker, M. (2021). Cyanobacteria and cyanotoxins in a changing environment: concepts, controversies, challenges. *Water*, *13*(18), 2463.

Collin, M. S., Venkatraman, S. K., Vijayakumar, N., Kanimozhi, V., Arbaaz, S. M., Stacey, R. G. S., Anusha, J., Choudhary, R., Lvov, V., & Tovar, G. I. (2022). Bioaccumulation of lead (Pb) and its effects on human: a review. *Journal of Hazardous Materials Advances*, *7*, 100094.

De Morais, M. G., & Costa, J. A. V. (2007). Biofixation of carbon dioxide by *Spirulina* sp. and *Scenedesmus obliquus* cultivated in a three-stage serial tubular photobioreactor. *Journal of Biotechnology*, *129*(3), 439–445.

Dineshkumar, R., Duraimurugan, M., Sharmiladevi, N., Lakshmi, L. P., Rasheeq, A. A., Arumugam, A., & Sampathkumar, P. (2020). Microalgal liquid biofertilizer and biostimulant effect on green gram (*Vigna radiata* L.) an experimental cultivation. *Biomass Conversion and Biorefinery*, *12*(8), 3007–3027.

Dineshkumar, R., Kumaravel, R., Gopalsamy, J., Sikder, M. N. A., & Sampathkumar, P. (2018). Microalgae as bio-fertilizers for rice growth and seed yield productivity. *Waste and Biomass Valorization*, *9*, 793–800.

Dineshkumar, R., Subramanian, J., Gopalsamy, J., Jayasingam, P., Arumugam, A., Kannadasan, S., & Sampathkumar, P. (2019). The impact of using microalgae as biofertilizer in maize (*Zea mays* L.). *Waste and Biomass Valorization*, *10*, 1101–1110.

Dineshkumar, R., Subramanian, J., & Sampathkumar, P. (2020). Prospective of *Chlorella vulgaris* to augment growth and yield parameters along with superior seed qualities in black gram, *Vigna mungo* (L.). *Waste and Biomass Valorization*, *11*, 1279–1287.

El Arroussi, H., Benhima, R., Elbaouchi, A., Sijilmassi, B., El Mernissi, N., Aafsar, A., Meftah-Kadmiri, I., Bendaou, N., & Smouni, A. (2018). *Dunaliella salina* exopolysaccharides: A promising biostimulant for salt stress tolerance in tomato (*Solanum lycopersicum*). *Journal of Applied Phycology*, *30*, 2929–2941.

El Arroussi, H., Elbaouchi, A., Benhima, R., Bendaou, N., Smouni, A., & Wahby, I. (2015). Halophilic microalgae *Dunaliella salina* extracts improve seed germination and seedling growth of *Triticum aestivum* L. under salt stress. *II World Congress on the Use of Biostimulants in Agriculture 1148*, 13–26.

Elarroussia, H., Elmernissia, N., Benhimaa, R., El Kadmiria, I. M., Bendaou, N., Smouni, A., & Wahbya, I. (2016). Microalgae polysaccharides a promising plant growth biostimulant. *Journal of Algal Biomass Utilization*, *7*(4), 55–63.

El-din, S. M., & Hassan, S. M. (2016). The promotive effect of different concentrations of marine algae on spinach plants (*Spinacia oleracea* L.). *Egyptian Journal of Horticulture*, *43*, 109–122.

Esquivel-Hernández, D. A., Ibarra-Garza, I. P., Rodriguez-Rodriguez, J., Cuéllar-Bermúdez, S. P., Rostro-Alanis, M. d J., Alemán-Nava, G. S., García-Pérez, J. S., & Parra-Saldívar, R. (2017). Green extraction technologies for high-value metabolites from algae: a review. *Biofuels, Bioprod Biorefin*, *11*, 215–231.

Fantozzi, F., Bartocci, P., & others. (2016). Carbon footprint as a tool to limit greenhouse gas emissions. *Greenhouse Gases*, *285287*.

Ferreira, A., Bastos, C. R. V, Marques-dos-Santos, C., Acién-Fernandez, F. G., & Gouveia, L. (2023). Algaeculture for agriculture: from past to future. *Frontiers in Agronomy*, *5*, 1064041.

Ferreira, A., Figueiredo, D., Ferreira, F., Ribeiro, B., Reis, A., da Silva, T. L., & Gouveia, L. (2022). Impact of high-pressure homogenization on the cell integrity of *Tetradesmus obliquus* and seed germination. *Molecules*, *27*(7), 2275.

Garcia-Gonzalez, J., & Sommerfeld, M. (2016). Biofertilizer and biostimulant properties of the microalga *Acutodesmus dimorphus*. *Journal of Applied Phycology*, *28*, 1051–1061.

Gasol, C. M., Salvia, J., Serra, J., Antón, A., Sevigne, E., Rieradevall, J., & Gabarrell, X. (2012). A life cycle assessment of biodiesel production from winter rape grown in Southern Europe. *Biomass and Bioenergy*, *40*, 71–81.

Ghosh, D., Ghorai, P., Debnath, S., Indrama, T., Kondi, V., & Tiwari, O. N. (2022). Algal biofertilizer towards green sustainable agriculture. In *New and Future Developments in Microbial Biotechnology and Bioengineering* (pp. 27–45). Elsevier.

Godlewska, K., Michalak, I., Pacyga, P., Baśladyńska, S., & Chojnacka, K. (2019). Potential applications of cyanobacteria: *Spirulina platensis* filtrates and homogenates in agriculture. *World Journal of Microbiology and Biotechnology*, *35*, 1–18.

Gonçalves, A. L. (2021). The use of microalgae and cyanobacteria in the improvement of agricultural practices: a review on their biofertilising, biostimulating and biopesticide roles. *Applied Sciences*, *11*(2), 871.

Grzesik, M., & Romanowska-Duda, Z. (2014). Improvements in germination, growth, and metabolic activity of corn seedlings by grain conditioning and root application with cyanobacteria and microalgae. *Polish Journal of Environmental Studies*, *23*(4), 1147–1153.

Grzesik, M., Romanowska-Duda, Z., & Kalaji, H. M. (2017). Effectiveness of cyanobacteria and green algae in enhancing the photosynthetic performance and growth of willow (*Salix viminalis* L.) plants under limited synthetic fertilizers application. *Photosynthetica*, *55*, 510–521.

Gupta, A. K. (2004). *The Complete Technology Book on Biofertilizers and Organic Farming*, Vol. 168 (p. 242). National Institute of Industrial Research Press.

Hernández-Herrera, R. M., Sánchez-Hernández, C. V., Palmeros-Suárez, P. A., Ocampo-Alvarez, H., Santacruz-Ruvalcaba, F., Meza-Canales, I. D., & Becerril-Espinosa, A. (2022). Seaweed extract improves growth and productivity of tomato plants under salinity stress. *Agronomy*, *12*(10), 2495.

Herrmann, L., & Lesueur, D. (2013). Challenges of formulation and quality of biofertilizers for successful inoculation. *Applied Microbiology and Biotechnology*, *97*(20), 8859–8873.

Hubenov, V., Ivanova, J., Nacheva, L., & Kabaivanova, L. (2023). Agricultural waste utilization for biomethane and algae-based fertilizer production for circular economy. *Bulgarian Journal of Agricultural Science*, *29*(6), 1022–1026.

Iriarte, A., Rieradevall, J., & Gabarrell, X. (2010). Life cycle assessment of sunflower and rapeseed as energy crops under Chilean conditions. *Journal of Cleaner Production*, *18*(4), 336–345.

Jaimes Duarte, I., Humberto Álvarez Hernández, S., Laura Ibañez, A., & Rodr\'\iguez Canto, A. (2018). Macroalgae as soil conditioners or growth promoters of *Pisum sativum* (L). *Annual Research & Review in Biology*, *27*(6), 1–8.

Javed, M. U., Mukhtar, H., Hayat, M. T., Rashid, U., Mumtaz, M. W., & Ngamcharussrivichai, C. (2022). Sustainable processing of algal biomass for a comprehensive biorefinery. *Journal of Biotechnology*, *352*(August 2021), 47–58. https://doi.org/10.1016/j.jbiotec.2022.05.009

Kadam, K. L. (2002). Environmental implications of power generation via coal-microalgae cofiring. *Energy*, *27*(10), 905–922.

Kapustka, L. A., & DuBois, J. D. (1987). Dinitrogen fixation by cyanobacteria and associative rhizosphere bacteria in the Arapaho Prairie in the Sand Hills of Nebraska. *American Journal of Botany*, *74*(1), 107–113.

Khan, A., Panthari, D., Sharma, R. S., Punetha, A., Singh, A. V., & Upadhayay, V. K. (2023). Biofertilizers: a microbial-assisted strategy to improve plant growth and soil health. In *Advanced Microbial Techniques in Agriculture, Environment, and Health Management* (pp. 97–118). Elsevier.

Kholssi, R., Marks, E. A. N., Miñón, J., Montero, O., Debdoubi, A., & Rad, C. (2019). Biofertilizing effect of *Chlorella sorokiniana* suspensions on wheat growth. *Journal of Plant Growth Regulation*, *38*, 644–649.

Kumar, K. S., Dahms, H.-U., Won, E.-J., Lee, J.-S., & Shin, K.-H. (2015). Microalgae–a promising tool for heavy metal remediation. *Ecotoxicology and Environmental Safety*, *113*, 329–352.

Kumar, R., Kumar, R., & Prakash, O. (2019). The impact of chemical fertilizers on our environment and ecosystem Chapter-5 the impact of chemical fertilizers on our environment and ecosystem. 2019. February.

Kuraganti, G., Edla, S., & Pallaval, V. B. (2020). Cyanobacteria as biofertilizers: current research, commercial aspects, and future challenges. In *Advances in Plant Microbiome and Sustainable Agriculture: Functional Annotation and Future Challenges*, 259–278. Springer.

Lehmann, J., & Joseph, S. (2015). Biochar for environmental management: an introduction. In *Biochar for Environmental Management* (pp. 1–13). Routledge.

Li, J., Hu, W., Lu, Z., Meng, F., Cong, R., Li, X., Ren, T., & Lu, J. (2022). Imbalance between nitrogen and potassium fertilization influences potassium deficiency symptoms in winter oilseed rape (*Brassica napus* L.) leaves. *The Crop Journal*, *10*(2), 565–576.

Li, Y., Horsman, M., Wu, N., Lan, C. Q., & Dubois-Calero, N. (2008). Biocatalysts and bioreactor design. *Biotechnology Progress*, *24*, 815–820.

Li, Z., Yuan, H., Yang, J., & Li, B. (2011). Optimization of the biomass production of oil algae *Chlorella minutissima* UTEX2341. *Bioresource Technology*, *102*(19), 9128–9134.

Li, Z., Zhang, R., Xia, S., Wang, L., Liu, C., Zhang, R., Fan, Z., Chen, F., & Liu, Y. (2019). Interactions between N, P and K fertilizers affect the environment and the yield and quality of satsumas. *Global Ecology and Conservation*, *19*, e00663.

Liu, C.-W., Sung, Y., Chen, B.-C., & Lai, H.-Y. (2014). Effects of nitrogen fertilizers on the growth and nitrate content of lettuce (*Lactuca sativa* L.). *International Journal of Environmental Research and Public Health*, *11*(4), 4427–4440.

Liu, X., Qu, J., Li, L., Zhang, A., Jufeng, Z., Zheng, J., & Pan, G. (2012). Can biochar amendment be an ecological engineering technology to depress N_2O emission in rice paddies?—A cross site field experiment from South China. *Ecological Engineering*, *42*, 168–173.

Mafat, I. H., Palla, S., & Surya, D. V. (2024). Machine learning and artificial intelligence for algal cultivation, harvesting techniques, wastewater treatment, nutrient recovery, and biofuel production and optimization. In *Value Added Products from Bioalgae Based Biorefineries: Opportunities and Challenges* (pp. 463–487). Springer.

Mahapatra, D. M., Chanakya, H. N., Joshi, N. V., Ramachandra, T. V., & Murthy, G. S. (2018). Algae-based biofertilizers: a biorefinery approach. In *Microorganisms for Green Revolution: Volume 2: Microbes for Sustainable Agro-Ecosystem* (pp. 177–196) Springer.

Mahapatra, D. M., Chanakya, H. N., & Ramachandra, T. V. (2014). Bioremediation and lipid synthesis through mixotrophic algal consortia in municipal wastewater. *Bioresource Technology*, *168*, 142–150.

Mahapatra, D. M., Satapathy, K. C., & Panda, B. (2022). Biofertilizers and nanofertilizers for sustainable agriculture: Phycoprospects and challenges. *Science of the Total Environment*, *803*, 149990. https://doi.org/https://doi.org/10.1016/j.scitotenv.2021.149990

Mancia, A., Chadwick, D. R., Waters, S. M., & Krol, D. J. (2022). Uncertainties in direct N_2O emissions from grazing ruminant excreta (EF3PRP) in national greenhouse gas inventories. *Science of the Total Environment*, *803*, 149935.

McKnight, D. M., & Morel, F. M. M. (1980). Copper complexation by siderophores from filamentous blue-green algae 1. *Limnology and Oceanography*, *25*(1), 62–71.

Michalak, I., Chojnacka, K., Dmytryk, A., Wilk, R., Gramza, M., & Rój, E. (2016). Evaluation of supercritical extracts of algae as biostimulants of plant growth in field trials. *Frontiers in Plant Science*, *7*, 1591.

Miranda, A. M., Hernandez-Tenorio, F., Villalta, F., Vargas, G. J., & Sáez, A. A. (2024). Advances in the development of biofertilizers and biostimulants from microalgae. *Biology*, *13*(3), 199.

Mógor, Á. F., Ördög, V., Lima, G. P. P., Molnár, Z., & Mógor, G. (2018). Biostimulant properties of cyanobacterial hydrolysate related to polyamines. *Journal of Applied Phycology*, *30*, 453–460.

Mona, S., Kaushik, A., & Kaushik, C. P. (2013). Prolonged hydrogen production by Nostoc in photobioreactor and multi-stage use of the biological waste for column biosorption of some dyes and metals. *Biomass and Bioenergy*, *54*, 27–35.

Mondal, M., Goswami, S., Ghosh, A., Oinam, G., Tiwari, O. N., Das, P., Gayen, K., Mandal, M. K., & Halder, G. N. (2017). Production of biodiesel from microalgae through biological carbon capture: a review. *3 Biotech*, *7*, 1–21.

Moreira, D., & Pires, J. C. M. (2016). Atmospheric CO_2 capture by algae: negative carbon dioxide emission path. *Bioresource Technology*, *215*, 371–379.

Moudrý Jr, J., & Moudrý, J. (2014). Environmental aspects of organic farming. In *Organic Agriculture towards Sustainability* (pp. 247–274). IntechOpen.

Omar, H. H., Abdullatif, B. M., Al-Kazan, M. M., & El-Gendy, A. M. (2015). Various applications of seaweed improves growth and biochemical constituents of *Zea mays* L. and *Helianthus annuus* L. *Journal of Plant Nutrition*, *38*(1), 28–40.

Ozlu, E., & Kumar, S. (2018). Response of soil organic carbon, pH, electrical conductivity, and water stable aggregates to long-term annual manure and inorganic fertilizer. *Soil Science Society of America Journal*, *82*(5), 1243–1251.

Pahalvi, H. N., Rafiya, L., Rashid, S., Nisar, B., & Kamili, A. N. (2021). Chemical fertilizers and their impact on soil health. In *Microbiota and Biofertilizers, Vol 2: Ecofriendly Tools for Reclamation of Degraded Soil Environs* (pp. 1–20) Springer.

Pan, S., Jeevanandam, J., & Danquah, M. K. (2019). Benefits of algal extracts in sustainable agriculture. In *Grand Challenges in Algae Biotechnology* (pp. 501–534) Springer.

Pankratz, S., Oyedun, A. O., & Kumar, A. (2019). Development of cost models of algae production in a cold climate using different production systems. *Biofuels, Bioproducts and Biorefining*, *13*(5), 1246–1260.

Park, J.-R., Jang, Y.-H., Kim, E.-G., Lee, G.-S., & Kim, K.-M. (2023). Nitrogen fertilization causes changes in agricultural characteristics and gas emissions in rice field. *Sustainability*, *15*(4), 3336.

Prasanna, R., Babu, S., Bidyarani, N., Kumar, A., Triveni, S., Monga, D., Mukherjee, A. K., Kranthi, S., Gokte-Narkhedkar, N., & Adak, A. (2015). Prospecting cyanobacteria-fortified composts as plant growth promoting and biocontrol agents in cotton. *Experimental Agriculture*, *51*(1), 42–65.

Puglisi, I., La Bella, E., Rovetto, E. I., Lo Piero, A. R., & Baglieri, A. (2020). Biostimulant effect and biochemical response in lettuce seedlings treated with a *Scenedesmus quadricauda* extract. *Plants*, *9*(1), 123.

Rahaman, M. S., Mise, N., & Ichihara, S. (2022). Arsenic contamination in food chain in Bangladesh: a review on health hazards, socioeconomic impacts and implications. *Hygiene and Environmental Health Advances*, *2*, 100004.

Raj, E. F. I., Appadurai, M., & Athiappan, K. (2022). Precision farming in modern agriculture. In *Smart Agriculture Automation Using Advanced Technologies: Data Analytics and Machine Learning, Cloud Architecture, Automation and IoT* (pp. 61–87). Springer.

Ramírez-López, C., Esparza-García, F. J., Ferrera-Cerrato, R., Alarcón, A., & Cañizares-Villanueva, R. O. (2019). Short-term effects of a photosynthetic microbial consortium and nitrogen fertilization on soil chemical properties, growth, and yield of wheat under greenhouse conditions. *Journal of Applied Phycology*, *31*(6), 3617–3624.

Ramya, S. S., Vijayanand, N., & Rathinavel, S. (2015). Foliar application of liquid biofertilizer of brown alga *Stoechospermum marginatum* on growth, biochemical and yield of *Solanum melongena*. *International Journal of Recycling of Organic Waste in Agriculture*, *4*, 167–173.

Rizwan, M., Mujtaba, G., Memon, S. A., Lee, K., & Rashid, N. (2018). Exploring the potential of microalgae for new biotechnology applications and beyond: a review. *Renewable and Sustainable Energy Reviews*, *92*, 394–404.

Rutkowski, K., & Łysiak, G. P. (2023). Effect of nitrogen fertilization on tree growth and nutrient content in soil and cherry leaves (*Prunus cerasus* L.). *Agriculture*, *13*(3), 578.

Saadaoui, I., Sedky, R., Rasheed, R., Bounnit, T., Almahmoud, A., Elshekh, A., Dalgamouni, T., al Jmal, K., Das, P., & Al Jabri, H. (2019). Assessment of the algae-based biofertilizer influence on date palm (*Phoenix dactylifera* L.) cultivation. *Journal of Applied Phycology*, *31*, 457–463.

Safinaz, A. F., & Ragaa, A. H. (2013). Effect of some red marine algae as biofertilizers on growth of maize (*Zea mayz* L.) plants. *International Food Research Journal*, *20*(4), 1629.

Saif, S., Abid, Z., Ashiq, M. F., Altaf, M., & Ashraf, R. S. (2021). Biofertilizer formulations. In *Biofertilizers: Study and Impact*, 211–256. Scrivener Publishing LLC.

Schmid, B. (2021). *Assessment of Microalgal Biomass as Potential Feedstock for Sustainable, Eco-Friendly Biostimulants and Biopesticides in Plant Production*. Universidade do Algarve (Portugal).

Schreiber, C., Schiedung, H., Harrison, L., Briese, C., Ackermann, B., Kant, J., Schrey, S. D., Hofmann, D., Singh, D., & Ebenhöh, O. (2018). Evaluating potential of green alga *Chlorella vulgaris* to accumulate phosphorus and to fertilize nutrient-poor soil substrates for crop plants. *Journal of Applied Phycology*, *30*, 2827–2836.

Selvan, S. T., Velramar, B., Chandrasekaran, R. K., Ramamurthy, D., & Natarajan, P. M. (2024). Phototrophic Carbon Capture Using Natural Microalgal Whole-Cell Support: An Eco-Technological Approach. In *Whole-Cell Biocatalysis* (pp. 515-554). Apple Academic Press.

Senthilnathan, P. R., & Ganczarczyk, J. J. (1990). Application of biomass carriers in activated sludge process. In *Wastewater Treatment by Immobilized Cells,* 103–141. CRC Press.

Shamim, A., Mahfooz, S., Hussain, A., & Farooqui, A. (2020). Ability of Al-acclimatized immobilized nostoc muscorum to combat abiotic stress and its potential as a biofertilizer. *Journal of Pure and Applied Microbiology*, *14*(2), 1377–1386.

Simranjit, K., Kanchan, A., Prasanna, R., Ranjan, K., Ramakrishnan, B., Singh, A. K., & Shivay, Y. S. (2019). Microbial inoculants as plant growth stimulating and soil nutrient availability enhancing options for cucumber under protected cultivation. *World Journal of Microbiology and Biotechnology*, *35*, 1–14.

Singh, J. S. (2014). Cyanobacteria: a vital bio-agent in eco-restoration of degraded lands and sustainable agriculture. *Climate Change and Environmental Sustainability*, *2*(2), 133–137.

Singh, J. S. (2015). *Plant–microbe interactions: A viable tool for agricultural sustainability Plant Microbes Symbiosis: Applied Facets*, NK Arora (Ed.). Springer, New Delhi/Heidelberg/New York/Dordrecht/London (2015). 384 pp., Price: 149.99€, ISBN: 9788132220671. Elsevier.

Singh, R., Parihar, P., Singh, M., Bajguz, A., Kumar, J., Singh, S., Singh, V. P., & Prasad, S. M. (2017). Uncovering potential applications of cyanobacteria and algal metabolites in biology, agriculture and medicine: current status and future prospects. *Frontiers in Microbiology*, *8*, 515.

Srivastava, S., Srivastava, S., Bist, V., Awasthi, S., Chauhan, R., Chaudhry, V., Singh, P. C., Dwivedi, S., Niranjan, A., & Agrawal, L. (2018). *Chlorella vulgaris* and *Pseudomonas putida* interaction modulates phosphate trafficking for reduced arsenic uptake in rice (*Oryza sativa* L.). *Journal of Hazardous Materials*, *351*, 177–187.

Su, M., Bastiaens, L., Verspreet, J., & Hayes, M. (2023). Applications of microalgae in foods, pharma and feeds and their use as fertilizers and biostimulants: legislation and regulatory aspects for consideration. *Foods*, *12*(20), 3878.

Suleiman, A. K. A., Lourenço, K. S., Clark, C., Luz, R. L., da Silva, G. H. R., Vet, L. E. M., Cantarella, H., Fernandes, T. V, & Kuramae, E. E. (2020). From toilet to agriculture: fertilization with microalgal biomass from wastewater impacts the soil and rhizosphere active microbiomes, greenhouse gas emissions and plant growth. *Resources, Conservation and Recycling*, *161*, 104924.

Sun, J., Li, W., Li, C., Chang, W., Zhang, S., Zeng, Y., Zeng, C., & Peng, M. (2020). Effect of different rates of nitrogen fertilization on crop yield, soil properties and leaf physiological attributes in banana under subtropical regions of China. *Frontiers in Plant Science*, *11*, 613760.

Sun, N., Thompson, R. B., Xu, J., Liao, S., Suo, L., Peng, Y., Chen, Q., Yang, J., Li, Y., & Zou, G. (2021). Arsenic and cadmium accumulation in soil as affected by continuous organic fertilizer application: implications for clean production. *Agronomy*, *11*(11), 2272.

Tayoh, L. N. (2020). Destruction of soil health and risk of food contamination by application of chemical fertilizer. In *Ecological and Practical Applications for Sustainable Agriculture*, 53–64. Springer.

Thomas, E. Y., Omueti, J. A. I., & Ogundayomi, O. (2012). The effect of phosphate fertilizer on heavy metal in soils and *Amaranthus caudatus*. *Agriculture and Biology Journal of North America*, *3*(4), 145–149.

Tian, S.-L., Khan, A., Zheng, W.-N., Song, L., Liu, J.-H., Wang, X.-Q., & Li, L. (2022). Effects of *Chlorella* extracts on growth of *Capsicum annuum* L. seedlings. *Scientific Reports*, *12*(1), 15455.

Uysal, O., Uysal, F. O., & Ekinci, K. (2015). Evaluation of microalgae as microbial fertilizer. *European Journal of Sustainable Development*, *4*(2), 77.

Vishwakarma, P., Pal, V., Nagar, J., & Rai, R. (2024). Biofertilizers and soil amendments from algae: Production and agronomic benefits. In *Value Added Products from Bioalgae Based Biorefineries: Opportunities and Challenges* (pp. 343–363). Springer.

Wang, N., Chang, Z.-Z., Xue, X.-M., Yu, J.-G., Shi, X.-X., Ma, L. Q., & Li, H.-B. (2017). Biochar decreases nitrogen oxide and enhances methane emissions via altering microbial community composition of anaerobic paddy soil. *Science of the Total Environment*, *581*, 689–696.

Wei, B., Yu, J., Cao, Z., Meng, M., Yang, L., & Chen, Q. (2020). The availability and accumulation of heavy metals in greenhouse soils associated with intensive fertilizer application. *International Journal of Environmental Research and Public Health*, *17*(15), 5359.

Xu, Z.-M., Li, Q.-S., Yang, P., Ye, H.-J., Chen, Z.-S., Guo, S.-H., Wang, L.-L., He, B.-Y., & Zeng, E. Y. (2017). Impact of osmoregulation on the differences in Cd accumulation between two contrasting edible amaranth cultivars grown on Cd-polluted saline soils. *Environmental Pollution*, *224*, 89–97.

Yan, L.-J., & Allen, D. C. (2021). Cadmium-induced kidney injury: oxidative damage as a unifying mechanism. *Biomolecules*, *11*(11), 1575.

Yang, Y., Chai, Z., Wang, Q., Chen, W., He, Z., & Jiang, S. (2015). Cultivation of seaweed *Gracilaria* in Chinese coastal waters and its contribution to environmental improvements. *Algal Research*, *9*, 236–244.

Zhang, A., Cui, L., Pan, G., Li, L., Hussain, Q., Zhang, X., Zheng, J., & Crowley, D. (2010). Effect of biochar amendment on yield and methane and nitrous oxide emissions from a rice paddy from Tai Lake plain, China. *Agriculture, Ecosystems \& Environment*, *139*(4), 469–475.

Zhang, C., Zeng, G., Huang, D., Lai, C., Chen, M., Cheng, M., Tang, W., Tang, L., Dong, H., Huang, B., & others. (2019). Biochar for environmental management: Mitigating greenhouse gas emissions, contaminant treatment, and potential negative impacts. *Chemical Engineering Journal*, *373*, 902–922.

Zhang, J., Song, X., Wei, H., Zhou, W., Peng, C., & Li, D. (2021). Effect of substituting nitrogen fertilizer with nitrogen-fixing cyanobacteria on yield in a double-rice cropping system in southern China. *Journal of Applied Phycology*, *33*, 2221–2232.

Zou, Y., Zeng, Q., Li, H., Liu, H., & Lu, Q. (2021). Emerging technologies of algae-based wastewater remediation for bio-fertilizer production: a promising pathway to sustainable agriculture. *Journal of Chemical Technology & Biotechnology*, *96*(3), 551–563.

Zrimec Berden, M., Reinhardt, R., Rossi, S., Ficara, E., Rodero, M. del R., & Muñoz, R. (2024). Algae and biogas plants: Digestate remediation and nutrient recycling with algal systems. *Algae Mediated Bioremediation: Industrial Prospectives*, *1*, 246–270.

8 Role of Biogas in Climate Change Resilience and Circular Economy

Muhammad Uzair Javed and Hamid Mukhtar

8.1 INTRODUCTION

The linear economic model which is currently in use does not align with the ideas of sustainable development. A linear economy is predicated on unsustainable practices such as the mining of raw materials for industry, the utilization of fossil fuels as energy sources, the continuous depletion of water and soil resources, and waste management systems that rely on landfills (LFs) (FAO, 2020). Given the threats posed by socioeconomic inequality, resource scarcity, and climate change, this economic system is thus out of date. Ineffective resource management and exploitation result in pollution, poverty, regional disparities, depletion of fossil fuels, rising energy consumption, local or regional climate change issues, and augmenting global environmental disputes (Kaza et al., 2018; Scheidel et al., 2020).

Climate change is a result of the rapid increase in population and industrialization, which has also led to the build-up of greenhouse gases (GHGs) in the atmosphere (Mikhaylov et al., 2020). According to the Intergovernmental Panel on Climate Change's (IPCC) Global Warming Special Study, global warming must be limited to 1.5°C to avoid detrimental effects on ecosystems and populations. This goal is still attainable, but it requires swift, significant, and sustained worldwide climate action to achieve net-zero global emissions by 2050 (Rhodes, 2019). The existing linear economy can be replaced with a circular economy model that addresses long-term economic growth, social equality, and environmental conservation. As the world is witnessing unheard-before levels of human expansion, extending human activity in every aspect of the environment, efforts must be made for sustainable and responsible anthropogenic activities (Ghosh et al., 2020).

Given the potential for recovering valuable raw materials and utilizing their energy potential, waste management and pollution control should be given special consideration while executing the goals of the circular economy. After many years of research and practical application, the prevailing belief is that waste-to-energy (WtE) is a good option for waste management and the development of energy resources since it offers several advantages (Obaideen et al., 2022). For several uses, WtE can serve as a semi-renewable energy source, replacing fossil fuels. Any method that creates fuel, fire, power, or other forms of energy from a waste feedstock is referred to as a WtE system (Rafiee et al., 2021).

In order to accomplish the goal of converting waste into electricity, biogas is a competitive resource. Biogas is widely regarded as a sustainable energy resource because of its plentiful, low-cost feedstocks and its diverse applications, which include fuel, electricity generation, heating, and as a raw material for producing sustainable chemicals such as carbon dioxide, hydrogen, and biofuels. One technique for obtaining energy from organic waste and generating biogas is anaerobic digestion (AD). In the last decade, the capacity for biogas-based power generation has grown

DOI: 10.1201/9781003685814-8

rapidly. Globally, the capacity increased by 90%, rising from 65 GW in 2010 to 120 GW in 2019 (Zupančič et al., 2022). The chapter outlines various feedstocks to illustrate the biogas generation paths. Its advantages and contributions to the current circular economy initiatives are also covered. The adoption of biogas might present several obstacles, including socio-cultural and infrastructure issues; however with more study and advancement, it can contribute further to the shifting of the linear economy to a circular economy, thereby becoming a cost-effective as well as an eco-friendly alternative to conventional energy resources.

8.2 BIOGAS

Bioenergy is the collective term for fuels derived from biological sources that are utilized in transportation, heat, power, and gas. It is the third most significant energy source in the world, nearly emission-neutral, and capable of advancing and balancing both existing and future energy systems (Mao et al., 2015). The main areas for usage include LF gas (LF gas), dairy waste, municipal solid waste, garden waste, agricultural waste, wastewater treatment facilities, and urban food waste. When organic waste decomposes in an oxygen-free environment, it generates a combination of gases, predominantly methane, and carbon dioxide, which is called biogas (Jameel et al., 2024). Biogas was once thought of as "poor man's fuel," but in the context of international energy planning, it has become one of the main alternatives (Abbasi et al., 2012). Biogas is now recognized as a sustainable energy carrier, produced from various sources such as LFs, AD of waste, wastewater treatment sludge, animal manure, and energy crops (Rafiee et al., 2021).

8.3 PRODUCTION PROCESS OF BIOGAS

Biogas can be produced in a variety of ways and environments. Its composition depends on several factors such as the process design and the nature of the substrate used to produce biogas. The various routes of biogas production are discussed in the following.

8.3.1 Anaerobic Digestion

AD is the term for the biological conversion process that takes place in an oxygen-free environment. It consists of four stages: hydrolysis, acidogenesis, acetogenesis, and methanogenesis. This biological mechanism is catalyzed by a multitude of microorganisms working together in the absence of oxygen. It is commonly recognized that AD is responsible for recycling carbon in a variety of settings, such as wetlands, manures, rice fields, and animal intestines. This method is also widely used on an industrial scale for valorization of organic wastes (Kougias & Angelidaki, 2018).

The different degradation processes are carried out by distinct consortia of bacteria, with varying environmental requirements and a partly syntrophic relationship (Angelidaki et al., 1993). The first attack on polymers and monomers is caused by hydrolyzing and fermenting microorganisms, which primarily provide acetate and hydrogen as well as different concentrations of volatile fatty acids like butyrate and propionate. Hydrolytic enzymes, such as cellulase, cellobiase, amylase, xylanase, lipase, and protease, are expelled by hydrolytic bacteria. The hydrolysis and fermentation of organic materials involve a varied community of microorganisms. The majority of bacteria, including *Bifidobacteria*, *Clostridium*, and *Bacteriocides*, are stringent anaerobes. In addition, several facultative anaerobes participate, including *Enterobacteriaceae* and *Streptococci* (Mignogna et al., 2023).

The longer-chain volatile fatty acids are transformed into acetate and hydrogen by acetogenic bacteria that exclusively produce hydrogen, as illustrated in Figure 8.1. The acetogenic bacteria that produce hydrogen are poorly understood. *Acetobacterium woodii* and *Clostridium aceticum* are the common homoacetogenic bacteria. The buildup of hydrogen has the ability to inhibit the metabolism of acetogenic bacteria. Therefore, acetogenic and H_2-producing bacteria require the

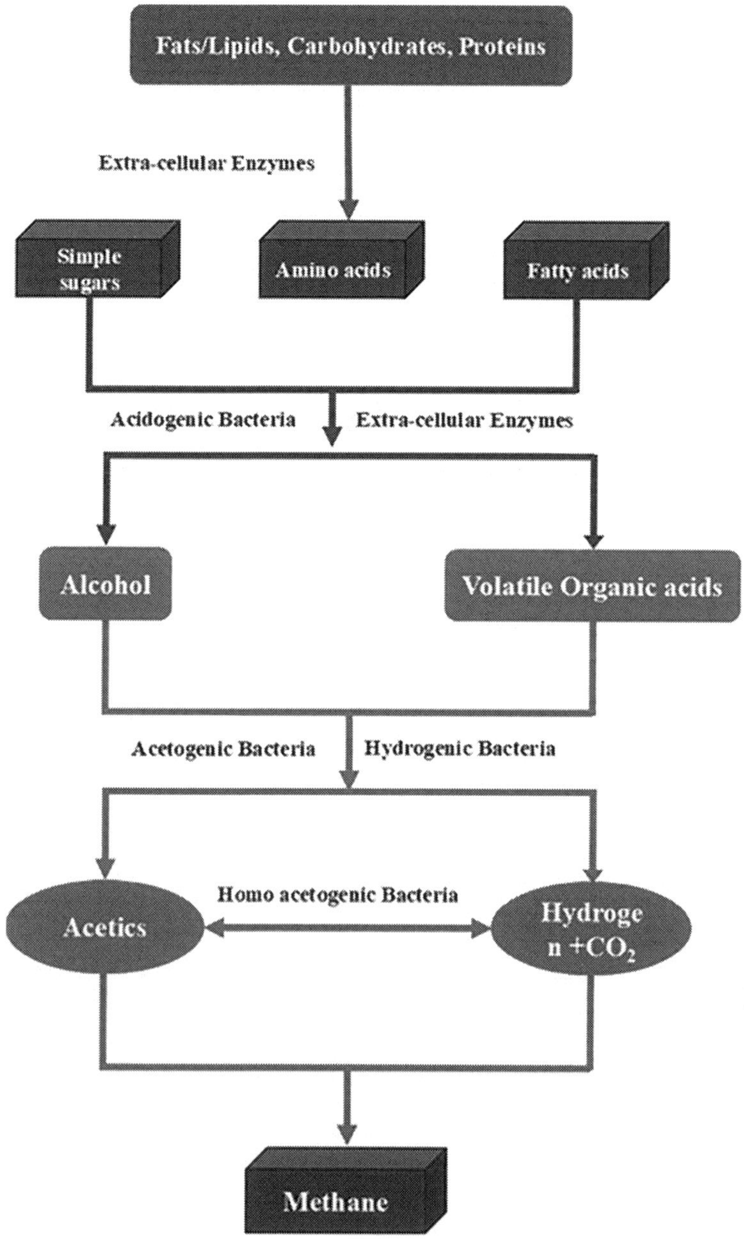

FIGURE 8.1 Process model of anaerobic digestion.

maintenance of an exceptionally low partial pressure of hydrogen. Despite the fact that many microbiological aspects of a methanogenic consortia's metabolic networks are yet unknown, recent research indicates that hydrogen may be a limiting substrate for methanogens (Bagi et al., 2007). This assumption stems from the observation that a higher presence of H_2-producing bacteria within the biogas-generating community leads to a daily increase in the production of biogas. At the final stage of the degradation process, two types of methanogenic bacteria convert carbon dioxide and acetate into methane. Compared to most other anaerobic bacteria, these bacteria are stringent anaerobic and need a lower redox potential for growth. Only a few species have been found to be able

TABLE 8.1
Classification of Feedstocks for Biogas Production, Along with the Necessary Pretreatment Methods, Conversion Processes, and Estimated Biogas Yield Potential

Origin	Feedstock Examples	Pretreatment Required	Conversion Process	Biogas Yield Potential	References
Agriculture	• Agricultural residues and waste • Energy crops • Livestock waste • Manure (cattle, pig, poultry) • Algae • Animal byproducts • Mosses • Lichens (Crustose lichens, foliose lichens) • Herbaceous woody crops • Oilseed crops • Harvest remains • Kelps • Bryophyta • Polytrichales.	• **Physical:** mechanical, thermal • **Biological:** partial composting, fungi, enzymatic • **Combined:** nitrogen extraction, extrusion	• Anaerobic digestion • Co-digestion with other organic wastes	200–500 Nm^3 biogas/ton of volatile solids	Chevalier et al. (2023), Kamusoko et al. (2019), Orlando and Borja (2020)
Municipal waste	• Community biowastes (garden wastes, food remains, excreta waste) • Sewage sludge • Organic fraction from municipal waste	• **Physical:** mechanical • **Biological** pre-composting • **Combined:** steam explosion	• Anaerobic digestion • Wet and dry digestion	100–400 Nm^3 biogas/ ton of total solids (TS)	Kamali et al. (2023), Karthikeyan et al. (2018), Mitraka et al. (2022)
Industry	• Industrial wastes and wastewater • Waste from food and beverage processing • Starch industry waste • Sugar industry residues • Slaughterhouses byproducts	• **Physical:** thermal, electrokinetic, ultrasound, microwave • **Combined:** steam explosion	• Anaerobic digestion • Dark fermentation	400–800 Nm^3 biogas/ton of chemical oxygen	Aliyu Salihu (2016), Anacleto et al. (2022), Ketsub et al. (2022)

to degrade acetate into CH_4 and CO_2, such as *Methanosarcina barkeri*, *Methanococcus mazei*, and *Methanothrix soehngenii*, though all methanogenic bacteria are capable of utilizing hydrogen to produce methane (Weiland, 2010).

8.3.2 Anaerobic Digestion in Landfills

A LF site is a designated area for the disposal of garbage, refuse, and other types of solid waste. Landfilling has been the most popular method of disposal as a result of the increasing amount of trash produced by hospitals, workplaces, residents, schools, and marketplaces. LFs are either buried or stacked in piles. Although LFs are the most economical way to dispose of waste materials, they cause environmental issues. The waste products that build up in the LFs are primarily responsible for the contamination of the soil, water, and air (Ghasemi Ghodrat et al., 2018). Furthermore, organic compounds affect future generations over time since they disintegrate slowly in an oxygen-free environment. There are five distinct types of landfilling: anaerobic, aerobic, semi-aerobic, and hybrid, LF as a deposit of inert waste (Pawłowska, 2014).

Complex biochemical conversion processes with many steps need to be developed in order to produce biogas from LFs. Anaerobic decomposition takes place when the oxygen is no longer available. A LF gas plant consists of two systems: a production system and a recovery system. To collect the generated gas, a recovery system may include membrane coverings, vertical perforated pipe wells, or horizontal perforated ditches or pipes. Biogas may be used as fuel for energy conversion processes, which offers a very interesting source of energy (Mavridis & Voudrias, 2021).

Novel AD technologies include membrane bioreactors (MBR), integrated high-rate anaerobic reactors (HRAR)-MBR, and HRAR (Singh et al., 2022). Depending on the source, biogas may include contaminants such as halogenated compounds, organic silicon species, nitrogen, sulfur compounds (H_2S, thiols, sulphides, and disulphides), and sulfur compounds, all of which are removed during upgrading. A conventional LF produces biogas which is a complex mixture of several ingredients. LF gas may consist of various components, including methane (35–65%), carbon dioxide (15–50%), halocarbons (20–200 ppm), nitrogen (5–40%), oxygen (0–5%), hydrogen (0–3%), carbon monoxide (0–3%), siloxanes (0–50 mg/m³), water vapor (0–5%), hydrogen sulfide (0–100 ppm), ammonia (0–5 ppm), and volatile organic compounds (VOC) (0–4,500 mg/m³) (Abanades et al., 2022; Rafiee et al., 2021).

8.3.3 Feedstocks

AD utilizes a diverse range of waste materials as substrates, categorized into three primary types of organic waste based on their origin (Atelge et al., 2020). Moreover, studies indicate that choosing one or more appropriate pretreatment methods can boost methane production or increase the AD rate, ultimately enhancing the performance of a biogas plant (Chandel et al., 2019). Table 8.2 lists the different feedstocks used to produce biogas along with their required pretreatment technologies.

8.3.3.1 Food Waste as Feedstock

Approximately 1.3×10^9 tons of various foods are disposed away as garbage globally each year. This type of waste can be produced in establishments including homes, grocery stores, restaurants, company cafeterias, and bars (Chew et al., 2021).Organic materials including proteins, carbohydrates, and lipids make up the majority of food waste. These materials may decompose into other materials

TABLE 8.2
Contributions of Biogas to Climate Change Mitigation, Resilience, and Circular Economy across Relevant SDGs

Sustainability Goals	Contribution of Biogas	Benefit
SDG 6: Clean Water and Sanitation	• Increasing the capacity of wastewater treatment to improve the quality of the water. • Providing decentralized wastewater treatment facilities in rural areas.	Environmental
SDG 7: Affordable and Clean Energy	• Increasing overall energy sustainability. • Improving the reliability and affordability of the energy supply.	Environmental Economic
SDG 9: Industry, Innovation, and Infrastructure	• Reducing material's carbon footprint. • Building sustainable infrastructure.	Environmental Economic
SDG 12: Responsible Consumption and Production	• Improving the effectiveness of the use of natural resources. • Improving the waste recycling process. • Providing a lower-emission energy source in order to reduce GHG emissions. • Reducing air and water pollution.	Environmental
SDG 13: Climate Action	• Reducing GHG emissions from the livestock industry. • Reducing GHG emissions from LFs.	Environmental

such as organic amino acids, glucose, and fatty acids (Mishra et al., 2021). AD of food waste produces very little contamination of solid waste as compared to other treatment processes such as gasification, pyrolysis, and incineration (Chew et al., 2021).

8.3.3.2 Industrial Feedstock

Many industries that process various raw materials produce significant quantities of residues, by-products, and waste, all of which can be utilized for biogas production. Waste may contain a variety of viruses, contaminants, and heavy metals, depending on the techniques that industries utilize in their production processes. In general, industrial dumps include waste from pulp and paper, petro-chemical, textile, and agro-industrial sectors, among other industries (Atelge et al., 2020). Biogas can be produced by utilizing these materials. An eco-friendly method of producing products, cutting down on the expense of disposing biowaste, and meeting environmental regulations is to produce biogas from various forms of biowaste (Yellezuome et al., 2024). The use of agro-industrial waste and its degradation procedure are affected by several factors, including its chemical composition, as well as its physical and thermal characteristics (Devi et al., 2022).

8.3.3.3 Activated Sludge from Wastewater Treatment Plants

The production of activated sludge results in large amounts of waste-activated sludge, which is then used in wastewater treatment plants. In 2020, approximately 9×10^6 metric tons of waste-activated sludge were generated and released throughout Europe. Due to its high organic content, waste-activated sludge has the potential to be used as a substrate for the production of biogas, an energy source. Biogas has the potential to meet the energy requirements of wastewater treatment facilities or be integrated into the architecture of current grids to generate power and heat. Numerous studies are currently focused on improving production of biogas by using activated sludge through several methods of pretreatment, including ultrasonication, enzymatic and thermal treatments, and the use of natural zeolites, among others (Tang et al., 2023).

8.3.3.4 Livestock Manure as a Feedstock

Organic materials found in livestock manure can be used as a feed in the procedure for producing biofuels (Wang et al., 2021). Only a small portion of the 1.4 billion tons of animal manure produced annually by farms in the EU and the UK is getting accumulated as biowaste (Köninger et al., 2021). Waste from plants, animals, and humans is an organic substance that decomposes naturally and has components that may be used to make biogas. AD can be performed with any type of moist organic waste. Generally, biomasses suitable for biogas production include animal wastes, plant materials like straw and forage, manure, and fodder residues, as well as household waste (Zareei, 2018).

8.3.3.5 Microalgae as Feedstock

Another feedstock used to produce biogas is microalgae. Algal biomass is a great feedstock for the AD process of producing biogas (Sher et al., 2024). Due to their high lipid and carbohydrate content, microalgae are considered to be a very efficient source for biogas production. In addition, they grow fast and are easy to cultivate, harvest, and transform into biogas. A study conducted by Ahmad et al. (2022) utilized *Spirulina platensis* microalgae as a feedstock to produce biogas. The results of this research indicated that the highest biogas yield achieved was 740 mL/gVS with a reaction time of 4 hours at 70°C, employing an alkaline pretreatment with 6% NaOH. Moreover, another study by Kumari et al. (2021) reported yields of 479 mL and 147 mL/gVS for biogas and methane, respectively, from untreated *Chlorella pyrenoidosa* biomass.

8.4 GLOBAL BIOGAS PRODUCERS

Recently, the production of biogas has experienced a substantial rise in adoption due to its environmental advantages and effectiveness as a "waste-to-energy" solution (Bhatt & Tao, 2020). As a matter of fact, it is seen to be essential to achieving the EU's ambitious targets for renewable energy (Capodaglio et al., 2016). The US and the EU have had particularly noteworthy progress, although development in Asia and other regions has been relatively slower (Scarlat et al., 2018). According to the statistics of 2012, Germany, the UK, and Italy produced 77% of all the biogas produced in the EU; other countries have added more biogas plants to diversify their energy sources. In 2012, the European Union produced 6.8% of its energy from biogas, with Germany having the largest share at over 19%. This was made possible due to institutional and legislative support, advancements in technology, and associations of small farmers (Szymańska & Lewandowska, 2015). According to the current reports, the EU's leading producers of biogas at the moment are Germany, France, the UK, the Czech Republic, and Italy (Nsair et al., 2020; Perea-Moreno et al., 2019).

8.5 POTENTIAL OF BIOGAS IN MITIGATING CLIMATE CHANGE AND ROLE IN THE CIRCULAR ECONOMY

8.5.1 Biogas as an Energy Carrier

The development of distributed generation, renewable energy, and energy efficiency measures are being promoted by industrialized nations in response to current energy crises and the possible impending depletion of fossil or non-renewable fuel supplies. This shift is not only essential for enhancing energy security but also aligns with global sustainable development goals, which now encompass energy efficiency and access to renewable energy. Achieving a transition to more efficient energy systems requires coordinated actions across all political levels, from local initiatives to global agreements (Zupančič et al., 2022). In this context, biogas emerges as a versatile resource that can be converted into various fuels. It can be upgraded to increase its methane content, combusted to generate electricity, gasified to produce syngas, or utilized through processes like methanotrophy and partial oxidation. By leveraging biogas, we can contribute to a more sustainable energy landscape while addressing pressing environmental challenges (Mignogna et al., 2023).

8.5.1.1 Domestic and Industrial Heating

Biogas presents a compelling substitute for natural gas as a means of generating electricity. There are several direct and indirect applications for raw biogas. The direct applications of biogas include cooking and lighting. However, it may also be utilized indirectly for a variety of systems when it undergoes chemical, physical, or biological treatment in order to enhance its quality or prepare it for future purposes (Kapoor, Ghosh, Kumar, et al., 2020). The most common, least expensive, and widely used technique for the combustion of biogas is the direct approach, which involves technology that is typically easy to maintain. This form of biogas utilization does not require H_2S removal and has a high moisture content. Moreover, it has been widely used globally for many years, particularly in rural areas. When burned, biogas produces a blue, clean flame and emits fewer pollutants compared to other energy sources (Kapoor, Ghosh, Tyagi, et al., 2020). In order to improve the future of biogas globally, physical biogas treatment techniques like upgrading and cleaning are frequently used (Wang et al., 2022).

8.5.1.2 Combined Heat and Power

Combined heat and power (CHP) is the term for the simultaneous production of useful heat and power in an integrated process using a common fuel or energy source. The energy source, the energy conversion procedure, and the demand side—which uses the generated heat and electricity—make up the CHP system. In addition to generating electricity, the power plant may provide process heat for industrial applications or heat for residential buildings in colder climates. CHP systems may employ a wide variety of energy sources. Biogas has gained acceptance as a renewable energy source globally due to its unique properties, setting it apart from other biomass resources (Hosseini et al., 2013). A significant portion of the biogas used in Europe (for example, 40% in 2011) is used by CHP plants in order to generate heat and power -. Combining the production of heat and electricity allows for more effective utilization of the energy source, which lowers CO_2 emissions and other environmental impacts. Engine-based systems are the most widely used kind of CHPs based on the AD of biogas. The building's hot water and heating requirements are met by the engine's waste heat. In this case, the system is configured to satisfy the boiler's needs in order to meet the building's heat demand. The system relies on the heat load, but any excess electricity generated must be capable of being fed into the electricity grid (Jonsson et al., 1999; Mago et al., 2009).

8.5.1.3 Vehicular Fuel

Biogas has the potential to yield several transportation fuels through various processes; the most probable alternatives are methanol, hydrogen, dimethyl ether (DME), compressed biogas (CBG), liquefied biogas (LBG), and Fischer–Tropsch (FT) fuels. While biogas is the same resource that can be used to create all of the fuels—CBG, LBG, hydrogen, methanol, DME, and FT—they differ in their strengths, limitations, and level of maturity. This unprocessed biogas could be used in the transportation network in a few different ways: it can be burned to generate electricity, upgraded to include more methane, gasified to create syngas, or partially oxidized to make methanol or methanotrophic acid as shown in Figure 8.2.

Raw biogas can be upgraded to produce biomethane, which is purer, provided that the methane content is at least 97%. The upgraded biomethane can be used directly as fuel in two forms: either compressed to 200 bar (CBG) or cooled to −162°C (LBG). It is frequently necessary to perform an extra polishing process after liquefying biomethane in order to further purify it. LBG often seems to have less of an effect on global warming as compared to CBG because of its lower methane slip. In several countries, a trend exists toward escalated biogas production into CBG (Gustafsson et al., 2020). Sweden is among such countries, as approximately 60% (1.3 TWh) of the biogas produced there in 2018 was converted to CBG, CBG has been used there since the 1990s (Klackenberg, 2019). Moreover, it was estimated that Denmark, which had previously concentrated on using biogas for heat and power, would upgrade 50% of its biogas production in 2018, a significant increase from the non-existent upgrading in 2012 (Sørensen & Jørgensen, 2022).

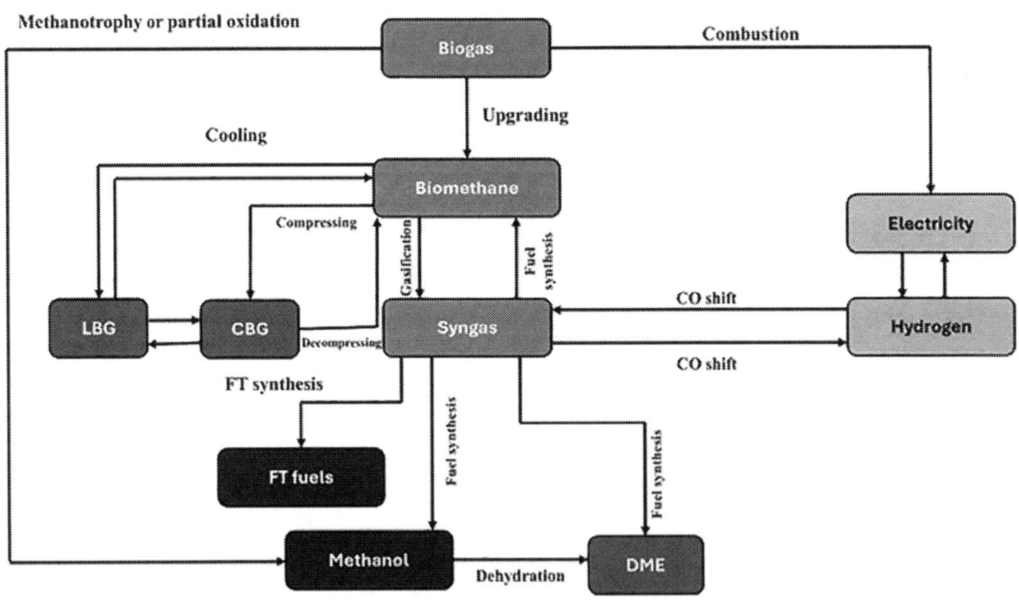

FIGURE 8.2 The different ways in which biogas can be used to produce fuels.

Through the process of gasification, biomethane may be utilized not only as fuel but also to create additional fuels. Syngas, a combination of hydrogen and carbon monoxide (CO), is produced during the gasification of biomethane. The syngas can then be transformed into other fuels. Hydrogen may be produced from syngas by changing its carbon dioxide concentration. Hydrogen generated from renewable resources often requires the usage of electricity and water.

In turn, the hydrogen can generate electricity. It also functions in reverse, producing hydrogen with electricity and water that may be transformed into syngas. Additionally, in fuel synthesis, syngas can be utilized to create methanol, which can then be used to create DME (Zinoviev et al., 2010). There is just one methanol plant in the Netherlands that produces more than 300 GWh of methanol annually and applies biogas as a resource. In addition to that facility, several factors influence the manufacture of methanol from various renewable sources, including some that are commercially established and others still in research and development (Hobson et al., 2018).

Moreover, syngas can also be used directly to create DME (Semelsberger et al., 2006), but methanol is the primary feedstock used in most of the DME production that is now underway. In a FT, syngas can also be used to create FT fuels like synthetic diesel. There are a few auto manufacturers, such as Isuzu Trucks, AB Volvo, Nissan Diesel, and Shanghai Diesel Co., that have focused on manufacturing heavy vehicles that are compatible with DME (Shah, 2017).

8.5.2 Environmental Benefits of Biogas

Biogas plays a significant role in addressing environmental challenges, particularly in mitigating climate change. The actual advantage of biogas and its demand is frequently seen to be its potential to control renewable energy output to account for fluctuations in wind energy. No matter how much energy it contains, biogas will always release CO_2, which is significantly less harmful to the environment than methane, even if it is flared. Reducing GHG emissions through the replacement of conventional fossil fuels is one of the primary environmental benefits (Cuéllar & Webber, 2008; Rafiee et al., 2021). A country's carbon credits increase as it adopts more renewable energy sources and embraces WtE initiatives. For example, India's largest bovine population could generate 335 million tons of cow dung each month, with approximately 110 million tons lost during collection and transportation. There is enough material to support 50 million small-scale biogas plants with the remaining 225 MT every month. There is a chance that his massive biogas production can lessen global warming (Thiruselvi et al., 2021). Biogas has other benefits for climate resilience beyond its ability to mitigate the consequences of global warming, which are discussed in more detail in the following section (Figure 8.3).

8.5.2.1 Deforestation Reduction

The 55% of the wood that is removed from forests for fuel accounts for 5% of global deforestation. Particularly in developing countries, biogas technology provides a sustainable energy source for families. Living in a rural region, this technology might be cost-effective and environmentally advantageous for the people. The implementation of household biogas systems can help reduce deforestation, safeguard forests, and enhance the outcomes of the Grain for Green Project (GGP) in the northwest regions of China. An 8 m^3 rural biogas digester in National Gas Grid Programme (NGGP) has the potential to save around 2,000 kg of wood yearly, which is the same as planting 0.23 ha of new forest (Deng et al., 2020). Moreover, the digester has the potential to hold up to 2,000 kg of straw yearly, which is comparable to 16.67–20 ha of natural grassland's grazing capacity, and may sustain eight to ten sheep annually (Qiu et al., 2005). Thus, the rural resident's requirement for firewood for lighting and cooking is reduced by biogas technology. Moreover, studies have reported a reduction in the amount of forest deforestation and degradation. Installing home biogas systems can protect forests, lessen deforestation, and improve the results of the GGP in the northwest of China (Tajebe, 2016).

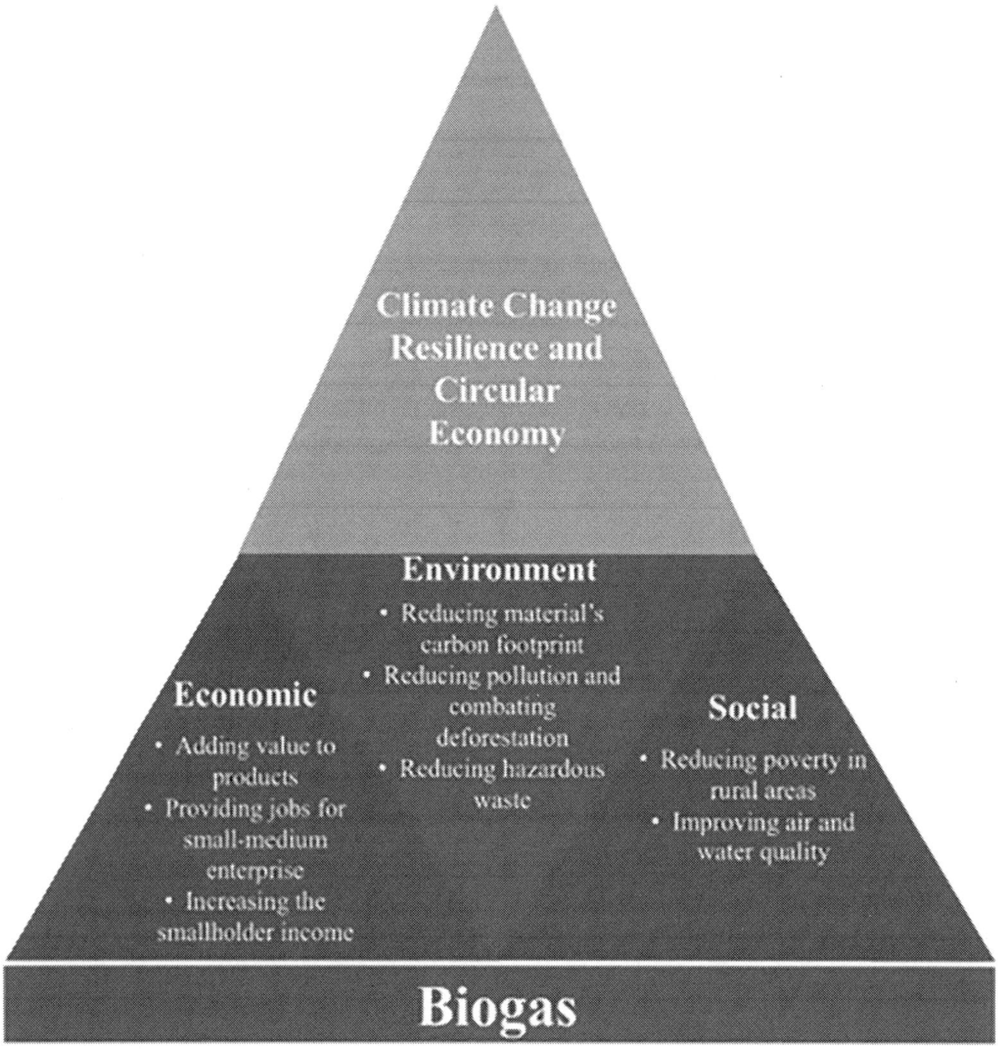

FIGURE 8.3 Role of biogas in climate change resilience and circular economy.

8.5.2.2 Agricultural Benefits

In a circular agricultural model, biogas projects contribute to enhancing the food chain and facilitate the reuse of agro-resources, aiming for multi-level energy utilization. The number of biogas projects might be influenced by legislative support, economic situations, agricultural practices, and animal husbandry. Rich and healthful food has been made available to people thanks to the growth of agricultural and animal husbandry production, as well as the high densities of crops and cattle. Biogas technology is a useful method for utilizing agricultural waste and reducing pollution in the environment (Sun et al., 2014). Biogas, the world's fourth-largest energy source, has the ability to mitigate two major issues that are causing the environment to deteriorate and the energy crisis (Chen et al., 2010a). Currently, the development of biogas is of great importance for countries all over the world.

In order to produce biogas, certain European countries are investigating the possibility of substituting energy cover crops with dedicated crops. It's true that energy cover crops can produce significant biomass without taking up space on the field from food crops. Moreover, there are several agricultural benefits associated with energy cover crops. Initially, in contrast to traditional cover crops, energy cover crops are intensively regulated to generate significantly more biomass, with yields of up to 16 t DM/ha, compared to less than 5 t DM/ha for regular cover crops. Second, digestate from cover crops contains more nitrogen than residue from cover crops as nitrogen is conserved during AD. However, depending on the application technique and storage conditions, ammonia is lost by volatilization, which might reduce the efficacy of using nitrogen (Launay et al., 2022). Third, between 43% and 80% of the initial carbon in crops is converted into biogas. Despite this, soil carbon storing levels may bear a resemblance to those of cover crops due to carbon stabilization during AD, with the roots and stubble of energy cover crops left in the soil. Fourth, energy cover crops provide various benefits, such as lowering nitrate leaching, boosting soil microbial action, and improving soil physical characteristics during fallow periods. However, they also pose certain challenges, including the compaction of soil, diminished groundwater recharge, the necessity for extra inputs (e.g., irrigation, fertilization, pesticides), and potentially lower yields for subsequent crops (Launay et al., 2022). Additionally, energy cover crops contribute to agricultural productivity by improving soil conditions through nutrient recovery, increasing yields with reasonably priced fertilizer, and recirculating essential nutrients such as phosphorus, nitrogen, and potassium during the digestion process (Obaideen et al., 2022).

8.5.2.3 Pollution Control and Waste Management

The production of biogas from farm use has emerged as a viable substitute, with the bonus of pollution control. Biogas offers superior environmental performance compared to other fuel and engine options for bus transport. The enhancement of biogas via chemical transformation is still a relatively underexplored field. Chemical industries are offered numerous opportunities of decreasing GHG emissions by adopting this domain (Sher et al., 2024). Municipal waste, generated by households, commercial establishments, and institutions such as schools and hospitals, serves as a valuable substrate for biogas production. Its widespread adoption in biogas systems aids in effective waste management reduces LF burden, lowers GHG emissions, and contributes to improved air quality (Bhattacharjee et al., 2013).

Industries like food, breweries, agro-industries, and fodder produce large volumes of organic waste, byproducts, and residues. Similarly, organic-loaded wastewater sludge is produced from biorefineries and thus requires it to be properly treated or disposal. Much of this industrial waste can be efficiently repurposed through biogas production, as it serves as a valuable feedstock for AD. Organic materials like pomace from winemaking, animal feed, brewery waste, sugar refinery by-products, and fruit processing residues are all viable sources for AD. Additionally, wastewater from dairies and slaughterhouse waste can be used to generate biogas, contributing to the reduction of air and water pollution (Deublein & Steinhauser, 2011).

8.5.3 Economic Development

In addition to its benefits for resistance against climate change, biogas can also impact positively to economic growth. Solid biomass fuels, including firewood, dried dung cakes, crop leftovers, heat and cooking straw, and other agricultural wastes, are thought to be used by about 3 billion people globally (Caubel et al., 2020). Mostly in underdeveloped countries, women and children bear the burden of gathering firewood, as well as the air pollution caused by cooking indoors. Employing AD and biogas as a domestic fuel could significantly augment the rural life's quality and is by now broadly implemented in countries such as Bangladesh and India. Thus, biogas has been regarded as a significant energy source, having proved to be critical for the economic growth (Korberg et al., 2020; Okoro & Sun, 2019).

Moreover, converting waste into biogas transforms it from a burden on government into an energy producing chance, while also generating new job opportunities for the community. For example, the biogas industry has produced approximately 23,000 full-time operating jobs and about 335,000 temporary construction jobs, despite the fact that biogas constituted less than 2.2% of all renewable energy in 2019 (Irena, 2019). Only in China, biogas provides employment for 209,000 people, and in the next years, biogas is expected to increase significantly. Comparing this amount of jobs to those in other renewable energy sources, 116,500 for wind, 99,000 for geothermal, 3,755,000 for solar, and 1,100 for tidal, it is adequate (Irena, 2019). If the slurry from biogas digesters is not properly utilized, it becomes an active breeding ground for disease-transmitting insects (Hamburg, 1989). Slurry mixed with plant-based materials can also serve as a base for vermicomposting. Additionally, the humic acid present in the slurry makes it an excellent soil conditioner (Sarkar et al., 2020). Numerous studies have explored the economic advantages of biogas (Kumar et al., 2021).

8.5.3.1 Balancing Rural Areas Earnings

In rural areas, diversifying income sources is essential for economic stability, especially in the face of fluctuating agricultural yields and limited industrial opportunities. One promising avenue for boosting rural earnings is the development of biogas plants, which not only provide energy but also offer financial benefits through byproducts like bio-fertilizer. The profitability of biogas plants, mainly from the cash excess from using biogas and biofertilizer in comparison to the cost of the plants, is referred to as financial viability in this context. When revenue exceeds expenses, profitability is attained. Typically, the choice with the best profitability is selected while making economic decisions (Sarker et al., 2020).

A suitable technology for rural communities is also the adoption of biogas energy (Bewket, 2012; Erdogdu, 2008). In rural areas with inadequate natural gas supplies, fuel wood, cow dung, and agricultural residues are used before liquified petroleum gas (LPG). Though, the supply of natural gas will run out in the future, making it more difficult for rural communities to expand sustainability (Hasan & Khan, 2012). In this case, sustainable and renewable energy sources are the ideal substitute for conventional fuel and energy sources. According to Propensity Score Matching (PSM) data, residents who use biogas technology generate more revenue than those who do not. The most likely reason for this outcome is that the homes that have adopted biogas generate biogas on a commercial scale, sell bio sludge, and give surplus gas to clients who use it in place of fossil fuels. Moreover, households that have switched to biogas save time by utilizing it is place of collecting biomass and burning wood (Moli et al., 2021).

Consequently, it contributes to an increase in the yearly household income. The bio sludge, which is frequently utilized as fertilizer and fish feed, also generates additional revenue for the biogas adopters. According to Hossain (2020), the low payback period and profitability of biogas plants make them economically viable to adopt. Studies by Sarker et al. (2020) and Kabir et al. (2012) have demonstrated that utilizing biogas in rural areas is economically feasible, regardless of plant size. Similar findings were reached by Abbas et al. (2017), Eshete et al. (2006), and Walekhwa et al. (2009), based on studies conducted in Pakistan, Ethiopia, and Uganda, respectively.

8.6 CHALLENGES TO BIOGAS IMPLEMENTATION

The implementation of biogas differs across regions based on factors such as the form, cost structure, and usage patterns, all of which are influenced by the country's level of development. In developing countries, challenges related to funding, policies, technical skills, sustainability, perceptions, and trainings are crucial factors that affect the effective implementation needed to realize the full potential of biogas. Considering these challenges, the following section identifies the key challenges to biogas implementation in developing countries.

8.6.1 Technical and Infrastructure Challenges

The lack of sustainable technology is still a major barrier to the use of biogas in developing countries in contrast to their developed counterparts. Developing countries are rich in biomass resources, but they are unable to utilize them to their full potential because they do not have the infrastructure to support the required technology. Moreover, the effectiveness of biogas implementation on how well users understand the technical aspects (Gebreegziabher et al., 2014).

The majority of biogas operators lack technical expertise in biogas production, which causes difficulty in integrate eco-agricultural technology and biogas technologies (Chen et al., 2010). For instance, the amount of biogas produced and the possibility of the biogas process failing altogether are influenced by the composition of the feedstock, which includes pollutants, organic and nutritional content, and possible inhibitors. Poor digester production and failure are also caused by inadequate knowledge of how to maintain these digesters, a lack of technical expertise to construct high-quality digesters and ineffective mixing capacity (Rupf et al., 2015). Process instability in biogas digesters occurs because of the fluctuating temperature, which adversely affect performance. Microorganisms involved in the biogas process become less active when temperatures deviate from their optimal range (Deng et al., 2006). This challenge is particularly pronounced in colder regions, where no efficient technology currently exists to regulate operating temperatures throughout winter.

Moreover, digesters lacking insulation cease functioning when temperatures drop below 15°C. Nepal serves as a notable example, where the combination of hilly terrain and low temperatures, often around 10°C, poses significant challenges for biogas production (Gautam et al., 2009; Surendra et al., 2011). A common issue in many developing countries is a shortage of water. One common method of producing biogas, called wet fermentation, needs a lot of water, which is problematic for poor countries where clean water is scarce. In addition, the high volume of biogas residue makes it difficult to transport and apply digestate to agricultural lands (Abdelkareem et al., 2019).

8.6.2 Socio-Cultural Barriers

Socio-cultural barriers also hinder the adoption of biogas. The lack of interest and engagement from the general population is a significant social obstacle. For example, Ghana's adoption of biogas is being hindered by a lack of technological expertise (Mengistu et al., 2015). Furthermore, the value of waste segregation, green, and sustainable development is not well recognized or understood in Malaysia by the public (Bong et al., 2017). The challenges faced by rural families in China are often attributed to the slow-growing rural economy, which has impacted their understanding of resource economics, environmental preservation, and health development (Chen et al., 2017). It is concerning that, despite farming households being involved in the adoption of new biogas technologies, they often do not feel responsible for their use, maintenance, or the ecological and economic benefits these technologies provide (Chen & Liu, 2017).

Biogas dissemination is also impacted by society specific factors, such as stigmatization. Some biogas projects were unsuccessful because they conflicted with local beliefs (Ghosh & Mandal, 2018). Family-sized biogas plants are disregarded since manure is utilized for cooking. The local population cannot accept biogas usage because of their cultural beliefs, as it is made from dung, manure, or other fecal waste (Giwa et al., 2017; Shane et al., 2015). The study of Gebreegziabher et al. (2014) highlights that certain religions in sub-Saharan Africa impose strict cleanliness rules, not only concerning human hygiene but also regarding the handling of animal excrement. For instance, in Zambia, women are prohibited from collecting cow dung due to traditional customs (Shane et al., 2015). Similarly, in Kenya, several households raised concerns about the "cleanliness" of biogas produced via certain types of waste (Sovacool et al., 2015). This can lead to challenges in managing slurry and feedstock, as a number of users are reluctant to perform the daily task of mixing dung, often viewing it as an unpleasant or dirty job (Khan et al., 2014). Similar to this, it is socially unacceptable to use human excrement in biogas plants when a house is surrounded by unhygienic conditions (Dabas et al., 2018). In rural Indian society, women, who are mostly subjected to the duty of cooking, often have a low social status and are unable to make decisions regarding the use of clean fuels (Mittal et al., 2018). Stigmatization also affects the export of biogas products. For example, Nevzorova and Kutcherov (2019) showed that products made from digestate constituting pig slurry cannot be sold to Islamic countries because of cultural and religious restrictions.

8.6.3 Biogas Impurities

Another challenge accompanying the usage of biogas is the existence of several contaminants, such as halogens, siloxanes, and sulfur compounds. The engine and other metallic components will corrode even after the biogas has been purified if these contaminants are still present. Eventually, the expense of maintaining or replacing the damaged parts would be increased (Khan et al., 2017; Qian et al., 2017). A significant fraction of biogas consists of carbon dioxide, which reduces the energy density of the fuel. Removing carbon dioxide, for instance through mechanization, requires additional equipment to enhance energy density. Besides carbon dioxide, biogas also contains several important impurities, including VOC, siloxanes, sulfur compounds, halogens, and ammonia (Wang et al., 2020; Xue et al., 2020).

8.6.4 Financial Challenges

One of the main challenges to the development of biogas in developing countries is the lack of funding. A DIY home-scale biodigester with a daily input capacity of 50 kg is currently estimated to cost approximately $1,500. This is a significant financial commitment for those with modest incomes, especially when you include the additional costs associated with the biogas compressor, storage tanks, operating expenses, purification, and maintenance (Morgan Jr et al., 2018). Mechanical pretreatment is sometimes mandatory before the digestion process in order to ensure efficient biogas production, but the cost can be high (Rodriguez et al., 2017). Achieving profitability from such an investment requires industrial-scale application, which demands costly equipment, often priced in the thousands of dollars. For example, a small industrial biodigester with a regular input scope of 2,000 kg is reported to be the costs varying between $500,000 and $1,000,000 (Morgan Jr et al., 2018). Additionally, a lack of technical skills also contributes to high maintenance costs. In developing countries, particularly in rural areas, it is quite challenging to secure commercial finance for the construction of biogas infrastructure (Nevzorova & Kutcherov, 2019).

8.6.5 Policy and Political Challenges

Most developing countries do not have a reliable system for collecting and sorting waste. Often, waste managers collect various types of waste and dispose of it in unregulated dumps or burn them openly. There is no defined policy for using renewable energy, and the environmental policy is relatively weak. One additional major obstacle impeding the progress of biogas implementation is the lack of government commitment and poor continuity of previous biogas program initiatives through successive governments (Akinbomi et al., 2014). Corruption poses another challenge, which raises capital as well as operational costs for implementing biogas and diminishes the return on investment (Taherzadeh & Rajendran, 2014).

8.7 FUTURE CONSIDERATIONS

As biogas technology continues to face various challenges in its implementation, there are key areas that require further exploration and development. Future considerations must focus on overcoming these obstacles to enhance the efficiency, scalability, and accessibility of biogas systems globally. The following steps can be taken in order to make the biogas industry more promising and sustainable in the long term.

Educating farmers, supplying bags made from flexible materials for storing biogas, insulating the top of the digester during winter, and providing flares are all strategies that could enhance the benefits of biogas (Bruun et al., 2014). Furthermore, integrating artificial intelligence into the biogas production process can optimize the operational parameters of the reactor, which may result in raising the yield of biogas and enhancing the process's economics (Davis et al., 2011). A novel design of the digester is required to maximize efficiency and minimize gas leakage. To increase the total efficiency of the digester, it is imperative to allow the utilization of easily accessible renewable energy sources, such as solar (Rafiee et al., 2021).

Producing higher value-added products from digestate, rather than traditional organic fertilizers, even in minute quantities but with greater worth, could enhance the economic viability of the biogas process (Mancini & Raggi, 2021). The bionutrients in the digestate can also enhance the economical prospects of the biogas production pathway. For example, studies have indicated that the digestate can biochemically prepare microbial proteins, such as volatile fatty acids (Fang et al., 2018; Vea et al., 2018).

Since the central targets of the Sustainable Development Goals (SDGs) are mutually interconnected, it is crucial for key stakeholders to focus on the nature of trade-offs as well as synergies between different goals. Through the consideration of the biogas plans in the realization of the overall concept of SDGs, the benefits of biogas products to the tendency of making a positive impact on environment space, employment creation, and poverty reduction can be well driven. This integrated approach will not only enhance the sustainability and stability of the biogas systems but also guarantee that they have a central function in supporting the universal goals of circular economy and resilient climate (Rafiee et al., 2021).

8.8 CONCLUSION

Biogas appears as a versatile approach in the context of circular economy perspective revealing large possibilities to solve existing environmental and energy problems. Biogas is an environmentally friendly form of energy since it helps in reducing the effects of climate change through the recycling and utilization of organic waste whilst at the same time helping conserve forest and agricultural resources. In future, it would act as the tool for equitable development due to its economic sustainability that compliments improvement of the livelihoods of people living in rural areas. In this respect, there are significant prospects for innovation and development in the biogas industry in spite of the numerous obstacles which were encountered in its application. In addition, integration of artificial intelligence and related technologies for optimally designing new reactors can improve the efficiency and minimize the leakage of digester. Therefore, for the next economic outlooks, it is crucial to promote biogas as one of the key tools to reach energy security and sustainable development goals while creating a new basis for the more sustainable economic model.

DECLARATION

The authors have not used any generative AI tools during the writing of this manuscript.

REFERENCES

Abanades, S., Abbaspour, H., Ahmadi, A., Das, B., Ehyaei, M. A., Esmaeilion, F., El Haj Assad, M., Hajilounezhad, T., Jamali, D. H., Hmida, A., & others. (2022). A critical review of biogas production and usage with legislations framework across the globe. *International Journal of Environmental Science and Technology*, 19(4):3377-3400

Abbas, T., Ali, G., Adil, S. A., Bashir, M. K., & Kamran, M. A. (2017). Economic analysis of biogas adoption technology by rural farmers: the case of Faisalabad district in Pakistan. *Renewable Energy*, *107*, 431–439.

Abbasi, T., Tauseef, S. M., Abbasi, S. A., Abbasi, T., Tauseef, S. M., & Abbasi, S. A. (2012). Capture of Biogas from Landfills. *Biogas Energy*, 145–169.

Abdelkareem, M. A., Tanveer, W. H., Sayed, E. T., Assad, M. E. H., Allagui, A., & Cha, S. W. (2019). On the technical challenges affecting the performance of direct internal reforming biogas solid oxide fuel cells. *Renewable and Sustainable Energy Reviews*, *101*, 361–375.

Ahmad, A., Banat, F., Alsafar, H., & Hasan, S. W. (2022). Algae biotechnology for industrial wastewater treatment, bioenergy production, and high-value bioproducts. *Science of the Total Environment*, *806*, 150585.

Akinbomi, J., Brandberg, T., Sanni, S. A., & Taherzadeh, M. J. (2014). Development and dissemination strategies for accelerating biogas production in Nigeria. *BioResources*, 9(3), 5707-5737

Aliyu Salihu, M. Z. A. (2016). Pretreatment methods of organic wastes for biogas production. *Journal of Applied Sciences*, *16*(3), 124–137.

Anacleto, T. M., Kozlowsky-Suzuki, B., Wilson, A. E., & Enrich-Prast, A. (2022). Comprehensive meta-analysis of pathways to increase biogas production in the textile industry. *Energies*, *15*(15). https://doi.org/10.3390/en15155574

Angelidaki, I., Ellegaard, L., & Ahring, B. K. (1993). A mathematical model for dynamic simulation of anaerobic digestion of complex substrates: focusing on ammonia inhibition. *Biotechnology and Bioengineering*, *42*(2), 159–166.

Atelge, M. R., Krisa, D., Kumar, G., Eskicioglu, C., Nguyen, D. D., Chang, S. W., Atabani, A. E., Al-Muhtaseb, A. H., & Unalan, S. (2020). Biogas production from organic waste: recent progress and perspectives. *Waste and Biomass Valorization*, *11*, 1019–1040.

Bagi, Z., Ács, N., Bálint, B., Horváth, L., Dobó, K., Perei, K. R., Rákhely, G., & Kovács, K. L. (2007). Biotechnological intensification of biogas production. *Applied Microbiology and Biotechnology*, *76*, 473–482.

Bewket, W. (2012). Climate change perceptions and adaptive responses of smallholder farmers in central highlands of Ethiopia. *International Journal of Environmental Studies*, *69*(3), 507–523.

Bhatt, A. H., & Tao, L. (2020). Economic perspectives of biogas production via anaerobic digestion. *Bioengineering*, *7*(3), 74.

Bhattacharjee, S., Miah, M. Y., & Sazzad, M. H. (2013). *Bio-fuel and Bio-fertilizer from municipal solid waste: conversion of waste management problem into an renewable energy generating solution*. LAP Lambert Academic Publishing.

Bong, C. P. C., Ho, W. S., Hashim, H., Lim, J. S., Ho, C. S., Tan, W. S. P., & Lee, C. T. (2017). Review on the renewable energy and solid waste management policies towards biogas development in Malaysia. *Renewable and Sustainable Energy Reviews*, *70*, 988–998.

Bruun, S., Jensen, L. S., Sommer, S., & others. (2014). Small-scale household biogas digesters: an option for global warming mitigation or a potential climate bomb? *Renewable and Sustainable Energy Reviews*, *33*, 736–741.

Capodaglio, A. G., Callegari, A., & Lopez, M. V. (2016). European framework for the diffusion of biogas uses: emerging technologies, acceptance, incentive strategies, and institutional-regulatory support. *Sustainability*, *8*(4), 298.

Caubel, J. J., Rapp, V. H., Chen, S. S., & Gadgil, A. J. (2020). Practical design considerations for secondary air injection in wood-burning cookstoves: an experimental study. *Development Engineering*, *5*, 100049.

Chandel, A. K., Albarelli, J. Q., Santos, D. T., Chundawat, S. P., Puri, M., & Meireles, M. A. A. (2019). Comparative analysis of key technologies for cellulosic ethanol production from Brazilian sugarcane bagasse at a commercial scale. *Biofuels, Bioproducts and Biorefining*, *13*(4), 994–1014.

Chen, L., Zhao, L. X., Dong, B. C., Wan, X. C., & Gao, X. X. (2010a). The status and trends of the development of biogas plants for crop straws in China. *Renewable Energy Resources*, *28*(3), 145–148.

Chen, Q., & Liu, T. (2017). Biogas system in rural China: upgrading from decentralized to centralized? *Renewable and Sustainable Energy Reviews*, *78*, 933–944.

Chen, Y., Hu, W., Chen, P., & Ruan, R. (2017). Household biogas CDM project development in rural China. *Renewable and Sustainable Energy Reviews*, *67*, 184–191.

Chen, Y., Yang, G., Sweeney, S., & Feng, Y. (2010b). Household biogas use in rural China: a study of opportunities and constraints. *Renewable and Sustainable Energy Reviews*, *14*(1), 545–549.

Chevalier, A., Evon, P., Monlau, F., Vandenbossche, V., & Sambusiti, C. (2023). Twin-screw extrusion mechanical pretreatment for enhancing biomethane production from agro-industrial, agricultural and catch crop biomasses. *Waste*, *1*(2), 497–514.

Chew, K. R., Leong, H. Y., Khoo, K. S., Vo, D.-V. N., Anjum, H., Chang, C.-K., & Show, P. L. (2021). Effects of anaerobic digestion of food waste on biogas production and environmental impacts: a review. *Environmental Chemistry Letters*, *19*(4), 2921–2939.

Cuéllar, A. D., & Webber, M. E. (2008). Cow power: the energy and emissions benefits of converting manure to biogas. *Environmental Research Letters*, *3*(3), 34002.

Dabas, J. P. S., Sharma, N., Dubey, S. K., Sharma, A., Singh, L., Dubey, A. V., & others. (2018). Utilizing cow dung and slurry for energy sufficiency of farms and households: experiences of field studies across three Indian states. *Indian Journal of Agricultural Sciences*, *88*(8), 1208–1213.

Davis, R., Aden, A., & Pienkos, P. T. (2011). Techno-economic analysis of autotrophic microalgae for fuel production. *Applied Energy*, *88*(10), 3524–3531. https://doi.org/10.1016/j.apenergy.2011.04.018

Deng, G. L., Li, K. L., Li, X., & Zhang, B. (2006). Role of biogas technology in new socialistic countryside. *Trans CSAE*, *22*(1), 61–64.

Deng, L., Liu, Y., & Wang, W. (2020). *Biogas Technology* (1st ed.). Springer Nature Singapore. https://doi.org/10.1007/978-981-15-4940-3

Deublein, D., & Steinhauser, A. (2011). *Biogas from waste and renewable resources: an introduction.* John Wiley & Sons.

Devi, M. K., Manikandan, S., Oviyapriya, M., Selvaraj, M., Assiri, M. A., Vickram, S., Subbaiya, R., Karmegam, N., Ravindran, B., Chang, S. W., & others. (2022). Recent advances in biogas production using agro-industrial waste: a comprehensive review outlook of techno-economic analysis. *Bioresource Technology*, *363*, 127871.

Erdogdu, E. (2008). An expose of bioenergy and its potential and utilization in Turkey. *Energy Policy*, *36*(6), 2182–2190.

Eshete, G., Sonder, K., & ter Heegde, F. (2006). Report on the feasibility study of a national programme for domestic biogas in Ethiopia. *SNV Netherlands Development Organization: Addis Ababa, Ethiopia.*

Fang, W., Zhang, P., Zhang, X., Zhu, X., van Lier, J. B., & Spanjers, H. (2018). White rot fungi pretreatment to advance volatile fatty acid production from solid-state fermentation of solid digestate: efficiency and mechanisms. *Energy*, *162*, 534–541.

FAO. (2020). *The state of food and agriculture 2020: overcoming water challenges in agriculture*. Food and Agriculture Organisation.

Gautam, R., Baral, S., & Herat, S. (2009). Biogas as a sustainable energy source in Nepal: present status and future challenges. *Renewable and Sustainable Energy Reviews*, *13*(1), 248–252.

Gebreegziabher, Z., Naik, L., Melamu, R., & Balana, B. B. (2014). Prospects and challenges for urban application of biogas installations in Sub-Saharan Africa. *Biomass and Bioenergy*, *70*, 130–140.

Ghasemi Ghodrat, A., Tabatabaei, M., Aghbashlo, M., & Mussatto, S. I. (2018). Waste management strategies; the state of the art. *Biogas: Fundamentals, Process, and Operation*, *6*, 1–33.

Ghosh, P., Shah, G., Sahota, S., Singh, L., & Vijay, V. K. (2020). Biogas production from waste: technical overview, progress, and challenges. *Bioreactors*, 89–104.

Ghosh, S. K., & Mandal, S. (2018). Evaluation of biogas as an alternative driving force of electrically operated vehicles: a case study. *International Journal of Engineering, IJE Transactions B: Applications*, *31*(5), 834–840.

Giwa, A., Alabi, A., Yusuf, A., & Olukan, T. (2017). A comprehensive review on biomass and solar energy for sustainable energy generation in Nigeria. *Renewable and Sustainable Energy Reviews*, *69*, 620–641.

Gustafsson, M., Cruz, I., Svensson, N., & Karlsson, M. (2020). Scenarios for upgrading and distribution of compressed and liquefied biogas—energy, environmental, and economic analysis. *Journal of Cleaner Production*, *256*, 120473.

Hamburg, R. A. (1989). Household cooking fuel hydrogen sulfide and sulfur dioxide emissions from stalks, coal and biogas. *Biomass*, *19*(3), 233–245.

Hasan, M. M., & Khan, M. F. (2012). A comparative study on installation of solar PV system for grid and non grid rural areas of Bangladesh. *2nd International Conference on the Developments in Renewable Energy Technology (ICDRET 2012),* 1–5.

Hobson, C., Márquez, C., & others. (2018). Renewable methanol report. *Methanol Institute, Singapore.*

Hossain, M. R. (2020). Can small-scale biogas projects mitigate the energy crisis of rural Bangladesh? A study with economic analysis. *International Journal of Sustainable Energy*, *39*(8), 744–760.

Hosseini, S. E., Andwari, A. M., Wahid, M. A., & Bagheri, G. (2013). A review on green energy potentials in Iran. *Renewable and Sustainable Energy Reviews*, *27*, 533–545.

Irena, I. (2019). *Renewable energy and jobs: annual review 2019*. International Renewable Energy Agency (IRENA).

Jameel, M. K., Mustafa, M. A., Ahmed, H. S., jassim Mohammed, A., Ghazy, H., Shakir, M. N., Lawas, A. M., khudhur Mohammed, S., Idan, A. H., Mahmoud, Z. H., & others. (2024). Biogas: production, properties, applications, economic and challenges: a review. *Results in Chemistry*, *7*, 101549.

Jonsson, M., Thorin, E., & Svedberg, G. (1999). *Gas engine bottoming cycles with ammonia-water mixtures as working fluid* (*PWR-Vol. 34*, Vol. 2, pp. 55–65). In *Proceedings of the 1999 International Joint Power Generation Conference* (Burlingame, CA). American Society of Mechanical Engineers. EDB-00:007147.

Kabir, H., Palash, M. S., & Bauer, S. (2012). Appraisal of domestic biogas plants in Bangladesh. *Bangladesh Journal of Agricultural Economics*, *35*, 71–89.

Kamali, M., Abdi, R., Rohani, A., Abdollahpour, S., & Ebrahimi, S. (2023). Enhancing biomethane production from OFMSW: the role of moderate temperature thermal pretreatment in anaerobic digestion. *Industrial & Engineering Chemistry Research*, *62*(46), 19471–19481. https://doi.org/10.1021/acs.iecr.3c02931

Kamusoko, R., Jingura, R. M., Parawira, W., & Sanyika, W. T. (2019). Comparison of pretreatment methods that enhance biomethane production from crop residues – a systematic review. *Biofuel Research Journal*, *6*(4), 1080–1089. doi: 10.18331/BRJ2019.6.4.4

Kapoor, R., Ghosh, P., Kumar, M., Sengupta, S., Gupta, A., Kumar, S. S., Vijay, V., Kumar, V., Vijay, V. K., & Pant, D. (2020). Valorization of agricultural waste for biogas based circular economy in India: a research outlook. *Bioresource Technology*, *304*, 123036.

Kapoor, R., Ghosh, P., Tyagi, B., Vijay, V. K., Vijay, V., Thakur, I. S., Kamyab, H., Nguyen, D. D., & Kumar, A. (2020). Advances in biogas valorization and utilization systems: a comprehensive review. *Journal of Cleaner Production*, *273*, 123052.

Karthikeyan, O. P., Trably, E., Mehariya, S., Bernet, N., Wong, J. W. C., & Carrere, H. (2018). Pretreatment of food waste for methane and hydrogen recovery: a review. *Bioresource Technology*, *249*, 1025–1039.

Kaza, S., Yao, L., Bhada-Tata, P., & Van Woerden, F. (2018). *What a waste 2.0: a global snapshot of solid waste management to 2050*. World Bank Publications.

Ketsub, N., Whatmore, P., Abbasabadi, M., Doherty, W. O. S., Kaparaju, P., O'Hara, I. M., & Zhang, Z. (2022). Effects of pretreatment methods on biomethane production kinetics and microbial community by solid state anaerobic digestion of sugarcane trash. *Bioresource Technology*, *352*, 127112. https://doi.org/https://doi.org/10.1016/j.biortech.2022.127112

Khan, E. U., Mainali, B., Martin, A., & Silveira, S. (2014). Techno-economic analysis of small scale biogas based polygeneration systems: Bangladesh case study. *Sustainable Energy Technologies and Assessments*, *7*, 68–78.

Khan, I. U., Othman, M. H. D., Hashim, H., Matsuura, T., Ismail, A. F., Rezaei-DashtArzhandi, M., & Azelee, I. W. (2017). Biogas as a renewable energy fuel – a review of biogas upgrading, utilisation and storage. *Energy Conversion and Management*, *150*, 277–294.

Klackenberg, L. (2019). Produktion och användning av biogas och rötrester 2018. *Bromma, Sweden: Statens Energimyndighet*.

Köninger, J., Lugato, E., Panagos, P., Kochupillai, M., Orgiazzi, A., & Briones, M. J. I. (2021). Manure management and soil biodiversity: towards more sustainable food systems in the EU. *Agricultural Systems*, *194*, 103251.

Korberg, A. D., Skov, I. R., & Mathiesen, B. V. (2020). The role of biogas and biogas-derived fuels in a 100% renewable energy system in Denmark. *Energy*, *199*, 117426.

Kougias, P. G., & Angelidaki, I. (2018). Biogas and its opportunities—a review. *Frontiers of Environmental Science \& Engineering*, *12*, 1–12.

Kumar, M., Dutta, S., You, S., Luo, G., Zhang, S., Show, P. L., Sawarkar, A. D., Singh, L., & Tsang, D. C. W. (2021). A critical review on biochar for enhancing biogas production from anaerobic digestion of food waste and sludge. *Journal of Cleaner Production*, *305*, 127143.

Kumari, P., Varma, A. K., Shankar, R., Thakur, L. S., & Mondal, P. (2021). Phycoremediation of wastewater by *Chlorella pyrenoidosa* and utilization of its biomass for biogas production. *Journal of Environmental Chemical Engineering*, *9*(1), 104974.

Launay, C., Houot, S., Frédéric, S., Girault, R., Levavasseur, F., Marsac, S., & Constantin, J. (2022). Incorporating energy cover crops for biogas production into agricultural systems: benefits and environmental impacts. A review. *Agronomy for Sustainable Development*, *42*(4), 57.

Mago, P. J., Chamra, L. M., & Hueffed, A. (2009). A review on energy, economical, and environmental benefits of the use of CHP systems for small commercial buildings for the North American climate. *International Journal of Energy Research*, *33*(14), 1252–1265.

Mancini, E., & Raggi, A. (2021). A review of circularity and sustainability in anaerobic digestion processes. *Journal of Environmental Management*, *291*, 112695.

Mao, C., Feng, Y., Wang, X., & Ren, G. (2015). Review on research achievements of biogas from anaerobic digestion. *Renewable and Sustainable Energy Reviews*, *45*, 540–555.

Mavridis, S., & Voudrias, E. A. (2021). Using biogas from municipal solid waste for energy production: comparison between anaerobic digestion and sanitary landfilling. *Energy Conversion and Management*, *247*, 114613.

Mengistu, M. G., Simane, B., Eshete, G., & Workneh, T. S. (2015). A review on biogas technology and its contributions to sustainable rural livelihood in Ethiopia. *Renewable and Sustainable Energy Reviews*, *48*, 306–316.

Mignogna, D., Ceci, P., Cafaro, C., Corazzi, G., & Avino, P. (2023). Production of biogas and biomethane as renewable energy sources: a review. *Applied Sciences*, *13*(18), 10219.

Mikhaylov, A., Moiseev, N., Aleshin, K., & Burkhardt, T. (2020). Global climate change and greenhouse effect. *Entrepreneurship and Sustainability Issues*, *7*(4), 2897.

Mishra, A., Kumar, M., Bolan, N. S., Kapley, A., Kumar, R., & Singh, L. (2021). Multidimensional approaches of biogas production and up-gradation: opportunities and challenges. *Bioresource Technology*, *338*, 125514.

Mitraka, G.-C., Kontogiannopoulos, K. N., Batsioula, M., Banias, G. F., Zouboulis, A. I., & Kougias, P. G. (2022). A comprehensive review on pretreatment methods for enhanced biogas production from sewage sludge. *Energies*, *15*(18). https://doi.org/10.3390/en15186536

Mittal, S., Ahlgren, E. O., & Shukla, P. R. (2018). Barriers to biogas dissemination in India: a review. *Energy Policy*, *112*, 361–370.

Moli, D. R. R., Hafeez, A. S. M. G., Majumder, S., Mitra, S., & Hasan, M. (2021). Does biogas technology adoption improve the livelihood and income level of rural people? *International Journal of Green Energy*, *18*(10), 1081–1090.

Morgan Jr, H. M., Xie, W., Liang, J., Mao, H., Lei, H., Ruan, R., & Bu, Q. (2018). A techno-economic evaluation of anaerobic biogas producing systems in developing countries. *Bioresource Technology*, *250*, 910–921.

Nevzorova, T., & Kutcherov, V. (2019). Barriers to the wider implementation of biogas as a source of energy: a state-of-the-art review. *Energy Strategy Reviews*, *26*, 100414.

Nsair, A., Cinar, S. O., Alassali, A., Qdais, H. A., & Kuchta, K. (2020). Operational parameters of biogas plants: a review and evaluation study. *Energies*, *13*(15). https://doi.org/10.3390/en13153761

Obaideen, K., Abdelkareem, M. A., Wilberforce, T., Elsaid, K., Sayed, E. T., Maghrabie, H. M., & Olabi, A. G. (2022). Biogas role in achievement of the sustainable development goals: evaluation, challenges, and guidelines. *Journal of the Taiwan Institute of Chemical Engineers*, *131*, 104207.

Okoro, O. V., & Sun, Z. (2019). Desulphurisation of biogas: a systematic qualitative and economic-based quantitative review of alternative strategies. *ChemEngineering*, *3*(3), 76.

Orlando, M.-Q., & Borja, V.-M. (2020). Pretreatment of animal manure biomass to improve biogas production: a review. *Energies*, *13*(14), 3573.

Pawłowska, M. (2014). *Mitigation of landfill gas emissions*. CRC Press.

Perea-Moreno, M.-A., Samerón-Manzano, E., & Perea-Moreno, A.-J. (2019). Biomass as renewable energy: worldwide research trends. *Sustainability*, *11*(3), 863.

Qian, Y., Sun, S., Ju, D., Shan, X., & Lu, X. (2017). Review of the state-of-the-art of biogas combustion mechanisms and applications in internal combustion engines. *Renewable and Sustainable Energy Reviews*, *69*, 50–58.

Qiu, L., Yang, G., & Bi, Y. (2005). Discussion on the conditions and countermeasures of developing marsh gas in rural areas of west China. *Agricultural Research in Arid Areas*, *23*, 200–204.

Rafiee, A., Khalilpour, K. R., Prest, J., & Skryabin, I. (2021). Biogas as an energy vector. *Biomass and Bioenergy*, *144*, 105935.

Rhodes, C. J. (2019). Only 12 years left to readjust for the 1.5-degree climate change option – says international panel on climate change report: current commentary. *Science Progress*, *102*(1), 73–87.

Rodriguez, C., Alaswad, A., El-Hassan, Z., & Olabi, A.-G. (2017). Mechanical pretreatment of waste paper for biogas production. *Waste Management*, *68*, 157–164.

Rupf, G. V., Bahri, P. A., de Boer, K., & McHenry, M. P. (2015). Barriers and opportunities of biogas dissemination in Sub-Saharan Africa and lessons learned from Rwanda, Tanzania, China, India, and Nepal. *Renewable and Sustainable Energy Reviews*, *52*, 468–476.

Sarkar, S., Skalicky, M., Hossain, A., Brestic, M., Saha, S., Garai, S., Ray, K., & Brahmachari, K. (2020). Management of crop residues for improving input use efficiency and agricultural sustainability. *Sustainability*, *12*(23), 9808.

Sarker, S. A., Wang, S., Adnan, K. M. M., & Sattar, M. N. (2020). Economic feasibility and determinants of biogas technology adoption: evidence from Bangladesh. *Renewable and Sustainable Energy Reviews*, *123*, 109766.

Scarlat, N., Dallemand, J.-F., & Fahl, F. (2018). Biogas: developments and perspectives in Europe. *Renewable Energy*, *129*, 457–472.

Scheidel, A., Del Bene, D., Liu, J., Navas, G., Mingorría, S., Demaria, F., Avila, S., Roy, B., Ertör, I., Temper, L., & others. (2020). Environmental conflicts and defenders: a global overview. *Global Environmental Change*, *63*, 102104.

Semelsberger, T. A., Borup, R. L., & Greene, H. L. (2006). Dimethyl ether (DME) as an alternative fuel. *Journal of Power Sources*, *156*(2), 497–511.

Shah, Y. T. (2017). *Chemical energy from natural and synthetic gas*. CRC Press.

Shane, A., Gheewala, S. H., & Kasali, G. (2015). Potential, barriers and prospects of biogas production in Zambia. *Journal of Sustainable Energy and Environment*, *6*, 21–26.

Sher, F., Smječanin, N., Hrnjić, H., Karadža, A., Omanović, R., Šehović, E., & Sulejmanović, J. (2024). Emerging technologies for biogas production: a critical review on recent progress, challenges and perspectives. *Process Safety and Environmental Protection*. 188, 834-859

Singh, R., Pareek, N., Kumar, R., & Vivekanand, V. (2022). Anaerobic biodigesters for the treatment of high-strength wastewater. In *Anaerobic biodigesters for human waste treatment* (pp. 75–94). Springer.

Sørensen, J. F. L., & Jørgensen, H. P. (2022). Rural development potential in the bioeconomy in developed countries: the case of biogas production in Denmark. *Sustainability*, *14*(17), 11077.

Sovacool, B. K., Kryman, M., & Smith, T. (2015). Scaling and commercializing mobile biogas systems in Kenya: a qualitative pilot study. *Renewable Energy*, *76*, 115–125.

Sun, H. S., Yang, D., & Wei, S. J. (2014). Analysis on mode and benefit of biogas project on livestock farm. *Advanced Materials Research*, *955*, 2644–2648.

Surendra, K. C., Khanal, S. K., Shrestha, P., & Lamsal, B. (2011). Current status of renewable energy in Nepal: opportunities and challenges. *Renewable and Sustainable Energy Reviews*, *15*(8), 4107–4117.

Szymańska, D., & Lewandowska, A. (2015). Biogas power plants in Poland—structure, capacity, and spatial distribution. *Sustainability*, *7*(12), 16801–16819.

Taherzadeh, M. J., & Rajendran, K. (2014). Factors affecting development of waste management: experiences from different cultures. In *Waste management and sustainable consumption* (pp. 67–87). Routledge.

Tajebe, L. (2016). Bio-gas technology adoption in rural Ethiopia: it's effect on the crisis of deforestation. *Journal of Energy Technologies and Policy*.6, 1-8.

Tang, S., Wang, Z., Lu, H., Si, B., Wang, C., & Jiang, W. (2023). Design of stage-separated anaerobic digestion: principles, applications, and prospects. *Renewable and Sustainable Energy Reviews*, *187*, 113702.

Thiruselvi, D., Kumar, P. S., Kumar, M. A., Lay, C.-H., Aathika, S., Mani, Y., Jagadiswary, D., Dhanasekaran, A., Shanmugam, P., Sivanesan, S., & others. (2021). A critical review on global trends in biogas scenario with its up-gradation techniques for fuel cell and future perspectives. *International Journal of Hydrogen Energy*, *46*(31), 16734–16750.

Vea, E. B., Romeo, D., & Thomsen, M. (2018). Biowaste valorisation in a future circular bioeconomy. *Procedia Cirp*, *69*, 591–596.

Walekhwa, P. N., Mugisha, J., & Drake, L. (2009). Biogas energy from family-sized digesters in Uganda: critical factors and policy implications. *Energy Policy*, *37*(7), 2754–2762.

Wang, D.-H., Zhu, M.-Y., Lian, S.-J., Zou, H., Fu, S.-F., & Guo, R.-B. (2022). Conversion of renewable biogas into single-cell protein using a combined microalga-and methane-oxidizing bacterial system. *ACS ES&T Engineering*, *2*(12), 2317–2325.

Wang, X., Yan, R., Zhao, Y., Cheng, S., Han, Y., Yang, S., Cai, D., Mang, H.-P., & Li, Z. (2020). Biogas standard system in China. *Renewable Energy*, *157*, 1265–1273.

Wang, Y., Zhang, Y., Li, J., Lin, J.-G., Zhang, N., & Cao, W. (2021). Biogas energy generated from livestock manure in China: current situation and future trends. *Journal of Environmental Management*, *297*, 113324.

Weiland, P. (2010). Biogas production: current state and perspectives. *Applied Microbiology and Biotechnology*, *85*, 849–860.

Xue, S., Song, J., Wang, X., Shang, Z., Sheng, C., Li, C., Zhu, Y., & Liu, J. (2020). A systematic comparison of biogas development and related policies between China and Europe and corresponding insights. *Renewable and Sustainable Energy Reviews*, *117*, 109474.

Yellezuome, D., Zhu, X., Liu, X., Liu, R., Sun, C., Abd-Alla, M. H., & Rasmey, A.-H. M. (2024). Effects of organic loading rate on hydrogen and methane production in a novel two-stage reactor system: performance, enzyme activity and microbial structure. *Chemical Engineering Journal*, *480*, 148055.

Zareei, S. (2018). Evaluation of biogas potential from livestock manures and rural wastes using GIS in Iran. *Renewable Energy*, *118*, 351–356.

Zinoviev, S., Müller-Langer, F., Das, P., Bertero, N., Fornasiero, P., Kaltschmitt, M., Centi, G., & Miertus, S. (2010). Next-generation biofuels: survey of emerging technologies and sustainability issues. *ChemSusChem*, *3*(10), 1106–1133.

Zupančič, M., Možic, V., Može, M., Cimerman, F., & Golobič, I. (2022). Current status and review of waste-to-biogas conversion for selected European countries and worldwide. *Sustainability*, *14*(3), 1823.

9 Algal Biomass Production and Carbon Sequestration

Afaan A. Malla, Fayaz A. Malla, Showkat Rashid, and Suhaib A. Bandh

9.1 INTRODUCTION

Algal biomass has attracted a lot of interest as a possible sustainable resource since it may be used to produce bioenergy and store carbon. A new solution to global energy and environmental concerns can be found in algae, thanks to their rapid growth rates and capacity to use CO_2 for photosynthesis. This chapter explores the creation of algal biomass and its function in reducing the impact of climate change by storing carbon. Integrating algae into sustainable practices is highlighted by a discussion of the latest breakthroughs and methods. The term "algal biomass" describes the organic matter created by the wide variety of photosynthetic organisms known as algae, which can range in size from very small microalgae to very large seaweeds. Freshwater, saltwater, and even sewer systems are all suitable habitats for algae. Because of their high lipid content and their ability to produce biofuels and biochemicals, microalgae have particularly attracted interest (Sarwer *et al.*, 2022). In addition to biofuels, algal biomass has many other potential uses, such as in the food, feed, and pharmaceutical industries. Biorefineries are one example of an emerging technology that has the potential to maximize the extraction of valuable metabolites from algae while simultaneously reducing operational costs and improving economic feasibility (Uma *et al.*, 2022).

Capturing and storing atmospheric CO_2 for future use in reducing global warming is known as carbon sequestration. In this process, algae play a key role by converting carbon dioxide gas into organic molecules through photosynthesis, which in turn lowers concentrations of greenhouse gases. Photobioreactors (PBRs) and other algal-based carbon sequestration technologies improve this capacity by creating an ideal setting for the growth of algae (Rambhiya *et al.*, 2021). Furthermore, there has been encouraging progress in combining algae with bioenergy carbon capture and storage, sometimes known as ABCS. Utilizing CO_2 emissions from bioenergy processes, this system integrates algae farming with bioenergy generation, hence promoting algal development. Sustainable biofuels and carbon sequestration are two goals that ABECCS could achieve at the same time, according to a techno-economic evaluation (Beal *et al.*, 2018).

The purpose of this chapter is to explain in detail how algae can produce biomass and how it could be used to store carbon and is aimed to investigate what aspects of algae's biology and ecology make them promising candidates for use in bioenergy generation and carbon sequestration. In addition, it investigates cutting-edge tactics and technology that aim to boost algal output and carbon capture effectiveness. To prove that algal systems can scale and remain sustainable, real-world examples and case studies are emphasized. Understanding the production processes, carbon capture capacities, and contributions to a sustainable bioeconomy of microalgal and macroalgal systems are the topics covered in this chapter. This chapter highlights the importance of algae as a potential answer to the energy and environmental problems of today by discussing these points.

DOI: 10.1201/9781003685814-9

9.2 BIOLOGICAL BASIS OF CARBON SEQUESTRATION IN ALGAE

The function of algae in carbon sequestration is fundamental to their ecological and industrial importance. Algae utilize their photosynthetic abilities to collect atmospheric CO_2, transforming it into organic matter and releasing oxygen in the process. Their swift growth rates, capacity to flourish in many conditions, and potential for incorporation into bioenergy and aquaculture systems provide them an asset in combating climate change.

9.2.1 Photosynthetic Pathways in Algae

Algae employ photosynthesis to transform sunlight into chemical energy, a process fundamental to carbon sequestration. In contrast to higher plants, algae exhibit several photosynthetic processes tailored to a broad spectrum of environmental conditions. The principal mechanism entails the Calvin–Benson cycle, in which CO_2 is assimilated into organic compounds through ribulose-1,5-bisphosphate carboxylase/oxygenase (Rubisco) (Murphy, 2024). Certain species, especially those in high-light settings, utilize C4-like mechanisms or the crassulacean acid metabolism (CAM) pathway to improve carbon fixation efficiency. These modifications enable algae to sustain photosynthetic activity in environments characterized by low CO_2 supply, elevated light intensity, or variable temperatures. This adaptability renders them very proficient at absorbing and sequestering carbon across various aquatic and terrestrial settings.

9.2.2 Carbon Uptake Mechanisms

The carbon absorption mechanisms of algae are a crucial aspect of their sequestration capability. Algae obtain inorganic carbon as CO_2 or bicarbonate (HCO_3^-) from their surroundings. Microalgae, including *Chlorella* and *Spirulina*, possess active transport mechanisms that concentrate CO_2 near the Rubisco activity site, hence reducing photorespiration and enhancing photosynthetic efficiency (Gurau *et al.*, 2025). Macroalgae mostly depend on passive diffusion and the usage of bicarbonate. Their extensive surface area and capacity to thrive in nutrient-dense marine ecosystems enhance their significant carbon sequestration capability. Species such as *Laminaria* can sequester CO_2 at rates far beyond those of terrestrial plants, rendering them optimal for extensive carbon capture projects (Lai *et al.*, 2022).

9.2.3 Comparative Efficiency of Microalgae and Macroalgae

Microalgae and macroalgae both play substantial roles in carbon sequestration, although their efficiency range owing to variations in physiology and ecological functions. Microalgae exhibit quick growth rates and great photosynthetic efficiency, rendering them appropriate for closed systems such as PBRs. Research indicates that microalgae may sequester CO_2 at rates of up to 1.8 kg of CO_2 per kg of biomass generated (Pal *et al.*, 2023). Macroalgae, despite their slower growth rates, facilitate long-term carbon sequestration via their integration into marine sediments and detrital cycles. Their substantial biomass production and capacity to colonize extensive oceanic regions enhance their sequestration potential. *Macrocystis pyrifera*, a species of kelp, can sequester up to 9.8 tons of CO_2 per ha each year, rendering it a significant asset in marine carbon offset initiatives (Murphy, 2024).

9.3 ALGAL BIOMASS PRODUCTION TECHNIQUES

9.3.1 Open Pond Systems

Open pond systems represent one of the most prevalent and economical techniques for algae cultivation. These systems utilize natural sunlight and atmospheric CO_2 for photosynthesis, rendering them economically beneficial for large-scale production. Open raceway ponds, a prevalent design, have shallow, oval-shaped channels equipped with paddle wheels to maintain continuous water circulation and nutrient homogenization (Zhang et al., 2024a. Nonetheless, the straightforward nature of open pond systems is counterbalanced by their vulnerability to infection from undesirable microbes and inadequate growth resulting from environmental variability. Notwithstanding these problems, these systems are especially conducive to species such as *Spirulina* and *Chlorella*, which flourish in alkaline or nutrient-specific environments that inhibit contamination (Mahmood *et al.*, 2022).

9.3.2 Photobioreactors

PBRs are sealed devices that create regulated conditions for algae cultivation, yielding superior productivity and enhanced biomass quality relative to open ponds. PBRs are available in multiple configurations, including as tubular, flat-panel, and bubble column systems, each specifically designed to enhance light exposure and CO_2 supply. Tubular PBRs are favored for their scalability and efficiency in light distribution (Novoveská *et al.*, 2023). PBRs exhibit reduced susceptibility to contamination and water evaporation, rendering them suitable for high-value product applications, including pharmaceuticals and nutraceuticals. Notwithstanding its benefits, the substantial capital and operational expenses of PBRs continue to impede widespread deployment. Innovative designs, such as Light Exchange Bubble-column systems, seek to enhance energy efficiency and cost-effectiveness (Falfushynska, 2024).

9.3.3 Hybrid Cultivation Methods

Hybrid systems integrate the advantages of open ponds and PBRs, mitigating the shortcomings of each method. Algae can be initially cultivated in open ponds to minimize expenses and subsequently transported to PBRs for the final growth phases to assure superior biomass quality. This integrated approach is particularly advantageous for large-scale biofuel production, as it balances cost efficiency and productivity (Saad *et al.*, 2019). In hybrid systems, wastewater can serve as a growth medium, offering a sustainable approach for nutrient recycling and CO_2 sequestration. These systems have demonstrated considerable potential in lowering operational expenses and environmental effects while improving the economic feasibility of algal biomass production (Ugwuanyi *et al.*, 2024).

9.3.4 Advances in Genetic and Metabolic Engineering

Genetic and metabolic engineering have transformed algal biomass production by facilitating the improvement of desirable characteristics, including lipid accumulation, photosynthetic efficiency, and stress resilience. Techniques such as CRISPR/Cas9 and RNA interference (RNAi) are progressively employed to alter algal genomes, yielding strains enhanced for biofuel production (Kumar *et al.*, 2018). The overexpression of crucial enzymes in lipid production has resulted in substantial enhancements in lipid content in species such as *Nannochloropsis*. Metabolic pathway engineering has been employed to augment carbohydrate production in microalgae for bioethanol applications. These improvements have significantly enhanced the viability of algae as a renewable energy source, providing a sustainable substitute for fossil fuels (Rathinavel *et al.*, 2024).

9.4 CARBON SEQUESTRATION POTENTIAL OF ALGAL SYSTEMS

9.4.1 Quantification of Carbon Fixation in Algal Cultures

Algal systems are exceptionally efficient in sequestering atmospheric CO_2 via photosynthesis, with rates far beyond those of terrestrial flora. Microalgae, including *Chlorella* and *Spirulina*, may sequester up to 1.8 kg of CO_2 per kg of dry biomass generated, rendering them suitable for scalable carbon capture systems (Xu *et al.*, 2023). This process is enhanced by their elevated photosynthetic efficiency and fast growth, allowing them to assimilate substantial quantities of CO_2 within a little period. The quantification of carbon fixation is generally accomplished by methods including gas exchange analysis and isotopic tagging. Brown algae, such as *Fucus vesiculosus*, excrete up to 0.3% of their biomass daily in the form of polysaccharides like fucoidan, hence augmenting dissolved organic carbon reservoirs in marine ecosystems (Buck-Wiese *et al.*, 2022). These indicators offer insights into the efficacy of algal systems for carbon sequestration.

9.4.2 Long-Term Storage Potential in Biomass

The enduring storage capacity of carbon in algal biomass relies on its transformation into stable organic or inorganic compounds. Macroalgae, including kelp, play a crucial role in carbon sequestration by integrating carbon into structural elements such as cellulose and alginates. When these elements are deposited in ocean sediments or transported to deep-sea habitats, they can stay sequestered for hundreds to thousands of years (Hurd *et al.*, 2022). Additionally, algae used into bioenergy carbon capture and storage (ABECCS) systems improve sequestration by transforming harvested biomass into biochar or bio-oil, both of which serve as stable carbon storage mediums. A study indicated that integrating algae production with bioenergy systems might trap up to 61 tons of CO_2 per ha yearly, illustrating the effectiveness of this method in addressing climate change (Beal *et al.*, 2018).

9.4.3 Contribution to Blue Carbon Ecosystems

Blue carbon ecosystems, which include mangroves, seagrasses, and macroalgal habitats, are essential to global carbon cycling. Macroalgae have been recognized as significant contributors to these ecosystems because of their broad covering and elevated production. Research conducted along the South American Pacific coast demonstrated that algal meadows, particularly those of *Chondracanthus chamissoi*, sequester substantial quantities of carbon, as evidenced by dry biomass measurements reflecting large carbon capture capabilities (Cerna-Arrue *et al.*, 2023). Furthermore, floating macroalgae in areas such as Greenland fjords facilitate carbon cycling by transferring biomass beyond coastal regions, thereby establishing enduring carbon sinks (Ager *et al.*, 2023). Notwithstanding these contributions, ambiguities persist concerning the accurate measurement of carbon fluxes in these systems, requiring additional research to enhance their effectiveness in climate change mitigation (Ross *et al.*, 2023).

9.5 APPLICATIONS OF ALGAL BIOMASS IN CLIMATE CHANGE MITIGATION

9.5.1 Biofuels and Renewable Energy Production

Algal biomass serves as a promising feedstock for biofuel production, providing considerable advantages compared to conventional energy sources. Microalgae, including *Chlorella* and *Spirulina*, are esteemed for their substantial lipid content, which can be transformed into biodiesel via transesterification. In contrast to traditional crops, algae do not compete with food supply chains and necessitate minimum land and freshwater for cultivation (Farghali *et al.*, 2023). Furthermore, macroalgae (seaweeds) like *Saccharina japonica* exhibit promise for bioethanol production owing to their abundant carbohydrate content. A comprehensive analysis reveals that algae-derived biofuels may diminish greenhouse gas emissions by as much as 68% in comparison to fossil fuels (Daneshvar *et al.*, 2022). Notwithstanding these benefits, issues including elevated culture and processing expenses must be resolved to improve the scalability of algal biofuels (Waqas *et al.*, 2024).

9.5.2 Bioplastics and Sustainable Materials

Algae-derived bioplastics offer an eco-friendly substitute for petroleum-based plastics, tackling worldwide issues related to plastic pollution. Polyhydroxyalkanoates (PHAs), sourced from algal biomass, are biodegradable plastics with mechanical qualities akin to those of conventional plastics (Irfan *et al.*, 2025). Microalgae farmed with industrial flue gas as a carbon source have effectively been employed to manufacture PHAs, providing a dual answer for CO_2 reduction and sustainable material generation. Seaweeds are being investigated for their polysaccharides, including agar and alginate, which are utilized as raw materials for biodegradable packaging and biomedical applications. A study indicated that the decomposition of seaweed-based bioplastics emits fewer greenhouse gases than conventional plastics, underscoring its environmental advantages (Zhang *et al.*, 2024). Nonetheless, enhancing the production of algal bioplastics necessitates additional research into economically viable technologies and market assimilation (Nawaz *et al.*, 2024).

9.5.3 Carbon Capture, Utilization, and Storage (CCUS)

Algal biomass is essential in carbon capture, utilization, and storage (CCUS) systems. Microalgae may directly sequester CO_2 from industrial emissions or the atmosphere and transform it into biomass via photosynthesis. This biomass can subsequently be converted into useful products, including biofuels, bioplastics, and charcoal, so establishing a circular carbon economy (Zieliński *et al.*, 2023). The biological carbon capture efficiency of algae has been shown to reach 90% in regulated PBR systems, indicating substantial promise for large-scale industrial applications. Furthermore, the integration of algae cultivation with wastewater treatment improves CO_2 sequestration by supplying vital nutrients for algal proliferation (Daneshvar *et al.*, 2022).

Innovative CCUS technologies, including bioenergy carbon capture and utilization (BECCU), utilize algal systems to capture and convert CO_2 emissions into energy and products. A techno-economic assessment revealed that seaweed-based CCUS systems might trap as much as 173 mt of carbon each year, so making a substantial contribution to global climate objectives (Farghali *et al.*, 2023).

9.6 ECONOMIC AND ENVIRONMENTAL ASPECTS

9.6.1 Cost–Benefit Analysis of Algal Production

The production of algal biomass presents considerable potential for mitigating global energy and environmental issues; nevertheless, it is still economically demanding. Microalgae production systems, whether in open ponds or PBRs, entail substantial expenses due to energy requirements for illumination, CO_2 supply, and thermal regulation. A rigorous assessment indicates that operational expenses, especially for closed systems, may account for as much as 80% of total costs (Moshood *et al.*, 2021). Nonetheless, progress in automation and hybrid farming techniques is progressively diminishing these expenses. Notwithstanding these costs, algal systems have other advantages, such as the generation of high-value bioproducts like biofuels, nutraceuticals, and bioplastics. Integrating algae growing with wastewater treatment diminishes nutrient input expenses and amplifies environmental advantages through resource recycling (Vance *et al.*, 2022). A life cycle analysis of microalgae-derived biofuel production indicated a possible return on investment of 1.5–2.3 times that of conventional fuels, contingent upon supportive regulatory frameworks for carbon credits and subsidies (Chowdhury *et al.*, 2025).

9.6.2 Environmental Benefits and Trade-Offs

The environmental benefits of algal biomass systems are many. Algae's remarkable photosynthetic efficiency facilitates substantial CO_2 sequestration, with certain strains able to fix as much as 1.8 kg of CO_2 for every kg of dry biomass generated (Daneshvar *et al.*, 2022). Moreover, algal systems mitigate eutrophication by sequestering surplus nutrients in aquatic environments, especially within integrated aquaculture and wastewater treatment frameworks. Nonetheless, environmental trade-offs must also be considered. Excessive energy usage for artificial lighting and agitation in PBRs may negate the carbon sequestration advantages of algae unless renewable energy sources are employed (Ansari *et al.*, 2023). Moreover, extensive algal culture may contend for land and water resources, particularly in arid areas, potentially resulting in conflicts over resource allocation (Mizik, 2021).

9.6.3 Socioeconomic Impacts of Large-Scale Algal Cultivation

Extensive algae culture has significant social advantages, such as employment generation and rural advancement. The creation of algal biorefineries can rejuvenate local economies by creating possibilities in biomass collecting, processing, and biofuel distribution (Brinkman *et al.*, 2019). Furthermore, the generation of high-value goods such as bioplastics and medicines from algae biomass facilitates income stream diversification. However, obstacles including substantial initial capital investments and a lack of skilled labor in underdeveloped nations may impede wider implementation. Collaboration among stakeholders and governmental policies that advocate for education, subsidies, and infrastructure enhancement are crucial for the expansion of algae technology (Khan *et al.*, 2021). Integrating algae with other bioeconomy sectors, like agriculture and aquaculture, improves resource efficiency and reduces trade-offs.

9.7 CHALLENGES IN ALGAL BIOMASS PRODUCTION

9.7.1 Technological Barriers

The cultivation of algal biomass has considerable technological obstacles that impede its widespread implementation. Effective growing techniques, especially for microalgae, necessitate sophisticated equipment like PBRs and raceway ponds, both of which entail substantial operational and maintenance expenses. Although PBRs provide regulated environments and enhanced production, they require substantial energy for illumination, temperature management, and CO_2 supply (Zhang *et al.*, 2024). In contrast, open systems are more economical yet susceptible to contamination and inefficiency. Emerging innovative technologies, such as digital twins—virtual representations of algae growth systems—are being developed to optimize factors like nutrient levels, temperature, and CO_2 inputs. Nonetheless, the incorporation of these technologies into current infrastructure poses a problem due to elevated prices and insufficient experience (Sheik *et al.*, 2024). Moreover, algae production for particular uses, including biofuels or bioplastics, necessitates strain-specific optimizations and sophisticated genetic engineering, which remain in the developmental phase.

9.7.2 Land, Water, and Resource Constraints

Extensive algal production requires substantial resources, such as land, water, and nutrients. Although algae do not directly compete with conventional crops on cultivable land, the creation of open ponds or extensive PBRs typically necessitates considerable spatial dimensions. In arid places, the scarcity of freshwater might present a significant obstacle; however, seawater and wastewater are progressively being investigated as alternatives (Rumin *et al.*, 2021). Nutrient inputs, especially nitrogen and phosphorus, are crucial for algal proliferation. Nonetheless, extensive utilization may result in eutrophication within aquatic environments if not meticulously regulated. Resource limitations also encompass energy requirements for culture, harvesting, and processing, hence exacerbating the environmental sustainability challenges of algal biomass production. Initiatives to combine algae production with wastewater treatment and carbon capture systems demonstrate potential in overcoming these limitations; nevertheless, their scalability is still restricted (Ugwuanyi *et al.*, 2024).

9.7.3 Regulatory and Policy Challenges

The regulatory framework for algal biomass production is sometimes disjointed and inconsistent, resulting in difficulties for industry participants. The categorization of algae-derived products—whether it is food, fuel, or pharmaceuticals—dictates the regulatory obligations, which may be intricate and interrelated in numerous areas. Microalgae-derived biofuels encounter rigorous emissions and safety regulations, frequently necessitating supplementary testing and certifications that elevate expenses (Makepa & Chihobo, 2024). Policy support is essential for promoting innovation and commercialization in the algal biomass industry. Subsidies, tax incentives, and grants for research and development can substantially diminish entry barriers. Nevertheless, conflicting policy frameworks and insufficient public knowledge impede the implementation of algae-based solutions, especially in low- and middle-income nations. Enhanced harmonization of international rules and the formulation of specific algal biomass strategies are essential to effectively tackle these difficulties (Samoraj *et al.*, 2024).

9.8 CONCLUSION

The production of algal biomass and carbon sequestration offer significant prospects for tackling global energy and environmental issues. Critical findings from this investigation underscore algae's viability as a sustainable resource for biofuels, bioplastics, and carbon capture systems, attributed to their quick growth, high photosynthetic efficiency, and capacity to thrive in various settings, including wastewater. Notwithstanding these advantages, considerable obstacles remain. Technological impediments, including the substantial expense of production equipment and the necessity for sophisticated genetic engineering, obstruct scaling. Moreover, limitations in resources like land, water, and nutrient availability, together with regulatory intricacies, hinder large-scale application. Nevertheless, the environmental advantages of algae, including the reduction of CO_2 emissions and the improvement of nutrient recycling, underscore their significance in a circular bioeconomy. Policymakers, researchers, and industry executives must work to surmount these obstacles. Investment in research, infrastructure, and supportive policies is crucial for reducing production costs, improving scalability, and harmonizing international rules. Moreover, enhancing public awareness and offering subsidies or incentives for algae technologies will expedite their adoption. The future of algal biomass production depends on the integration of novel solutions, such as digital twins and hybrid systems, with sustainable practices. By collaboratively tackling these problems, stakeholders may realize the complete potential of algae, advancing global efforts in energy security, climate change mitigation, and sustainable development.

DECLARATION

The authors have not used any generative AI tools during the writing of this manuscript.

REFERENCES

Ager, T.G., Krause-Jensen, D., Olesen, B., Carlson, D.F., Winding, M.H.S. & Sejr, M.K. (2023) Macroalgal habitats support a sustained flux of floating biomass but limited carbon export beyond a Greenland fjord. *Science of the Total Environment*, 872, 162224. DOI: 10.1016/j.scitotenv.2023.162224, PubMed: 36804986.

Ansari, A., Wuryandani, S., Pranesti, A., Telaumbanua, M., Ngadisih, T., Hardiansyah, M.Y., Alam, T., Supriyanta, Martini, T. & Taryono (2023) Optimizing water-energy-food nexus: Achieving economic prosperity and environmental sustainability in agriculture. *Frontiers in Sustainable Food Systems*, 7. Available at https://app.dimensions.ai/details/publication/pub.1163679747. DOI: 10.3389/fsufs.2023.1207197

Beal, C.M., Archibald, I., Huntley, M.E., Greene, C.H. & Johnson, Z.I. (2018) Integrating algae with bioenergy carbon capture and storage (ABECCS) increases sustainability. *Earth's Future*, 6, 524–542. Available at https://app.dimensions.ai/details/publication/pub.1101164248. DOI: 10.1002/2017EF000704

Brinkman, M.L.J., Wicke, B., Faaij, A.P.C. & van der Hilst, F. (2019) Projecting socio-economic impacts of bioenergy: Current status and limitations of ex-ante quantification methods. *Renewable and Sustainable Energy Reviews*, 115. Available at https://app.dimensions.ai/details/publication/pub.1121059765. DOI: 10.1016/j.rser.2019.109352

Buck-Wiese, H., Andskog, M.A., Nguyen, N.P., Bligh, M., Asmala, E., Vidal-Melgosa, S., Liebeke, M., Gustafsson, C. & Hehemann, J.H. (2022) Fucoid brown algae inject fucoidan carbon into the ocean. Proceedings of the National Academy of Sciences of the USA. Available at https://app.dimensions.ai/details/publication/pub.1154084818

Cerna-Arrue, A., Aponte, H. & Torres-Galarza, S. (2023) Carbon capture in Latin American algal meadows: A case study on South American Pacific coast. *Research Square*. Available at https://app.dimensions.ai/details/publication/pub.1164913880

Chowdhury, P., Mahi, N.A., Yeassin, R., Chowdhury, N.U.R. & Farrok, O. (2025) Biomass to biofuel: Impacts and mitigation of environmental, health, and socioeconomic challenges. *Energy Conversion and Management: X*, 25. Available at https://app.dimensions.ai/details/publication/pub.1184426801. DOI: 10.1016/j.ecmx.2025.100889

Daneshvar, E., Wicker, R.J., Show, P.L. & Bhatnagar, A. (2022) Biologically mediated carbon capture and utilization by microalgae towards sustainable CO_2 biofixation and biomass valorization – A review. *Chemical Engineering Journal*, 427. Available at https://app.dimensions.ai/details/publication/pub.1138951445. DOI: 10.1016/j.cej.2021.130884.

Falfushynska, H. (2024) Advancements and prospects in algal biofuel production: A comprehensive review. *Phycology*, 4, 548–575. Available at https://app.dimensions.ai/details/publication/pub.1181493593. DOI: 10.3390/phycology4040030

Farghali, M., Mohamed, I.M.A., Osman, A.I. & Rooney, D.W. (2023) Seaweed for climate mitigation, wastewater treatment, bioenergy, bioplastic, biochar, food, pharmaceuticals, and cosmetics: A review. *Environmental Chemistry Letters*, 21, 97–152. Available at https://app.dimensions.ai/details/publication/pub.1151708263. DOI: 10.1007/s10311-022-01520-y, PubMed: 36245550.

Gurau, S., Imran, M. & Ray, R.L. (2025) Algae: A cutting-edge solution for enhancing soil health and accelerating carbon sequestration – A review. *Environmental Technology and Innovation*, 37. Available at https://app.dimensions.ai/details/publication/pub.1183656549. DOI: 10.1016/j.eti.2024.103980

Hurd, C.L., Law, C.S., Bach, L.T., Britton, D., Hovenden, M., Paine, E.R., Raven, J.A., Tamsitt, V. & Boyd, P.W. (2022) Forensic carbon accounting: Assessing the role of seaweeds for carbon sequestration. *Journal of Phycology*, 58, 347–363. Available at https://app.dimensions.ai/details/publication/pub.1146248871. DOI: 10.1111/jpy.13249, PubMed: 35286717.

Irfan, F., Tasnim, N., Razzak, S.A. & Uddin, S. (2025) Advances in capturing carbon and producing Bioplastics through the microalgal approach towards a sustainable future: A review. *Journal of Cleaner Production*, 486. Available at https://app.dimensions.ai/details/publication/pub.1183820364. DOI: 10.1016/j.jclepro.2024.144576

Khan, N., Sudhakar, K. & Mamat, R. (2021) Role of biofuels in energy transition, green economy, and carbon neutrality. *Sustainability*, 13. Available at https://app.dimensions.ai/details/publication/pub.1143152359. DOI: 10.3390/su132212374

Kumar, P., Kumar, D., Nehra, P. & Sharma, P.K. (2018) Green algae biomass cultivation, harvesting and genetic modifications for enhanced cellular lipids. *Springer Nature*. Available at https://app.dimensions.ai/details/publication/pub.1101041799

Lai, Q., Ma, J., He, F., Zhang, A., Pei, D. & Yu, M. (2022) Current and future potential of shellfish and algae mariculture carbon sinks in China. *International Journal of Environmental Research and Public Health*, 19, 8873. Available at https://app.dimensions.ai/details/publication/pub.1149668092. DOI: 10.3390/ijerph19148873, PubMed: 35886723.

Mahmood, T., Hussain, N., Shahbaz, A., Mulla, S.I., Iqbal, H.M.N. & Bilal, M. (2022) Sustainable production of biofuels from the algae-derived biomass. *Bioprocess Engineering*. Available at https://app.dimensions.ai/details/publication/pub.1152461362

Makepa, D.C. & Chihobo, C.H. (2024) Barriers to commercial deployment of biorefineries: A multi-faceted review of obstacles across the innovation chain. *Heliyon*, 10, e32649. Available at https://app.dimensions.ai/details/publication/pub.1172603763. DOI: 10.1016/j.heliyon.2024.e32649, PubMed: 39183827.

Mizik, T. (2021) Economic aspects and sustainability of ethanol production—A systematic literature review. *Energies*, 14. Available at https://app.dimensions.ai/details/publication/pub.1141438026. DOI: 10.3390/en14196137

Moshood, T.D., Nawanir, G. & Mahmud, F. (2021) Microalgae biofuels production: A systematic review on socioeconomic prospects of microalgae biofuels and policy implications. *Environmental Challenges*, 5. Available at https://app.dimensions.ai/details/publication/pub.1139451663. DOI: 10.1016/j.envc.2021.100207

Murphy, D.J. (2024) Biological carbon sequestration: From deep history to the present day. *Earth*, 5, 195–213. Available at https://app.dimensions.ai/details/publication/pub.1172469735. DOI: 10.3390/earth5020010

Nawaz, T., Gu, L., Hu, Z., Fahad, S., Saud, S. & Zhou, R. (2024) Advancements in synthetic biology for enhancing cyanobacterial capabilities in sustainable plastic production: A green horizon perspective. *Fuels*, 5, 394–438. Available at https://app.dimensions.ai/details/publication/pub.1175151346. DOI: 10.3390/fuels5030023

Novoveská, L., Nielsen, S.L., Eroldoğan, O.T., Haznedaroglu, B.Z., Rinkevich, B., Fazi, S., Robbens, J., Vasquez, M. & Einarsson, H. (2023) Overview and challenges of large-scale cultivation of photosynthetic microalgae and cyanobacteria. *Marine Drugs*, 21, 445. Available at https://app.dimensions.ai/details/publication/pub.1163209682. DOI: 10.3390/md21080445, PubMed: 37623726.

Pal, A., Srivastava, I., Sahu, S. & Singh, K.R. (2023) Algae's promise to sequester carbon sheds new insight on changing climate. *International Journal of Current Science Research and Review*, 6. Available at https://app.dimensions.ai/details/publication/pub.1158215094. DOI: 10.47191/ijcsrr/V6-i5-28

Rambhiya, S.J., Magar, C.S. & Deodhar, M.A. (2021) Using seawater-based Na_2CO_3 medium for scrubbing the CO_2 released from Bio-CNG plant for enhanced biomass production of *Pseudanabaena limnetica*. *Discover Applied Sciences*. Available at https://app.dimensions.ai/details/publication/pub.1135111167

Rathinavel, L., Singh, S., Rai, P.K., Chandra, N., Jothinathan, D., Gaffar, I., Pandey, A.K., Choure, K., Waoo, A.A., Joo, J.C. & Pandey, A. (2024) Sustainable microalgal biomass for efficient and scalable green energy solutions: Fueling tomorrow. *Fuels*, 5, 868–894. Available at https://app.dimensions.ai/details/publication/pub.1183139896. DOI: 10.3390/fuels5040049

Ross, F.W.R., Boyd, P.W., Filbee-Dexter, K., Watanabe, K., Ortega, A., Krause-Jensen, D., Lovelock, C., Sondak, C.F.A., Bach, L.T., Duarte, C.M., Serrano, O., Beardall, J., Tarbuck, P. & Macreadie, P.I. (2023) Potential role of seaweeds in climate change mitigation. *Science of the Total Environment*, 885, 163699. Available at https://app.dimensions.ai/details/publication/pub.1157774225. DOI: 10.1016/j.scitotenv.2023.163699, PubMed: 37149169.

Rumin, J., Junior, R.G.O., Bérard, J.B. & Picot, L. (2021) Improving microalgae research and marketing in the European Atlantic Area: Analysis of major gaps and barriers limiting sector development. *Marine Drugs*, 19, 319. Available at https://app.dimensions.ai/details/publication/pub.1138495071. DOI: 10.3390/md19060319, PubMed: 34070907.

Saad, M.G., Dosoky, N.S., Zoromba, M.S. & Shafik, H.M. (2019) Algal biofuels: Current status and key challenges. *Energies*, 12. Available at https://app.dimensions.ai/details/publication/pub.1114981879. DOI: 10.3390/en12101920

Samoraj, M., Çalış, D., Trzaska, K., Mironiuk, M. & Chojnacka, K. (2024) Advancements in algal biorefineries for sustainable agriculture: Biofuels, high-value products, and environmental solutions. *Biocatalysis and Agricultural Biotechnology*, 58. Available at https://app.dimensions.ai/details/publication/pub.1171534337. DOI: 10.1016/j.bcab.2024.103224

Sarwer, A., Hamed, S.M., Osman, A.I., Jamil, F., Al-Muhtaseb, A.H., Alhajeri, N.S. & Rooney, D.W. (2022) Algal biomass valorization for biofuel production and carbon sequestration: A review. *Environmental Chemistry Letters*, 20, 2797–2851. Available at https://app.dimensions.ai/details/publication/pub.1148692579. DOI: 10.1007/s10311-022-01458-1

Sheik, A.G., Kumar, A., Ansari, F.A., Raj, V., Peleato, N.M., Patan, A.K., Kumari, S. & Bux, F. (2024) Reinvigorating algal cultivation for biomass production with digital twin technology – A smart sustainable infrastructure. *Algal Research*, 84. Available at https://app.dimensions.ai/details/publication/pub.1181765205. DOI: 10.1016/j.algal.2024.103779

Ugwuanyi, E.D., Nwokediegwu, Z.Q.S., Dada, M.A., Majemite, M.T. & Obaigbena, A. (2024) The role of algae-based wastewater treatment systems: A comprehensive review. *World Journal of Advanced Research and Reviews*. Available at https://app.dimensions.ai/details/publication/pub.1168961320

Uma, V.S., Usmani, Z., Sharma, M., Diwan, D., Sharma, M., Guo, M., Tuohy, M.G., Makatsoris, C., Zhao, X., Thakur, V.K. & Gupta, V.K. (2022) Valorisation of algal biomass to value-added metabolites: Emerging trends and opportunities. *Phytochemistry Reviews: Proceedings of the Phytochemical Society of Europe*, 1–26. Available at https://app.dimensions.ai/details/publication/pub.1145968724. DOI: 10.1007/s11101-022-09805-4, PubMed: 35250414.

Vance, C., Sweeney, J. & Murphy, F. (2022) Space, time, and sustainability: The status and future of life cycle assessment frameworks for novel biorefinery systems. *Renewable and Sustainable Energy Reviews*, 159. Available at https://app.dimensions.ai/details/publication/pub.1145695638. DOI: 10.1016/j.rser.2022.112259

Waqas, M.A., Hashemi, F., Mogensen, L. & Knudsen, M.T. (2024) Environmental performance of seaweed cultivation and use in different industries: A systematic review. *Sustainable Production and Consumption*, 48, 123–142. Available at https://app.dimensions.ai/details/publication/pub.1171576967. DOI: 10.1016/j.spc.2024.05.001

Xu, P., Li, J., Qian, J., Wang, B., Liu, J., Xu, R., Chen, P. & Zhou, W. (2023) Recent advances in CO2 fixation by microalgae and its potential contribution to carbon neutrality. *Chemosphere*, 319, 137987. Available at https://app.dimensions.ai/details/publication/pub.1154932524. DOI: 10.1016/j.chemosphere.2023.137987, PubMed: 36720412.

Zhang, J., Cran, M., Gao, L., Xie, Z. & Gray, S. (2024a) Contribution of seaweed farming to the mitigation of greenhouse gas emissions and microplastics pollution. *Algal Research*, 82. Available at https://app.dimensions.ai/details/publication/pub.1173973413. DOI: 10.1016/j.algal.2024.103623

Zhang, Q., Guan, Y., Zhang, Z., Dong, S., Yuan, T., Ruan, Z. & Chen, M. (2024b) Sustainable microalgae cultivation: A comprehensive review of open and enclosed systems for biofuel and high-value compound production. *E3S Web of Conferences*, 577. Available at https://app.dimensions.ai/details/publication/pub.1181262060. DOI: 10.1051/e3sconf/202457701008

Zieliński, M., Dębowski, M., Kazimierowicz, J. & Świca, I. (2023) Microalgal carbon dioxide (CO_2) capture and utilization from the European Union perspective. *Energies*, 16. Available at https://app.dimensions.ai/details/publication/pub.1155051585. DOI: 10.3390/en16031446

10 Algae-Based Carbon Credits

Fayaz A. Malla, Afaan A. Malla, Suhaib A. Bandh, and Showkat Rashid

10.1 INTRODUCTION

The application of algae in carbon credit systems is a novel approach to combating climate change. Algae, due to their fast proliferation and high carbon sequestration efficiency, present a promising avenue for enterprises seeking to mitigate greenhouse gas (GHG) emissions. Recent studies emphasise the incorporation of algal systems into comprehensive bioenergy and decarbonisation strategies, illustrating their ability to reduce CO_2 emissions while promoting circular economy principles (Luong *et al.*, 2022). Carbon credits function as an economic instrument to promote GHG reductions, enabling businesses to mitigate emissions by investing in verified carbon reduction initiatives. These credits have become essential in international climate initiatives, facilitating adaptability in achieving emission objectives. Algae-based systems effectively conform to this framework by directly sequestering CO_2 from industrial emissions and transforming it into biomass, which can then be processed into fuels, bioplastics, or other valuable commodities (Beal *et al.*, 2018). Bioenergy carbon capture and storage (BECCS) systems that incorporate algae (ABECCS) exhibit improved sustainability by eliminating competition for arable land and freshwater, while offering carbon-neutral or potentially carbon-negative solutions.

Algae are proficient in carbon sequestration owing to their elevated photosynthetic efficiency and capacity to thrive in many conditions, including wastewater and non-arable soil. Microalgae have been examined for their capacity to sequester CO_2 emissions from industrial sources, either within wastewater treatment systems or as independent culture platforms (Arora *et al.*, 2021). Moreover, biochar produced from algal biomass by pyrolysis functions as a long-term carbon sequestration medium, underscoring the versatile role of algal systems (Alazaiza *et al.*, 2023). Algae-derived biofuel production systems are becoming seen as a feasible substitute for fossil fuels. These systems not only employ CO_2 as a feedstock but also provide a means for the manufacture of sustainable aviation fuel. Streamlining algae-to-fuel conversion methods utilising techniques such as hydroprocessed esters and fatty acids (HEFA) technology significantly improves the economic and ecological feasibility of algal biofuels (Atnoorkar *et al.*, 2024).

The capacity of algae to decarbonise industrial processes is attributed to their function as biofactories for carbon capture and use. Industrial symbiosis, in which CO_2 emissions are channelled to algal growing facilities, illustrates a sustainable method for emission management (Beal *et al.*, 2018). Algae farming combined with eucalyptus forests in BECCS systems can produce renewable energy while sequestering atmospheric CO_2, exemplifying a comprehensive approach to sustainability. Furthermore, algae-based systems tackle the dual issues of nutrient contamination and GHG emissions. Nutrient-dense waste streams, a notable environmental issue, can be employed to cultivate algae, hence diminishing the reliance on chemical fertilisers and alleviating eutrophication dangers (Greenhalgh & Faeth, 2001). Innovations in algae processing, including supercritical water

 DOI: 10.1201/9781003685814-10

gasification and hydrothermal liquefaction, offer methods for generating bio-crude with reduced lifecycle emissions. These methods facilitate the transformation of algal biomass into high-energy-density fuels while minimising carbon emissions into the atmosphere (Onigbajumo *et al.*, 2021).

10.2 MECHANISMS OF CARBON CAPTURE AND SEQUESTRATION BY ALGAE

Algal carbon capture and sequestration (CCS) constitutes a viable approach to reduce atmospheric CO_2 concentrations. Algae employ photosynthesis and diverse metabolic processes to effectively sequester carbon, providing adaptable applications across numerous industries.

10.2.1 Algae Physiology and Carbon Uptake Processes

Algae, especially microalgae, utilise very effective carbon absorption processes that leverage their photosynthetic ability. They employ carbon-concentrating mechanisms (CCMs) to actively sequester CO_2 and transform it into biomass, which can be processed into biofuels, bioplastics, or other value-added products (Barry *et al.*, 2015). Certain species, including *Chlamydomonas reinhardtii*, have physiological modifications that improve their capacity to assimilate CO_2, even in nutrient-deficient conditions. Mutations have been engineered to enhance nutrition utilisation while preserving swift carbon absorption (Gopal & Sudhakar, 2016). The algae's cellular structure enhances these activities, optimising light absorption and photosynthetic efficiency. Moreover, research on algal consortia, encompassing cyanobacteria and green algae, underscores their collaborative capacity to absorb carbon under diverse settings, including variable pH and CO_2 concentrations. The consortia method has demonstrated considerable potential for optimising carbon capture while generating superior biomass (Nath *et al.*, 2019).

10.2.2 Microalgae vs. Macroalgae in Carbon Capture Efficiency

Microalgae and macroalgae vary in their methods and efficiency for carbon sequestration. Microalgae are unicellular organisms characterised by rapid growth rates and elevated surface-to-volume ratios, rendering them suitable for high-density cultures in photobioreactors (PBRs). These technologies optimise the distribution of light, nutrients, and CO_2, hence greatly improving biomass output (Barry *et al.*, 2015). Macroalgae, conversely, flourish in marine habitats, sequestering CO_2 dissolved in seawater. Marine calcifiers, such as coccolithophores, sequester carbon by producing calcium carbonate structures, hence facilitating long-term carbon storage in marine sediments (Moore *et al.*, 2022). Although their growth rates are inferior to those of microalgae, macroalgae necessitate less energy inputs and may be cultivated extensively in open water, offering a cost-efficient carbon capture and storage solution. Recent research indicates the integration of microalgae with wastewater treatment or desalination systems to sequester CO_2 while concurrently mitigating nutrient contamination. This dual-purpose technique enhances the environmental advantages of microalgae carbon capture and storage systems (Viswanaathan *et al.*, 2022).

10.2.3 Role of Algal Biomass in Long-Term Carbon Storage

Algal biomass is essential for long-term carbon sequestration. The collected biomass can be converted into biochar, biofuels, or preserved in geological formations, so preventing the re-release of trapped carbon into the atmosphere. The high-temperature pyrolysis of algal biomass yields biochar, a stable form of carbon utilised in agriculture and environmental remediation (Barry *et al.*, 2015). In marine ecosystems, calcifying algae facilitate long-term carbon sequestration by producing calcium carbonate, which descends to the ocean floor and persists for geological timeframes. This natural process highlights the significance of macroalgae in addressing climate change on a substantial scale

(Moore *et al.*, 2022). Moreover, advancements in algal bioprocessing have augmented the prospects for biofuel production. These processes guarantee the efficient utilisation of carbon sequestered during algae development, hence diminishing dependence on fossil fuels and lowering net carbon emissions (Gilani *et al.*, 2024).

10.3 TECHNOLOGY AND METHODS FOR ALGAE CULTIVATION

10.3.1 Open Pond Systems

Open pond systems (Figure 10.1) represent one of the oldest and most economical techniques for algae cultivation. These systems, generally superficial manmade ponds, harness sunlight and atmospheric CO_2 for photosynthesis. Open ponds are cost-effective to build and sustain, rendering them appropriate for extensive activities. Nonetheless, they encounter obstacles including contamination, evaporation, and suboptimal light utilisation. Research indicates that open pond systems can attain significant biomass output when tailored to certain strains and local environmental circumstances. Algae cultivated in wastewater-fed open ponds not only save water treatment expenses but also effectively sequester CO_2 (Lammers *et al.*, 2017). Notwithstanding their constraints, open ponds continue to be a feasible alternative for areas with abundant sunlight and minimal land expenses.

10.3.2 Photobioreactors

PBRs (Figure 10.2) exemplify a sophisticated and regulated technology for algae cultivation. These enclosed systems provide meticulous control of light, temperature, CO_2 concentration, and nutrient levels, thereby optimising productivity. PBRs are available in multiple configurations, including as

FIGURE 10.1 Open pond systems for algae cultivation.

FIGURE 10.2 Photobioreactors (PBRs) for algae cultivation.

flat-panel, tubular, and column reactors, each tailored for certain algal strains and operating objectives. Flat-panel PBRs are optimal for high-density cultures owing to their superior light distribution and gas exchange capabilities (In-na *et al.*, 2022). Tubular PBRs are extensively utilised in industrial applications for biofuel generation and carbon capture, as they enable continuous growing under optimal circumstances. The principal disadvantage of PBRs is their elevated capital and operational expenses. Nonetheless, progress in materials science and automation is rendering PBR systems progressively more economical and scalable. Moreover, their enhanced productivity and resistance to contamination render them essential for high-value applications such as pharmaceutical-grade algae production (Li *et al.*, 2021).

10.3.3 Advances in Genetic Engineering for High Carbon Uptake Algae

Genetic engineering is revolutionising algae culture by improving carbon capture efficiency and biomass production. Researchers have engineered algal strains that can rapidly fix CO_2 under various settings by altering metabolic pathways and including carbon-concentrating mechanisms. Genetic alterations in microalgae, specifically *C. reinhardtii*, have enhanced their photosynthetic efficiency and nutritional utilisation. These strains can flourish in nutrient-deficient conditions while sustaining elevated biomass productivity (Gopal & Sudhakar, 2016). Moreover, genetically modified algae with improved lipid biosynthesis pathways are being refined for biofuel generation, hence diminishing dependence on fossil fuels (Neeti *et al.*, 2023). The creation of self-flocculating algae presents a significant opportunity, since it streamlines harvesting procedures and diminishes costs related to downstream processing. These advancements are essential for enhancing algae-based carbon capture technology (Li *et al.*, 2021).

FIGURE 10.3 Industrial integration of algal systems capturing CO_2 emissions.

10.3.4 Integration of Algal Systems with Industrial Carbon Emissions

The integration of algal systems with industrial carbon emissions (Figure 10.3) provides a sustainable method for diminishing GHG emissions while producing value-added products. Algae may directly absorb CO_2 from flue gases, collecting pollutants at their origin. This integration has been examined in sectors including electricity generation, breweries, and wastewater treatment plants. A study on CO_2 collection from brewery fermentation with algal biocomposites revealed the viability of attaining high capture efficiencies in compact systems (In-na *et al.*, 2022). Algae cultivation in produced water from the oil and gas sector has the twin advantages of bioremediation and carbon sequestration (Alsarayreh *et al.*, 2022). Integrated systems tackle environmental issues while generating economic opportunities through the production of biofuels, bioplastics, and feedstock. Nonetheless, issues pertaining to system scalability, energy demands, and economic feasibility persist as critical domains for continued investigation.

10.4 ECONOMIC AND ENVIRONMENTAL BENEFITS OF ALGAE-BASED CARBON CREDITS

Algae-derived carbon credits present an innovative strategy for mitigating climate change while producing economic and environmental advantages. These systems utilise algae's capacity to sequester and store carbon while concurrently facilitating the production of biofuels, bioplastics, and other value-added products.

10.4.1 Generating Revenue through Carbon Markets

Algae-derived carbon credits offer financial incentives through the monetisation of CO_2 sequestration. These credits can be traded in voluntary or regulated carbon markets, providing revenue prospects for enterprises implementing sustainable practices. The sale of these credits is affected by the carbon sequestration efficiency of algae, which can be up to ten times greater than that of terrestrial plants, owing to algae's rapid growth and superior photosynthetic efficiency (Beal *et al.*, 2016). Algae farming combined with industrial emissions can produce carbon offsets that qualify for marketable credits. Research indicates that algae can sequester up to 90% of generated CO_2 in optimally constructed systems, greatly improving the economic feasibility of these technologies (Somers *et al.*, 2021). Moreover, carbon pricing mechanisms such as cap-and-trade systems generate market opportunities for algae-based credits, enabling enterprises to adhere to emissions limitations economically while facilitating the transition to a low-carbon economy (Cruce *et al.*, 2021).

10.4.2 Co-Benefits: Biofuel, Bioplastics, and Fertilisers

Algae's adaptability not just encompasses carbon sequestration but also provides economic benefits through the manufacture of biofuels, bioplastics, and fertilisers. Algal biofuels, such as biodiesel and bioethanol, have gained viability owing to technological progress and governmental incentives. These biofuels are renewable and generate considerably reduced GHG emissions in comparison to fossil fuels (Hise *et al.*, 2016). The bioplastics sector also utilises algae as a feedstock. Algal biomass can be converted into biodegradable polymers, diminishing dependence on petroleum-derived products and mitigating plastic pollution. Additionally, the remaining biomass from algae production functions as an organic fertiliser abundant in nitrogen and phosphorus, hence promoting sustainable agricultural practices (Pereira *et al.*, 2021). Economic assessments indicate that co-products such as biofuels and bioplastics substantially enhance the financial viability of algae-based systems by diversifying income streams and mitigating production costs (Liang *et al.*, 2023).

10.4.3 Algae's Role in Enhancing Ecosystem Services

The production of algae enhances ecosystem services such as water purification, habitat restoration, and climate regulation. Integrating algae into wastewater treatment systems eliminates undesirable nutrients like nitrogen and phosphorus, hence mitigating the risk of eutrophication in aquatic habitats (Greenhalgh & Faeth, 2001). Marine algae, including seaweed, are essential to "blue carbon" ecosystems by capturing carbon in coastal sediments. These ecosystems not only sequester carbon but also safeguard shorelines from erosion and augment biodiversity (Sovacool *et al.*, 2024). Furthermore, cultivating algae in degraded or arid regions can rejuvenate soil fertility and mitigate desertification, demonstrating its capacity to provide terrestrial ecosystem services. Algae-based systems provide environmental advantages by diminishing industrial emissions, fostering circular economies, and advancing global climate objectives (Cruce *et al.*, 2021).

10.5 CHALLENGES IN SCALING ALGAE-BASED CARBON CREDIT SYSTEMS

10.5.1 High Initial Costs and Operational Challenges

The advancement and expansion of algae-based carbon credit systems face significant economic obstacles, chiefly marked by excessively high initial investments and operational expenses that fundamentally impede their broad adoption. Lee *et al.* (2017) projected that the establishment of large-scale algae production facilities may necessitate an investment of \$5–\$10 million per hectare, with expenses including specialised equipment such as PBRs, nutrient delivery systems, and sophisticated carbon capture technology. The considerable economic difficulties are exacerbated by high operational costs, especially energy usage, which Brennan and Owende (2010) noted can represent 30–40% of overall operational expenses. The technological intricacies of moving from laboratory-scale to commercial-scale production present further challenges, as Chisti (2013) emphasised the essential difficulties in sustaining uniform algal productivity and carbon sequestration efficiency at larger operational scales. Furthermore, the complex technological prerequisites require precision-controlled environments with ideal lighting, temperature, and nutrient management, necessitating sophisticated engineering solutions that inherently elevate infrastructure and maintenance costs, thereby establishing significant economic obstacles to the widespread implementation of algae-based carbon credit systems.

10.5.2 Land, Water, and Nutrient Requirements

The spatial and resource requirements of algae-based carbon credit systems pose complex environmental and logistical obstacles that greatly hinder their large-scale deployment. projected that generating adequate algal biomass for significant carbon sequestration would necessitate vast land expanses, potentially leading to conflicts with current agricultural, residential, and conservation land utilisations. Water resource management is a significant limitation, as Gerbens-Leenes *et al.* (2014) illustrate that water usage for algal biomass production can vary from 100 to 500 litres per kilogramme of dry biomass, contingent upon growing methods and environmental factors. Nutrient requirements hamper scaling, as emphasised the considerable demands for nitrogen and phosphorus that may compete with agricultural fertiliser markets and present significant problems over sustainable nutrient sources. Mitigation strategies have arisen, encompassing the use of saline or brackish water sources, the creation of closed-loop water recycling systems, and the incorporation of algae cultivation with wastewater treatment facilities; however, these methods necessitate advanced technological solutions and significant initial investments to address the inherent resource challenges linked to large-scale algae-based carbon credit systems.

10.5.3 Regulatory and Policy Barriers

The emerging and intricate framework of algae-based carbon credit systems faces considerable regulatory and policy obstacles that significantly hinder their broad acceptance and market incorporation. Wang *et al.* (2020) identified significant inconsistencies in carbon accounting methodologies, resulting in considerable uncertainty for prospective investors and carbon market participants, while current carbon credit regulations often inadequately address the distinct characteristics of algae-based sequestration. Thompson *et al.* (2016) contended that robust policy frameworks are crucial for fostering investment and technological innovation; nevertheless, existing regulatory landscapes exhibit considerable deficiencies in acknowledging and facilitating algae-based carbon sequestration methods. These policy impediments arise from ambiguous certification standards, inconsistent international carbon trading protocols, and insufficient acknowledgement of algae-based methods in current carbon markets, thereby establishing significant institutional barriers that hinder the seamless transition of these innovative carbon sequestration technologies from experimental phases to extensive commercial deployment. The lack of strong, standardised regulatory processes deters potential investors and obstructs the establishment of a unified, global framework for evaluating and trading carbon credits produced by algae-based sequestration systems.

10.6 FUTURE PERSPECTIVES AND RESEARCH DIRECTIONS

10.6.1 Innovations in Algal Technology for Enhanced Carbon Sequestration

The domain of algal carbon sequestration has experienced significant technological progress that has the potential to transform carbon capture methodologies, tackling essential environmental issues using cutting-edge biotechnological techniques. Genetic engineering and synthetic biology have become crucial technology for improving algal carbon sequestration, as researchers create innovative genetic modification methods to enhance photosynthetic efficiency and carbon absorption processes. Martinez *et al.* (2020) revealed that targeted genetic alterations can enhance carbon fixation rates by as much as 35%, so substantially augmenting the overall carbon sequestration capacity of microalgal species. Innovative PBR designs have revolutionised cultivation methodologies, as demonstrated by Xu *et al.* (2019), who presented advanced closed-loop systems that optimise carbon capture efficiency and reduce resource utilisation. These modern technologies integrate complex environmental control systems, comprising precise nutrition supply, optimised light spectrum regulation, and sophisticated CO_2 injection mechanisms that establish optimal conditions for improved carbon sequestration. Innovative nanotechnology methods have augmented the capabilities of algal carbon capture, as demonstrated by Zhang and Wang (2021), who created nano-enhanced growth substrates that raise CO_2 absorption rates and cellular metabolic efficiency. These technological innovations exemplify a multidisciplinary strategy for carbon sequestration, amalgamating biotechnology, materials science, and environmental engineering to develop more efficient and scalable carbon capture solutions that confront the urgent challenges of climate change mitigation.

10.6.2 Synergies with Circular Economy Models

The use of algae-based carbon sequestration technologies inside circular economy frameworks signifies a revolutionary method for sustainable resource management, establishing interconnected systems that optimise both environmental and economic value. Chen *et al.* (2022) emphasised the capacity of algae growing systems to fulfil many economic and environmental roles concurrently, such as carbon sequestration, wastewater remediation, biofuel generation, and nutrient reclamation. This comprehensive strategy illustrates how algae can serve as a versatile biorefinery platform, transforming waste streams into valuable resources while concurrently mitigating carbon emissions. Integration of wastewater treatment presents a promising synergy, by algae-based systems can concurrently eliminate nutrients from industrial and municipal wastewater while producing biomass for diverse industrial applications. The economic model of these integrated systems contests conventional linear production methods, establishing closed-loop processes that reduce waste and optimise resource use. Strategies of industrial symbiosis have proven highly effective, as evidenced by Wang *et al.* (2020), who documented successful implementations of co-locating algae growing with industrial facilities, leveraging waste CO_2 and nutrients to establish a mutually advantageous ecosystem. Circular economy strategies improve carbon sequestration efficiency and generate economic opportunities by converting waste streams into valuable biomass products, such as biofuels, animal feed, pharmaceutical compounds, and high-value nutraceuticals, thus forming a holistic sustainable development strategy that simultaneously tackles various environmental and economic issues.

10.6.3 Potential of Artificial Intelligence and Blockchain in Monitoring Algal Carbon Credits

The integration of artificial intelligence (AI) and blockchain technology offers unparalleled prospects to transform the monitoring, verification, and trading of algae-based carbon credits, tackling essential issues of transparency, accountability, and market efficiency. Li *et al.* (2023) illustrated the capacity of machine learning algorithms to enhance algal growth parameters in real time, hence increasing carbon sequestration efficiency via predictive modelling and adaptive management techniques. These AI solutions facilitate accurate environmental monitoring by employing sophisticated sensor networks and predictive analytics to oversee intricate biological systems with unparalleled precision. Blockchain technology serves as an essential framework for guaranteeing the authenticity and traceability of carbon credit generation, as noted by Thompson *et al* (2016), who emphasise its capacity to provide immutable and transparent records of carbon sequestration activities. The decentralised architecture of blockchain systems facilitates real-time validation of carbon capture parameters, hence mitigating the risk of fraudulent operations and furnishing investors with reliable, immediate data on carbon sequestration efficacy. Smart contract technologies augment the dependability of carbon credit trading by autonomously completing transactions according to predetermined performance criteria, so establishing a more efficient and transparent market mechanism. Kumar *et al.* (2021) illustrated that AI-enhanced blockchain solutions could diminish verification expenses by as much as 40% while augmenting the general dependability of carbon credit accounting. These technical advancements tackle essential issues in carbon markets, offering advanced instruments for the monitoring, reporting, and verification of carbon sequestration activities with unparalleled accuracy and transparency.

10.7 CONCLUSION

The incorporation of algae into carbon credit systems offers a promising and complex approach to mitigating climate change. Algae's exceptional photosynthetic efficiency and capacity to flourish in various conditions, such as wastewater and industrial contexts, establish them as effective carbon capture agents. Their swift proliferation facilitates substantial CO_2 sequestration, surpassing that of numerous terrestrial plants, and the resultant biomass can be transformed into valuable commodities such as biofuels, bioplastics, and fertilisers, thereby augmenting the economic feasibility of these systems. Moreover, algae production can enhance ecosystem services by treating wastewater and rehabilitating damaged lands, exemplifying a comprehensive strategy for environmental sustainability. The processes of carbon capture by algae are varied, with microalgae thriving in high-density, regulated systems, while macroalgae play a role in long-term carbon sequestration in marine ecosystems. Technological innovations in cultivation techniques, ranging from economical open ponds to sophisticated PBRs, are enhancing the scalability and efficiency of algae cultivation. Genetic engineering is augmenting the carbon absorption potential of algae, while the incorporation of algal systems with industrial emissions provides a sustainable method for emission control. Nonetheless, considerable obstacles persist in the expansion of algae-based carbon credit schemes. Substantial initial expenses, together with operating intricacies and the demand for land, water, and nutrients, present considerable challenges. Moreover, discrepancies in regulations and policies regarding carbon accounting and market processes hinder investment and widespread adoption. Confronting these issues is essential for actualising the complete potential of algae-derived carbon offsets. Future study should concentrate on advancements in algal technology, including genetic engineering and sophisticated bioreactor designs, to optimise carbon sequestration. Integrating algae cultivation with wastewater treatment and industrial waste streams is essential for developing resource-efficient systems through synergies with circular economy models. The utilisation of AI and blockchain technologies for the monitoring and verification of carbon credits will improve transparency and reliability in carbon markets. Algae-based carbon credits present a distinctive avenue for climate change mitigation, promoting economic development and environmental restoration; nonetheless, addressing existing hurdles through technological advancement and policy endorsement is essential for their extensive implementation.

DECLARATION

The authors have not used any generative AI tools during the writing of this manuscript.

REFERENCES

Alazaiza, M. Y. D., Albahnasawi, A., Eyvaz, M., Al Maskari, T., Nassani, D. E., Abu Amr, S. S., Abujazar, M. S. S., & Bashir, M. J. K. (2023). An overview of green bioprocessing of algae-derived biochar and biopolymers: Synthesis, preparation, and potential applications. *Energies*, 16(2), 791. https://doi.org/10.3390/en16020791

Alsarayreh, M., Almomani, F., Khraisheh, M., Nasser, M. S., & Soliman, Y. (2022). Biological-based produced water treatment using microalgae: Challenges and efficiency. *Sustainability*, 14(1), 499. https://doi.org/10.3390/su14010499

Arora, P., Chance, R. R., Hendrix, H., Realff, M. J., Thomas, V. M., & Yuan, Y. (2021). Greenhouse gas impact of algal bio-crude production for a range of CO_2 supply scenarios. *Applied Sciences*, 11(24), 11931. https://doi.org/10.3390/app112411931

Atnoorkar, S., Wiatrowski, M., Newes, E., Davis, R., & Peterson, S. (2024). Algae to HEFA: Economics and potential deployment in the United States. *Biofuels, Bioproducts and Biorefining*. https://doi.org/10.1002/bbb.2623

Barry, A. N., Starkenburg, S. R., & Sayre, R. T. (2015). Strategies for optimizing algal biology for enhanced biomass production. *Frontiers in Energy Research*, 3, 1. https://doi.org/10.3389/fenrg.2015.00001

Beal, C. M., Archibald, I., Huntley, M. E., Greene, C. H., & Johnson, Z. I. (2018). Integrating algae with bioenergy carbon capture and storage (ABECCS) increases sustainability. *Earth's Future*, 6(4), 524–542. https://doi.org/10.1002/2017ef000704

Beal, C. M., Davidson, F. T., Webber, M. E., & Quinn, J. C. (2016). Flare gas recovery for algal protein production. *Algal Research*, 19, 316–323. https://doi.org/10.1016/j.algal.2016.09.022

Brennan, L., & Owende, P. (2010). Biofuels from microalgae—A review of technologies for production, processing, and extractions of biofuels and co-products. *Renewable and Sustainable Energy Reviews*, 14(2), 557–577. https://doi.org/10.1016/j.rser.2009.10.003

Chen, L., Zhang, W., & Liu, H. (2022). Algae-based biorefinery models: Integration of carbon capture and resource recovery. *Renewable and Sustainable Energy Reviews*, 156, 112009.

Chisti, Y. (2013). Constraints to commercialization of algal fuels. *Journal of Biotechnology*, 167(3), 201–213. https://doi.org/10.1016/j.jbiotec.2013.07.004

Cruce, J. R., Beattie, A., Chen, P., Quiroz, D., Somers, M., Compton, S., DeRose, K., Beckstrom, B., & Quinn, J. C. (2021). Driving toward sustainable algal fuels: A harmonization of techno-economic and life cycle assessments. *Algal Research*, 54, 102169. https://doi.org/10.1016/j.algal.2020.102169

Gerbens-Leenes, W., Xu, L., De Masilla, A., & Mollers, C. (2014). Water footprint of biogas and bio-electricity from algae. *Water Resources and Industry*, 7, 40–52. https://doi.org/10.1016/j.wri.2014.09.002

Gilani, I. E., Hosseini, H., Al-Ghouti, M. A., Saadaoui, I., & Sayadi, S. (2024). Microalgal-based desalination brine remediation: Achievements, challenges, and future research trends. *Environmental Technology & Innovation*, 29, 103592. https://doi.org/10.1016/j.eti.2024.103592

Gopal, N., & Sudhakar, K. (2016). Ultraviolet modification of *Chlamydomonas reinhardtii* for carbon capture. *Research and Reports in Biology*, 7, 1–10. https://doi.org/10.2147/rrb.s98536

Greenhalgh, S., & Faeth, P. (2001). A potential integrated water quality strategy for the Mississippi River Basin and the Gulf of Mexico. *The Scientific World Journal*, 1, 927–936. https://doi.org/10.1100/tsw.2001.354

Hise, A. M., Characklis, G. W., Kern, J., Gerlach, R., Viamajala, S., Gardner, R. D., & Vadlamani, A. (2016). Evaluating the relative impacts of operational and financial factors on the competitiveness of an algal biofuel production facility. *Bioresource Technology*, 220, 329–338. https://doi.org/10.1016/j.biortech.2016.08.050

In-na, P., Byrne, F., Caldwell, G. S., & Lee, J. G. M. (2022). Techno-economic analysis of living biocomposites for carbon capture from breweries. *Algal Research*, 64, 102781. https://doi.org/10.1016/j.algal.2022.102781

Kumar, A., Singh, P., & Ravi, S. (2021). Blockchain and AI technologies in carbon credit markets: A comprehensive analysis. *Journal of Cleaner Production*, 312, 127725. https://doi.org/10.1016/j.jclepro.2021.127725

Lammers, P. J., Huesemann, M., Boeing, W., Anderson, D. B., Arnold, R. G., Bai, X., ..., & Olivares, J. A. (2017). Review of the cultivation program within the National Alliance for Advanced Biofuels and Bioproducts. *Algal Research*, 22, 246–266. https://doi.org/10.1016/j.algal.2016.11.021

Lee, O. K., Paik, S. M., Lee, K., Hong, S. H., & Kim, D. H. (2017). Advances and challenges of algal biomass in the microalgae production pipeline. *Process Biochemistry*, 56, 23–37. https://doi.org/10.1016/j.procbio.2016.10.016

Li, T., Hu, J., & Zhu, L. (2021). Self-flocculation as an efficient method to harvest microalgae: A mini-review. *Water*, 13(18), 2585. https://doi.org/10.3390/w13182585

Liang, C., Gracida-Alvarez, U. R., Hawkins, T. R., & Dunn, J. B. (2023). Life-cycle assessment of biochemicals with clear near-term market potential. *ACS Sustainable Chemistry & Engineering*, 11(5), 1569–1581. https://doi.org/10.1021/acssuschemeng.2c05764

Luong, N. N., Aditya, L., Vu, H. P., Johir, A. H., Bennar, L., Ralph, P., Hoang, N. B., Zdarta, J., & Nghiem, L. D. (2022). Nutrient removal by algae-based wastewater treatment. *Current Pollution Reports*, 8(3), 213–224. https://doi.org/10.1007/s40726-022-00230-x

Martinez, R., Garcia, J., & Rodriguez, M. (2020). Genetic engineering approaches for enhanced carbon fixation in microalgae. *Algal Research*, 47, 101876. https://doi.org/10.1016/j.algal.2020.101876

Moore, D., Heilweck, M., Burton Fears, W., Petros, P., Squires, S. J., Tamburini, E., & Waldron, R. P. (2022). An assessment of the potential value for climate remediation of ocean calcifiers in sequestration of atmospheric carbon. ScienceOpen Preprints. https://doi.org/10.14293/s2199-1006.1.sor-.ppgszey.v1

Nath, A., Tiwari, P. K., Rai, A. K., & Sundaram, S. (2019). Evaluation of carbon capture in competent microalgal consortium for enhanced biomass, lipid, and carbohydrate production. *3 Biotech*, 9, 381. https://doi.org/10.1007/s13205-019-1910-6

Neeti, K., Gaurav, K., & Singh, R. (2023). The potential of algae biofuel as a renewable and sustainable bioresource. *Engineering Proceedings*, 26(1), 14716. https://doi.org/10.3390/ecp2023-14716

Onigbajumo, A., Taghipour, A., Ramirez, J., Will, G., Ong, T. C., Couperthwaite, S., Steinberg, T., & Rainey, T. (2021). Techno-economic assessment of solar thermal and alternative energy integration in supercritical water gasification of microalgae. *Energy Conversion and Management*, 228, 113807. https://doi.org/10.1016/j.enconman.2020.113807

Pereira, S. A., Kimpara, J. M., & Valenti, W. C. (2021). Sustainability of the seaweed *Hypnea pseudomusciformis* farming in the tropical Southwestern Atlantic. *Ecological Indicators*, 122, 107101. https://doi.org/10.1016/j.ecolind.2020.107101

Rodriguez, J., Reynoso, C. E., Prieto, M. A., & Arnaiz, E. (2018). Nutrient recovery from wastewater using microalgae. *Algal Research*, 35, 223–238. https://doi.org/10.1016/j.algal.2018.08.020

Somers, M. D., Chen, P., Clippinger, J., Cruce, J. R., Davis, R., Lammers, P. J., & Quinn, J. C. (2021). Techno-economic and life-cycle assessment of fuel production from mixotrophic *Galdieria sulphuraria* microalgae on hydrolysate. *Algal Research*, 60, 102419. https://doi.org/10.1016/j.algal.2021.102419

Sovacool, B. K., Baum, C. M., Low, S., & Fritz, L. (2024). The sociotechnical dynamics of blue carbon management: Testing typologies of ideographs, innovation, and co-impacts for marine carbon removal. *Environmental Science & Policy*, 155, 103730. https://doi.org/10.1016/j.envsci.2024.103730

Thompson, P. B., Millstone, E., & Raffensperger, C. (2016). Governing the transition to sustainable agriculture. *Agriculture and Agricultural Science Procedia*, 8, 246–254. https://doi.org/10.1016/j.aaspro.2016.02.099

Viswanaathan, S., Perumal, P. K., & Sundaram, S. (2022). Integrated approach for carbon sequestration and wastewater treatment using algal–bacterial consortia: Opportunities and challenges. *Sustainability*, 14(3), 1075. https://doi.org/10.3390/su14031075

Wang, H., Liu, X., & Chen, P. (2020). Industrial symbiosis and algal biotechnology: Integrated carbon capture strategies. *Bioresource Technology*, 298, 122521. https://doi.org/10.1016/j.biortech.2019.122521

Xu, L., Chen, Z., & Li, D. (2019). Advanced photobioreactor design for optimized carbon sequestration. *Biotechnology Advances*, 37(5), 701–718. https://doi.org/10.1016/j.biotechadv.2019.02.002

Zhang, M., & Wang, R. (2021). Nanotechnology applications in microalgal carbon capture systems. *Nano Energy*, 81, 105638. https://doi.org/10.1016/j.nanoen.2020.105638

11 Algal Bioplastics for Climate-Resilient Packing

Karan Kapoor, Saumya Bharti, Sudarshan Sahu, and Shailendra Kumar Arya

11.1 INTRODUCTION

11.1.1 Plastic Pollution and Its Significant Detrimental Impacts

Rampant increase in plastic production and poor waste management practices make plastic pollution a very serious challenge for the environment (Elkaliny et al., 2024). Over time, production rapidly escalated, with more than 368 million metric tons of plastic being produced annually by 2019, mainly of single-use plastics that can last in the environment for centuries. Even when recycled, we only manage a small portion of plastic waste – the remainder finds its way to landfills, incinerators, or natural ecosystems (Elkaliny et al., 2024). Many of these regions are developing countries, and much of the mismanagement results in extensive plastic leakage into marine and terrestrial environments. Marine littering, primarily of plastic, is widespread, and it has been estimated that about 23 million metric tons of plastic enter the oceans annually (Bergmann et al., 2022), accumulating in large garbage patches and causing ingestion and entanglement in marine life. Land-based pollution also presents a serious risk, as plastics make their way into soils and freshwater systems, which in turn affects wildlife and plant health (Kumar et al., 2021). Microplastics, which have been detected in locations ranging from the deep sea to the Arctic (Bergmann et al., 2022), have been associated with potential adverse health effects in humans: they can be ingested through drinking water, food, and the air. Plastic pollution takes a heavy toll on the economy in a number of ways, including clean-up operations, slowdowns in tourism and the fishery industry, and the loss of biodiversity (IUCN, 2023). A comprehensive strategy to combat the crisis needs to include the following: decreasing plastic materials' production and use, enhancing waste management infrastructure, increasing public consciousness of the problems associated with plastic pollution, and strengthening policies and international agreements (Kibria et al., 2023). By transitioning to sustainable practices and fostering global cooperation, we can mitigate the environmental and health impacts of plastic pollution and work toward a more sustainable future.

11.1.2 History of Bioplastic

Bioplastics have a long history that has seen several important breakthroughs and innovations over the years. The idea that plastics may be made from renewable resources dates back to the early 1900s. To create bio-based polymers, scientists started experimenting in the 1920s with substances like cellulose and starch (Rafiqah et al., 2021). In the 1970s, the first bioplastics sold commercially appeared. One prominent example was polylactic acid (PLA), which was developed from corn starch and offered as a compostable and biodegradable substitute for conventional polymers (Teixeira et al., 2023). In the 1980s and 1990s, biopolymer technology saw substantial breakthroughs.

DOI: 10.1201/9781003685814-11

Scholars investigated diverse sustainable resources and devised methodologies to transform them into biodegradable polymers. Additionally, new biopolymers were introduced, including polyhydroxyalkanoates (PHAs) (Gahlawat et al., 2020). Bioplastics began to draw notice as more environmentally friendly substitutes for fossil plastics by the early 2000s. Numerous industries, such as packaging, automotive, agricultural, and consumer products, find use for them. Businesses started manufacturing bioplastic goods on a bigger scale (Siracusa & Blanco, 2020). Bioplastics have seen an increase in innovation and interest in recent years. To enhance the functionality, environmental effect, and biodegradability of bioplastics, scientists and businesses are investigating novel materials and methods. It is being developed to create bioplastics from waste materials, bacteria, and algae (Siracusa & Blanco, 2020). With continuous research and development aimed at enhancing their qualities, cutting costs, and broadening their uses, the future of bioplastics is bright. The objective is to develop more environmentally friendly and sustainable substitutes for traditional polymers.

11.1.3 Bioplastics and Their Importance

In this era of growing environmental hazards, environmentally friendly alternatives to conventional plastics have garnered attention on a global scale. Bioplastics are materials derived from biological sources, contrasting with traditional plastics that are based on fossil fuels. These materials are often derived from renewable sources such as algal biomass and plant-based components like sugar, starch, or oil. The production of bioplastics may have a lower carbon footprint compared to fossil fuel-based plastics, making them environmentally appealing. There is no net increase in carbon dioxide gas emission upon biodegradation of bioplastics (Nanda et al., 2022). Biodegradable bioplastics have the ability to decompose into natural compounds such as water, carbon dioxide, and biomass when exposed to certain conditions (Coppola et al., 2023).

11.2 CLASSIFICATION OF BIOPLASTICS

Based on their content, source, and degree of biodegradability, bioplastics can be categorized. Renewable resources like plants, crops, or algal biomasses are the source of bio-based or renewable resource-based bioplastics. Bio-based polymers derived from corn, potatoes, sugarcane, or wood pulp cellulose are used to make them (Shah et al., 2021; Pathak et al., 2014). Biodegradable or compostable bioplastics are another type. These polymers are made to disintegrate into smaller parts when exposed to certain elements, such as heat, moisture, or microbes, or when they undergo natural processes. A particular kind of biodegradable plastic known as compostable bioplastics decomposes into organic matter in a composting setting without producing any hazardous byproducts (Pathak et al., 2014; Dróżdż et al., 2022). It's critical to understand that not all plastics, whether fossil or biodegradable, are equally biodegradable. There is a difference between the terms "bioplastic" and "biodegradable." For sustainable plastic consumption and waste management, it is essential to comprehend the unique properties and EOL (End of Life) alternatives for each form of bioplastic (Rahman & Bhoi, 2021).

11.3 SOURCES OF BIOPLASTIC

11.3.1 Plants

Plant-based natural biopolymers have significant promise as feedstocks for the manufacturing of bioplastics (Coppola et al., 2023). These consist of sawdust, sugarcane bagasse, cassava, wheat starch, corn starch, and vegetable fats and oils, among others. To create bioplastics, these biopolymers' lipids or sugars are fermented or chemically altered. Moreover, cellulose and gluten, two naturally occurring polysaccharides, can be converted into bioplastics (Coppola et al., 2023). It has been demonstrated that adding starchy elements to sugarcane bagasse improves the bioplastics' tensile and mechanical qualities (Hamin et al., 2023). Fibers from coconut husks can be used to support bioplastics made from cassava starch (Babalola & Olorunnisola, 2019). It has been demonstrated that the modification of cassava starch with glycerol, vinegar, and water leads to the creation of bioplastic sheets (Mojibayo et al., 2020). It has been documented that plant cellulose can be used as a bio-based feedstock to create bioplastics (Sharif Hossain et al., 2018; Aguilar et al., 2019; Azmin et al., 2020). There have also been published studies on the production of lipid-based bioplastics using castor, olive, linseed, palm, and soybean oils. Bioplastics derived from plant oil exhibit thermostability and superior elongation and tensile strength (Liu & Zhu, 2014).

11.3.2 Animals

An affordable and sustainable source for the production of biodegradable plastics is animal byproducts from tannery industries, such as hides, skins, and tallow (Maroušek et al., 2015). Additionally, because of their advantageous functional characteristics, animal proteins like collagen, gelatine, and keratin are appropriate for the creation of biodegradable bioplastics (Khodaei et al., 2021). It has been shown that myofibrillar proteins derived from fish co-products may create plastic films with remarkable mechanical qualities, low water vapor permeability, and high transparency, which makes them appropriate for food packaging (Vasconcelos da Silva Pereira et al., 2019). A significant component of the exoskeleton of arthropods, such as shrimp, crabs, and crustaceans, is chitin. Chitosan is created when the acetyl group from chitin is removed. It has been demonstrated that chitosan-based biodegradable plastic films have strong mechanical, transparent, and antibacterial qualities (Khodaei et al., 2021).

11.3.3 Microorganisms

Microorganisms are essential to the future of sustainable plastic production, offering a pathway to reduce dependence on fossil fuels and mitigate environmental pollution. A variety of microorganisms (including bacteria, fungi, yeasts, and microalgae) produce and store bioplastics as carbon and energy sources. The extent of production of the polymers is determined by the microbial and substrate type, the physiology of the organisms, as well as the availability of appropriate nutritional and physicochemical parameters (Albuquerque et al., 2011). Bacteria such as *Citrobacter*, *Pseudomonas*, *Enterobacter*, *Klebsiella*, *Rhizobium*, *Azotobacter*, and *Alcaligenes* have been reported as being viable for bioplastic production (Moreno-Bayona et al., 2019; Simó-Cabrera et al., 2021; de Melo et al., 2023). In addition, some bacteria and yeasts (such as *Saccharomyces cerevisiae*, *Candida krusei*, *Rhodotorula glutinis*, *Ralstonia eutropha*, and *Kluyveromyces africans*) that utilize polyphosphate complexes in their membrane transport produce low-molecular weight polyhydroxybutyrate (PHB) (Portugal-Nunes et al., 2017; Soto et al., 2019). Microalgae such as *Spirulina* sp., *Enteromorpha crinita*, *Chlorella* sp., *Laminaria japonica*, and *Ulva armoricana* have been reported as suitable for bioplastic synthesis (Rahman & Miller, 2017).

11.4 MICROALGAE: A POTENTIAL SOURCE FOR PRODUCING BIOPLASTICS

Although there are several natural resources available for producing bioplastics, many of them contribute to the global food crisis, posing a significant issue. The food industry may face challenges when bioplastics are made from these resources. In contrast, microalgae have emerged as a promising alternative for bioplastic production. Microalgae are particularly well-suited for this purpose due to their rapid growth, minimal nutrient requirements, and the ease with which they can be cultivated. Moreover, bioplastics derived from microalgae offer superior mechanical and tensile strength compared to those from other sources. The production process involves the intracellular synthesis of biopolymers like PHAs and PHB within the microalgae (Rakhmi Sefrienda et al.,

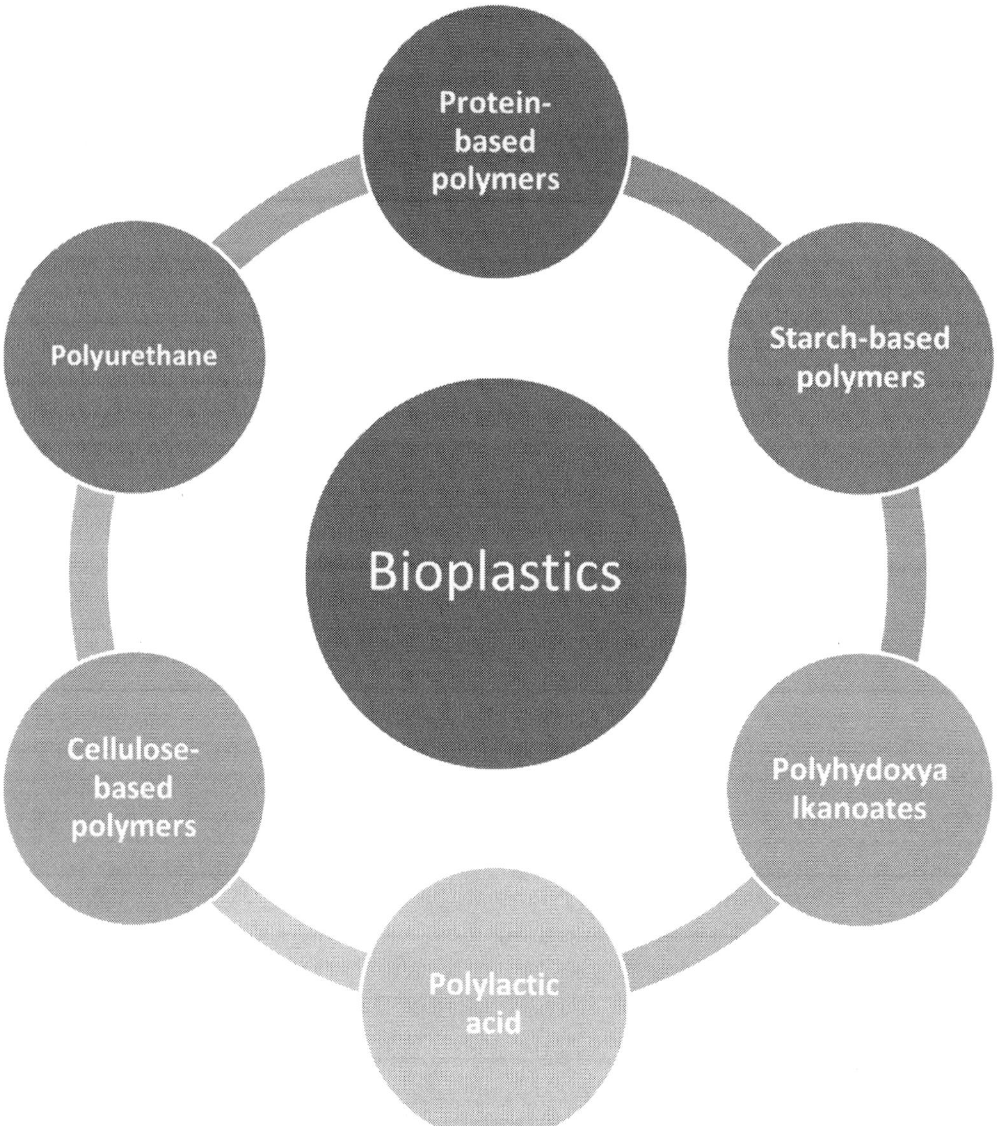

FIGURE 11.1 Schematic diagram depicting different microalgae-derived polymers. (Adetunji & Erasmus, 2024.)

2024). Microalgae represent one of the most viable options for bioplastic production due to their fast growth and efficient synthesis, along with their favorable material properties (Figure 11.1).

11.5 SIGNIFICANCE OF ALGAL BIOPLASTICS

Algal bioplastics present a unique opportunity to replace currently used plastics because they promise to solve some of the most critical environmental and economic problems. Derived from rapidly renewable algae, these bioplastics decrease the demand on finite fossil fuels while dramatically decreasing the negative impacts of GHGs (greenhouse gases) due to the carbon sequestration ability of algae during its growth phase. Reduced biodegradability of algal bioplastics is an advantage in tackling one of the

TABLE 11.1
Comparison between Bioplastics Derived from Algae and Conventional Plastics

S. No.	Properties	Algal Bioplastic	Conventional Plastics
1	**Molecular weight**	20–95 kDa	10–300 kDa
2	**Thermal analysis**	10–30°C	80–100°C
3	**Degree of crystallinity**	92–95%	10–80%
4	**Degradation duration**	65–90 days	Half-life of 5–250 years in landfills, compost, soil conditions
5	**Transparency**	Transparent	Transparent
6	**Fermentation**	1–4 weeks	Not applicable
7	**Water absorption**	Low	Low
8	**Cost consideration**	3.8 US $/Kg	1.6 US $/Kg
9	**Toxicity and safety**	Generally safe	Potential hazards
10	**Renewability of resources**	Renewable	Non-renewable
11	**Recyclability**	Less chance of recyclability	Recyclable
12	**Density**	1.397 g/cm^3	0.96 g/cm^3

biggest woes of the current world, the plastic pollution problem; more so since most of the conventional plastics take centuries to break down in ecosystems. From the economical aspect, it is apparent that algal bioplastics offer a cheap solution to bioplastic production and opportunity to grow in seas or ocean, which requires salty water and non-arable land so there will be no competition with the food industry. From a technological perspective, it can be engineered to possess mechanical properties equivalent to conventional plastic materials and in some cases even better than them besides being genetically and metabolically engineered. This makes them appropriate for various uses, such as packaging material for medical applications or uses. One of the major benefits of algal bioplastics is a relatively low level of incorporation of toxic additives compared to the use of traditional plastics. One of the global expectations is their incorporation of circular economy where they employ waste resources such as wastewater and CO_2 in production as indicators of resource efficiency and sustainability. Overall, algal bioplastics represent a critical innovation in developing climate-resilient, sustainable materials that support environmental health, economic viability, and societal well-being. Table 11.1 provides a comprehensive comparison between bioplastics produced from algae and conventional plastics (A & G, 2024).

11.6 ALGAE AS A RAW MATERIAL FOR BIOPLASTICS

Algae, a diverse group of photosynthetic organisms, are categorized into two primary groups based on their size and complexity: microalgae and macroalgae. Both types play crucial roles in the production of bioplastics (Dang et al., 2022). They produce a high concentration of metabolites, which can be converted into a variety of value-added products, opening up a wide range of potential across industries. Bioplastics are a category of products derived from algal metabolites, with notable examples including PLA and PHAs. While these metabolites are naturally produced by algal cells, their production can be significantly enhanced through the addition of specific chemicals and modifications in culture conditions (Sreenikethanam & Bajhaiya, 2022).

11.6.1 Microalgae vs. Macroalgae

Microalgae are microscopic, single-celled organism that thrives in various aquatic environments including freshwater and marine systems (Hannon et al., 2010). They possess high growth rates and can be cultivated in controlled environment, such as photobioreactor (PBR). Microalgae are rich in lipids and carbohydrates, which are essential for producing biopolymers like PHAs and PLAs. Their fast growth rate allows for efficient biomass production, reducing the time required to generate raw materials for bioplastics.

Seaweeds (macroalgae) are a type of photosynthetic organism that can be found in freshwater and marine habitats. They are divided into three classes based on color: brown (Phaeophyta), red (Rhodophyta), and green (Chlorophyta). Numerous sectors, including pharmaceuticals, cosmetics, fertilizers, biofuels, wastewater treatment, paper making, aquaculture, and plastics, have found uses for seaweeds (Ashour et al., 2022; Barakat et al., 2022; Huo et al., 2018). Compared to other feedstocks, macroalgae have unique advantages for the synthesis of bioplastics (Ji & Gao, 2021). Several aspects emphasize their usefulness as a sustainable and renewable supply of bioplastic. For example, macroalgae have an amazing growth rate and produce a substantial quantity of biomass per unit area compared to land-based plants, making bioplastic manufacture more cost-effective (Kostas et al., 2021).

11.6.2 Common Species Utilized in Bioplastic Production

11.6.2.1 *Spirulina* sp.

Bioplastics can be efficiently produced from *Spirulina* species without the use of binders or solvents. Compared to petroleum-based plastics like polystyrene, the bioplastics derived from *Spirulina* exhibit superior tensile properties. These polymers offer excellent tensile strength, elongation, and flexibility (Iyer et al., 2023). Under photoautotrophic conditions, *Spirulina* can synthesize biopolymers such as PHAs and PHB. Additionally, *Spirulina* can serve as a filler or reinforcing fiber in bioplastic composites, enhancing the mechanical properties of the resulting materials (Cheah et al., 2023).

11.6.2.2 *Chlorella* sp.

Chlorella species are a potential source for bioplastic production. Bioplastics derived from *Chlorella* typically exhibit lower melting points; however, this can be improved by incorporating a compatibilizer into the formulation. Additionally, the tensile strength and mechanical properties of *Chlorella*-PVA bioplastics can be enhanced through the pretreatment process using an ultrasonic homogenizer. *Chlorella* is also utilized as a filler to improve the mechanical properties of Polyvinyl Chloride (PVC) (Cheah et al., 2023). Compared to cellulose-based polymers, *Chlorella* biomass is more easily biodegradable. Furthermore, fungal mycelium can be added as a sorbent in the bioplastic manufacturing process. Utilizing *Chlorella* species presents a cost-effective and sustainable approach (Senko et al., 2023).

11.6.3 Factors Influencing Algal Production

Algal production is influenced by numerous factors that play a critical role in determining the growth rate, biomass yield, and overall cultivation efficiency. Table 11.2 shows the factor that influences algal production.

11.6.4 Cultivation Methods of Algae

Microalgae are highly potential for industrial raw materials because of their rapid growth and great photosynthetic efficiency, which enable them to store large amounts of bioproducts in their cells (Randrianarison & Ashraf, 2017). Furthermore, cultivation of microalgae does not require fertile soil, a huge quantity of freshwater, herbicides, and pesticides as compared to other crops, and hence will not compete for resources (Khan et al., 2018). Additionally, microalgae reduce atmospheric CO_2 through photosynthesis, helping to combat climate change. However, challenges

TABLE 11.2
Factors Influencing Algal Production

Factor	Role/Importance	Effects if Not Managed	Management	References
Light	Essential for photosynthesis. Intensity, duration, and wavelength directly affect productivity.	Poor photosynthesis, hindered growth.	Ensure sufficient sunlight or artificial lighting. Regulate light intensity, duration, and wavelength.	Huesemann et al. (2013), Jacob-Lopes et al. (2009), Mata et al. (2010), Ye et al. (2012)
Nutrient availability	Key nutrients such as nitrogen, phosphorus, and CO_2 are required for proper growth.	Growth limitations or unwanted contamination.	Monitor and balance nutrient levels carefully.	Juneja et al. (2013)
Temperature	Algae have specific temperature ranges for optimal growth.	Inhibited growth, reduced biomass yield, potential cellular damage.	Maintain optimal temperature range suited to the algal species (optimal temp – 20–30°C).	Béchet et al. (2017), Singh and Singh (2015)
pH Levels	Affects cellular functions and enzyme activity.	Growth inhibition, disrupted metabolism.	Keep pH within the ideal range to support cell metabolism.	Juneja et al. (2013), Lam and Lee (2012)
Mixing and aeration	Ensures even distribution of light and nutrients.	Cells settle at the bottom, reduced access to light and nutrients, clumping.	Ensure proper aeration to maintain uniform distribution of resources.	Show et al. (2017), Zeng et al. (2011)
Carbon dioxide (CO_2)	Plays a crucial role in photosynthesis. Higher levels can boost growth and biomass production.	Excessive CO_2 buildup can negatively affect culture.	Regulate CO_2 concentration to maximize photosynthesis without harmful effects.	Cheirsilp and Torpee (2012), Ye et al. (2012)
Overall goal	To maximize algal production for various applications such as bioplastics, biofuels, and other valuable products.	Inefficient algal production, lower yields.	Effective management of all factors is essential for successful cultivation.	Khan et al. (2018)

remain, such as low biomass production and high harvesting costs due to their small cell size (Tan et al., 2020).

Improving microalgae growth rates and optimizing harvesting technologies are essential to addressing these issues. While microalgae are easy to grow in controlled lab conditions, large-scale production remains difficult. An ideal cultivation system should provide sufficient light, efficient gas exchange, low contamination risk, simple operation, and cost-effectiveness (Jegathese & Farid, 2014). Algal cultivation is a critical step in the production of algal bioplastics, with different methods employed for microalgae and macroalgae due to their distinct biological characteristics and applications. Cultivation methods generally fall into two categories: open ponds and PBRs.

11.6.4.1 Open Pond

Open pond cultivation is one of the oldest and simplest methods for growing microalgae at a large scale, valued for its low construction, maintenance, and operational costs. Its benefits include scalability, low energy consumption, and ease of operation (Costa & de Morais, 2014). Open ponds can be human-made, like circular or raceway ponds, or they can be natural bodies of water like lakes. Sometimes, open ponds are used for cultivation, like tanks. However, challenges include low cell concentrations, the need for efficient harvesting, and environmental factors such as rainwater runoff, bank erosion, and contamination by protozoa or bacteria. Risks of contamination can be decreased by cultivating microalgae in extremely saline or alkaline environments. However, in open systems, managing growth parameters like light and temperature is still challenging (Stark & O'Gara, 2012).

The ancient Aztecs were among the first to harvest microalgae, particularly *Spirulina* from Lake Texcoco (Hamed, 2016). The first artificial ponds used for large-scale farming were circular ponds, yet they had drawbacks such as high energy consumption and expensive construction (Shen et al., 2009). Raceway ponds, which use closed-loop channels and paddlewheels to circulate biomass, are thought to be more energy-efficient and have been successfully implemented by firms such as Sapphire Energy in the United States, which generated 520 metric tons of dry biomass without any technical concerns (White & Ryan, 2015).

11.6.4.2 Photobioreactor

A PBR is an enclosed system for growing phototrophs such as microalgae in a controlled environment that prevents contamination and maximizes land use. It produces more biomass because it provides superior metabolic and nutritional efficiency as compared to open ponds. However, PBRs' high price and scaling issues prevent them from being used in large-scale production (Gupta et al., 2015). Tubular PBRs are made of transparent tubes that are intended to capture sunlight, but they have poor mass transmission. They are commercially utilized in Germany and Israel to cultivate *Haematococcus* and *Chlorella* (Mishra et al., 2019; Torzillo & Chini Zittelli, 2015). Although vertical column PBRs, which employ air spargers for gas exchange, have higher mass transfer efficiency, they are more expensive, more prone to light exposure, and more difficult to clean (Mohan et al., 2019; Huang et al., 2017). Experimental reactors have been developed, such as a 40-liter *Chlorella* reactor in China (Huo et al., 2018). The maximum photosynthetic efficiency is provided by flat-plate PBRs, which have a rectangular form but can stress microalgae cells. Their performance has been enhanced by innovations like twin-layer designs. This technique is used by Algamo in the Czech Republic to produce astaxanthin for commercial purposes (Vo et al., 2019).

11.7 PRODUCTION OF BIOPLASTICS BY MICROALGAE

The production of bioplastics from microalgae is a complex yet intriguing process that leverages the potential of these microorganisms to create eco-friendly materials. Bioplastics can be produced using microalgal biomass either directly or indirectly through different methods. In the direct method, the entire microalgal biomass is used as a base material for bioplastic synthesis. This process is eco-friendly, cost-effective, and recyclable. It also reduces carbon emissions and downstream processing expenses, while producing clear and transparent films (Jumaidin et al., 2018). In the indirect method, the whole or spent microalgal biomass is fermented to generate precursors for bioplastic production. This method eliminates the need for biomass pretreatment, though it is more time-consuming and yields lower output (Sudhakar et al., 2024). Additionally, microalgal biomass can be blended with other materials like starch, cellulose, petrochemical plastics, or bioplastics to extend the lifespan and improve the physical and mechanical properties of the resulting bioplastic (Chia et al., 2020). Figure 11.2 shows the steps involving in bioplastic production by microalgae (A & G, 2024).

Incorporating plasticizers (e.g., glycerol) and compatibilizers (such as maleic anhydride, diethyl succinate) into biopolymer mixtures significantly improves the flexibility, processability, tensile strength, thermal stability, miscibility, and elongation of the resulting bioplastics (Devadas et al., 2021). The success of bioplastic formation depends on factors such as the number of hydroxyl (OH) groups in the plasticizer, its compatibility with the biopolymer, as well as its concentration and type

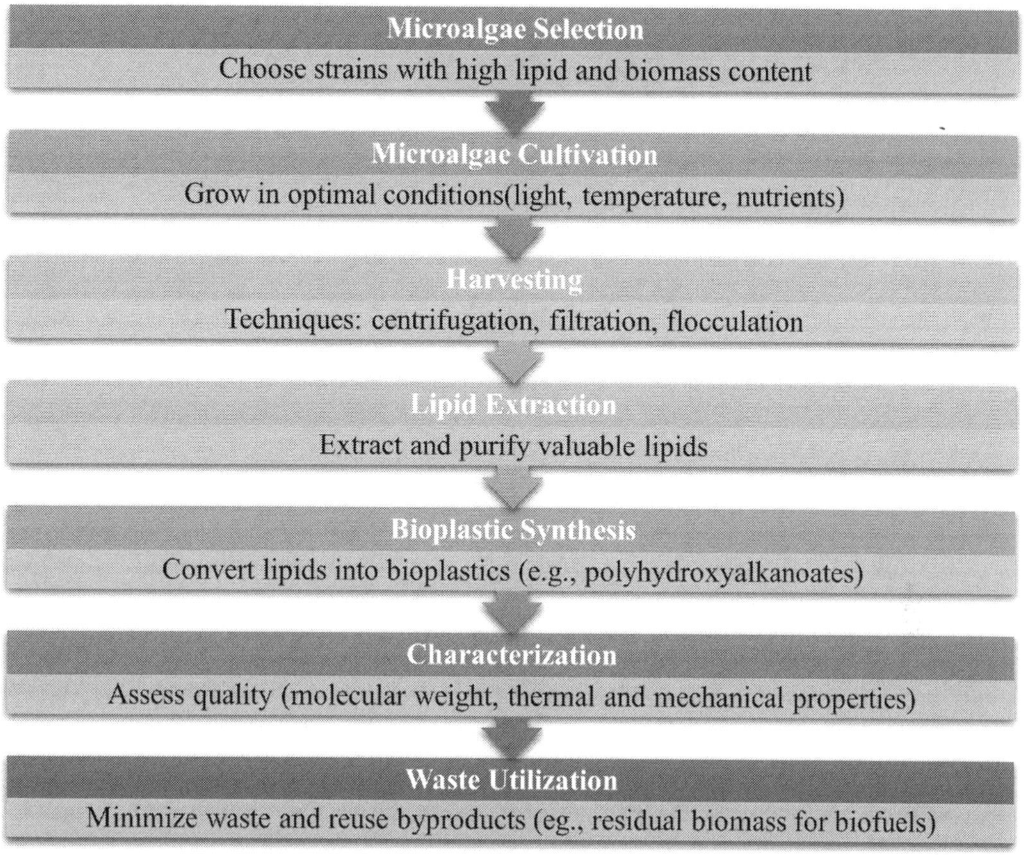

FIGURE 11.2 Production process of bioplastic by microalgae.

(Nanda & Bharadvaja, 2022). Compatibilizers also enhance the interfacial bonding between synthetic polymers and algae-based biopolymers (Dianursanti & Khalis, 2018). Additionally, surface modifiers like gluten can be used to improve phase morphology, performance, miscibility, and thermoplastic properties of algae-based bioplastics (Devadas et al., 2021; Zhu et al., 2017). Following these steps, the bioplastics are molded or extruded under heat and pressure to prepare them for their intended application. Microalgae-polymer blends can be produced using various techniques, including compression molding, solvent casting, twin-screw extrusion, or injection molding. Figure 11.3 shows the conversion of microalgal biomass to bioplastics by different techniques.

11.8 APPLICATIONS OF ALGAL BIOPLASTICS

Bioplastics are a sustainable alternative to conventional plastics due to their renewable nature, biodegradability, and minimal carbon footprint. These attributes make them a promising option for reducing environmental impact (Chia et al., 2020; Coppola et al., 2023). Figure 11.4 and Table 11.3 provide examples of various applications of algal bioplastics.

Algal bioplastics are finding diverse applications across various industries due to their sustainable and biodegradable nature. In the packaging industry, they are increasingly used for creating biodegradable containers, films, and wraps, helping to reduce reliance on conventional plastics and mitigate plastic pollution. The food industry benefits from algal bioplastics through the development of both edible and non-edible films are used to manage or prevent reactions occurring inside

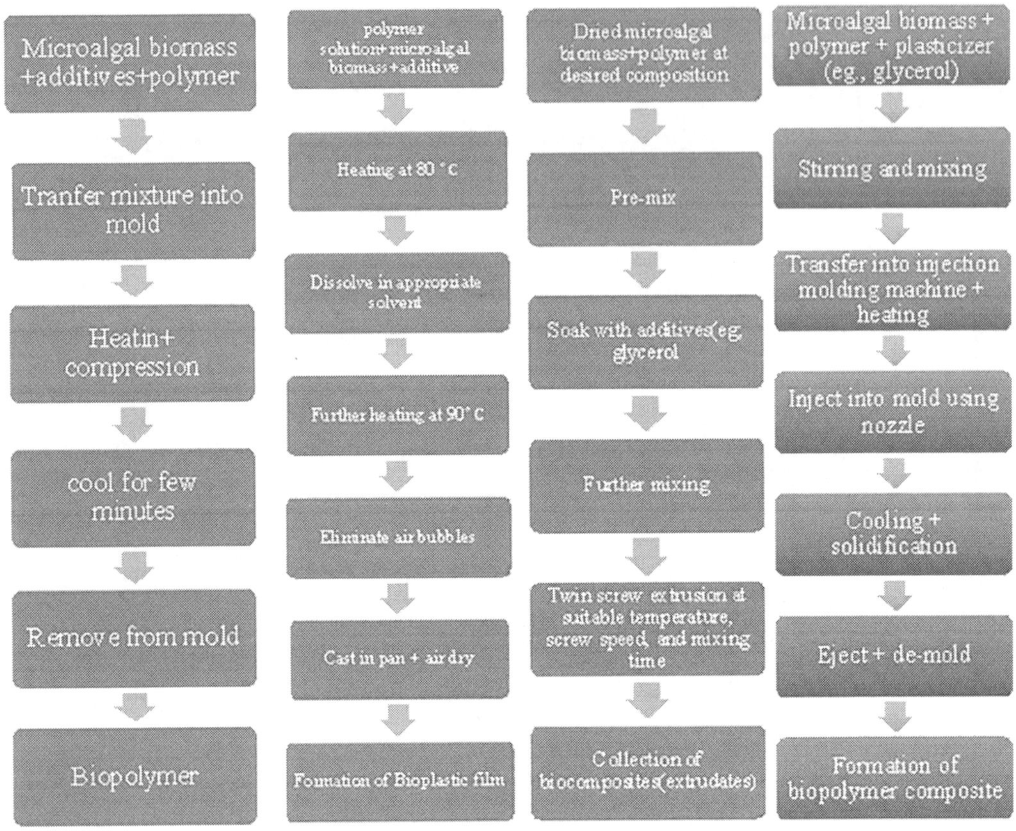

FIGURE 11.3 Schematic diagram showing conversion of microalgal biomass to bioplastics by different techniques: compression molding (Ciapponi et al., 2019), solvent casting (Abdo & Ali, 2019), twin-screw extrusion (Mathiot et al., 2019), and injection molding technique (Torres et al., 2015).

the package. These films serve as a thin barrier, limiting the movement of moisture, oxygen, and solutes, thereby helping to reduce food spoilage. Edible films offer an environmental advantage as they can be eaten along with the packaged food, while non-edible films can break down naturally, making them biodegradable. Algae-based bioplastic films have attracted attention in food packaging due to their flexibility, microwave resistance, and strong antimicrobial and antioxidant properties. A key goal of these films is to extend the shelf life of food products (Yap et al., 2023). In agriculture, these bioplastics are used to produce biodegradable mulch films and seed coatings, enhancing crop protection and soil health without contributing to long-term plastic waste. The medical and pharmaceutical sectors utilize algal bioplastics for making biocompatible materials, such as drug delivery systems, wound dressings, and medical implants, capitalizing on their non-toxic and bioactive properties. Additionally, the textile industry is exploring the use of algal bioplastics to create sustainable fibers and fabrics, offering an eco-friendly alternative to synthetic materials.

11.9 ADVANTAGES AND LIMITATIONS

Bioplastics offer several significant advantages over traditional petroleum-based plastics, making them an attractive alternative for sustainable development. One of the primary benefits is their **reduced environmental impact** (Nandakumar et al., 2021). Bioplastics are often derived from renewable resources, such as corn starch, sugarcane, or algae, which decreases dependence on

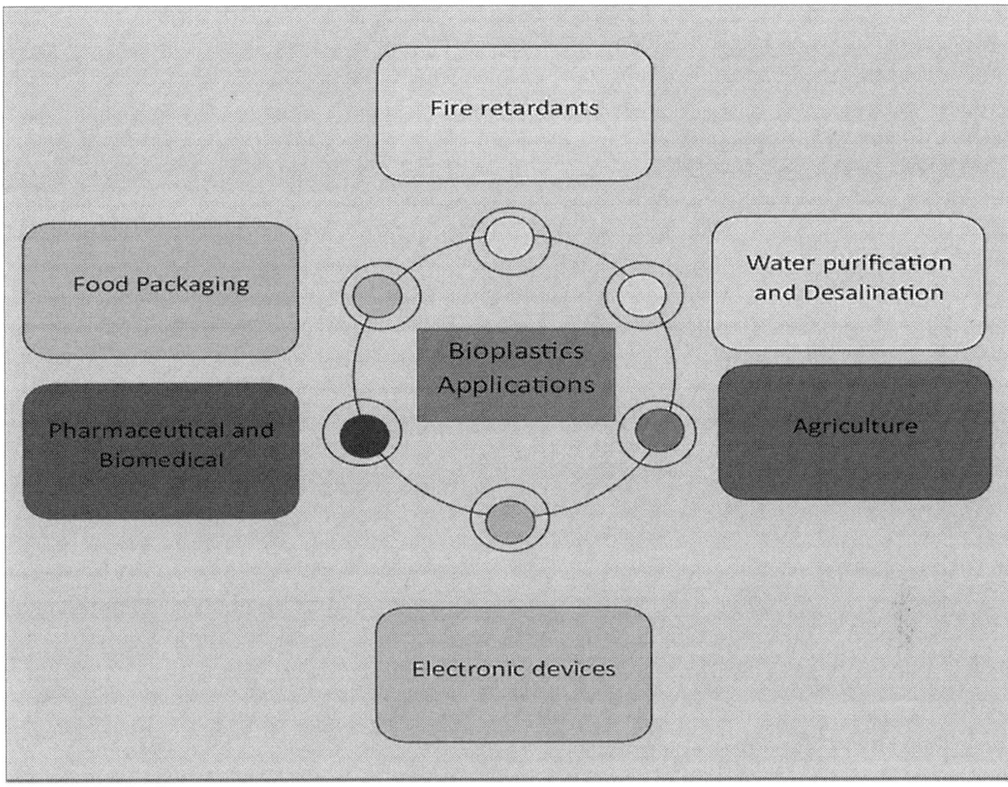

FIGURE 11.4 Different applications of algal bioplastic. (Elkaliny et al., 2024.)

TABLE 11.3
Application of Bioplastics in Different Industries

Industry	Application	References
Packaging industry	Biodegradable containers, films, and wraps.	Ibrahim et al. (2021), Jariyasakoolroj et al. (2020)
Food industry	Edible film, coatings for extending shelf life.	Zhao et al. (2023)
Agriculture industry	Biodegradable mulch film and seed coating.	Varghese et al. (2022), Iriani et al. (2019)
Medical and pharmaceutical industry	Drug delivery systems, wound dressings, implants.	Kai et al. (2019), Narancic et al. (2020)
Textile industry	Sustainable fibers.	Friedrich (2021)
Construction industry	Doors, frames, window, insulation.	Stabnikov et al. (2015), Verma et al. (2022)

finite fossil fuels and **lowers GHG emissions** during production (Islam et al., 2024). Additionally, many bioplastics are designed to be biodegradable or compostable, allowing them to break down more quickly and safely in natural environments, thereby **reducing plastic pollution and waste in landfills and oceans** (Nandakumar et al., 2021). This biodegradability also lessens the **long-term environmental impact** associated with traditional plastics, which can persist for hundreds of years (Narancic et al., 2020). Bioplastics also **contribute to carbon sequestration**; as plants grow, they absorb carbon dioxide from the atmosphere, effectively offsetting some emissions (Wang & Hao, 2008). Furthermore, the development and use of bioplastics can stimulate economic growth by fostering innovation in the green technology sector and creating new agricultural and manufacturing **jobs**. Overall, bioplastics represent a pivotal shift toward a more sustainable and eco-friendly future, with the potential to significantly mitigate the adverse effects of conventional plastics. Despite their advantages, bioplastics face several **challenges** that hinder their widespread adoption and impact. One of the main challenges is **cost competitiveness**; bioplastics are often more expensive to produce than traditional plastics due to the high costs associated with raw materials and processing technologies. Additionally, the **infrastructure** for recycling and composting bioplastics is not yet as developed or widespread as that for conventional plastics, leading to potential issues with waste management and the effectiveness of biodegradation in real-world conditions (Ghasemlou et al., 2024). There is also confusion among consumers and manufacturers about the differences between biodegradable, compostable, and recyclable bioplastics, which can lead to improper disposal and reduced environmental benefits. Lastly, the performance characteristics of some bioplastics, such as heat resistance and durability, may not yet match those of conventional plastics, limiting their application in certain industries. Addressing these challenges requires continued research and innovation to improve the cost-effectiveness, infrastructure, and performance of bioplastics while ensuring that their production and use are genuinely sustainable. Figure 11.5 shows the challenges and limitations of bioplastics.

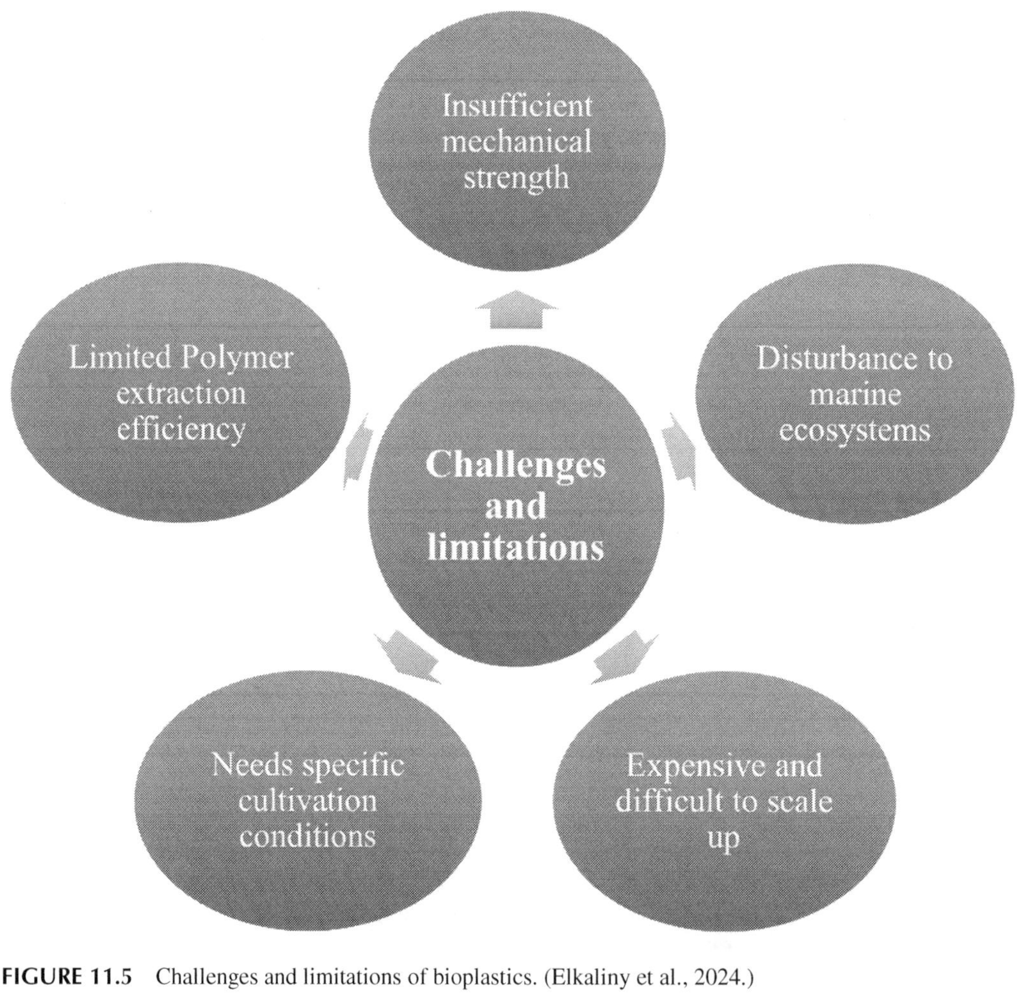

FIGURE 11.5 Challenges and limitations of bioplastics. (Elkaliny et al., 2024.)

11.10 CONCLUSION AND FUTURE PROSPECT

The widespread adoption of bioplastics represents a pivotal step toward achieving a more sustainable and environmentally conscious future. As we navigate the complexities of the modern world, the imperative to address plastic pollution and reduce our reliance on finite fossil fuel resources has never been more pressing. Bioplastics offer a compelling solution by presenting a viable alternative that combines innovation with environmental responsibility. In conclusion, embracing bioplastics entails recognizing and appreciating their potential to mitigate the environmental impact associated with traditional plastics. While challenges such as land use, water usage, and waste management need careful consideration, ongoing research and development in bioplastic technology are continually addressing these issues. As consumers, industries, and policymakers unite in their commitment to sustainable practices, embracing bioplastics emerges not merely as an option but as a collective responsibility. By supporting and investing in the development of bio-packaging, we contribute to a future where the lifecycle of materials aligns harmoniously with the delicate balance of our planet, paving the way for a more sustainable and resilient tomorrow.

DECLARATION

The authors have not used any generative AI tools during the writing of this manuscript.

REFERENCES

A, Anli Dino., & G, Kishore. (2024). Microalgae: An emerging source of bioplastics production. *Discover Environment*, *2*(1), 10. https://doi.org/10.1007/s44274-024-00038-0

Abdo, S. M., & Ali, G. H. (2019). Analysis of polyhydroxybutrate and bioplastic production from microalgae. *Bulletin of the National Research Centre*, *43*(1), 97. https://doi.org/10.1186/s42269-019-0135-5

Adetunji, A., & Erasmus, M. (2024). Green synthesis of bioplastics from microalgae: A state-of-the-art review. *Polymers*, *16*(10), 1322. https://doi.org/10.3390/polym16101322

Aguilar, N. M., Arteaga-Cardona, F., de Anda Reyes, M. E., Gervacio-Arciniega, J. J., & Salazar-Kuri, U. (2019). Magnetic bioplastics based on isolated cellulose from cotton and sugarcane bagasse. *Materials Chemistry and Physics*, *238*, 121921. https://doi.org/10.1016/j.matchemphys.2019.121921

Albuquerque, M. G. E., Martino, V., Pollet, E., Avérous, L., & Reis, M. A. M. (2011). Mixed culture polyhydroxyalkanoate (PHA) production from volatile fatty acid (VFA)-rich streams: Effect of substrate composition and feeding regime on PHA productivity, composition and properties. *Journal of Biotechnology*, *151*(1), 66–76. https://doi.org/10.1016/j.jbiotec.2010.10.070

Ashour, M., Al-Souti, A. S., Hassan, S. M., Ammar, G. A. G., Goda, A. M. A.-S., El-Shenody, R., Abomohra, A. E.-F., El-Haroun, E., & Elshobary, M. E. (2022). Commercial seaweed liquid extract as strawberry biostimulants and bioethanol production. *Life*, *13*(1), 85. https://doi.org/10.3390/life13010085

Azmin, Siti Nuurul Huda Mohammad., Hayat, Najah Aliah binti Mohd, N. A., & Nor, Mohd Shukri Mat. (2020). Development and characterization of food packaging bioplastic film from cocoa pod husk cellulose incorporated with sugarcane bagasse fibre. *Journal of Bioresources and Bioproducts*, *5*(4), 248–255. https://doi.org/10.1016/j.jobab.2020.10.003

Babalola, O. A., & Olorunnisola, A. O. (2019). Evaluation of coconut (*Cocos nucifera*) husk fibre as a potential reinforcing material for bioplastic production, 195–200. https://doi.org/10.21741/9781644900178-14

Barakat, K. M., Ismail, M. M., Abou El Hassayeb, H. E., El Sersy, N. A., & Elshobary, M. E. (2022). Chemical characterization and biological activities of ulvan extracted from *Ulva fasciata* (Chlorophyta). *Rendiconti Lincei. Scienze Fisiche e Naturali*, *33*(4), 829–841. https://doi.org/10.1007/s12210-022-01103-7

Béchet, Q., Laviale, M., Arsapin, N., Bonnefond, H., & Bernard, O. (2017). Modeling the impact of high temperatures on microalgal viability and photosynthetic activity. *Biotechnology for Biofuels*, *10*(1), 136. https://doi.org/10.1186/s13068-017-0823-z

Bergmann, M., Collard, F., Fabres, J., Gabrielsen, G. W., Provencher, J. F., Rochman, C. M., van Sebille, E., & Tekman, M. B. (2022). Plastic pollution in the Arctic. *Nature Reviews Earth & Environment*, *33*(5), 323–337. Springer Nature. https://doi.org/10.1038/s43017-022-00279-8

Cheah, W. Y., Er, A. C., Aiyub, K., Mohd Yasin, N. H., Ngan, S. L., Chew, K. W., Khoo, K. S., Ling, T. C., Juan, J. C., Ma, Z., & Show, P. L. (2023). Current status and perspectives of algae-based bioplastics: A reviewed potential for sustainability. *Algal Research*, *71*, 103078. https://doi.org/10.1016/j.algal.2023.103078

Cheirsilp, B., & Torpee, S. (2012). Enhanced growth and lipid production of microalgae under mixotrophic culture condition: Effect of light intensity, glucose concentration and fed-batch cultivation. *Bioresource Technology*, *110*, 510–516. https://doi.org/10.1016/j.biortech.2012.01.125

Chia, W. Y., Ying Tang, D. Y., Khoo, K. S., Kay Lup, A. N., & Chew, K. W. (2020). Nature's fight against plastic pollution: Algae for plastic biodegradation and bioplastics production. *Environmental Science and Ecotechnology*, *4*, 100065. https://doi.org/10.1016/j.ese.2020.100065

Ciapponi, R., Turri, S., & Levi, M. (2019). Mechanical reinforcement by microalgal biofiller in novel thermoplastic biocompounds from plasticized gluten. *Materials*, *12*(9), 1476. https://doi.org/10.3390/ma12091476

Coppola, G., Gaudio, M. T., Lopresto, C. G., Calabro, V., Curcio, S., & Chakraborty, S. (2023). Correction to: Bioplastic from Renewable Biomass: A Facile Solution for a Greener Environment. *Earth Systems and Environment*, *7*(2), 581–581. https://doi.org/10.1007/s41748-021-00280-z

Costa, J. A. V., & de Morais, M. G. (2014). An Open Pond System for Microalgal Cultivation. In *Biofuels from Algae* (pp. 1–22). Elsevier. https://doi.org/10.1016/B978-0-444-59558-4.00001-2

Dang, B.-T., Bui, X.-T., Tran, D. P. H., Hao Ngo, H., Nghiem, L. D., Hoang, T.-K.-D., Nguyen, P.-T., Nguyen, H. H., Vo, T.-K.-Q., Lin, C., Yi Andrew Lin, K., & Varjani, S. (2022). Current application of algae derivatives for bioplastic production: A review. *Bioresource Technology*, *347*, 126698. https://doi.org/10.1016/j.biortech.2022.126698

de Melo, R. N., de Souza Hassemer, G., Steffens, J., Junges, A., & Valduga, E. (2023). Recent updates to microbial production and recovery of polyhydroxyalkanoates. *3 Biotech*, *13*(6), 204. https://doi.org/10.1007/s13205-023-03633-9

Devadas, V. V., Khoo, K. S., Chia, W. Y., Chew, K. W., Munawaroh, H. S. H., Lam, M.-K., Lim, J.-W., Ho, Y.-C., Lee, K. T., & Show, P. L. (2021). Algae biopolymer towards sustainable circular economy. *Bioresource Technology*, *325*, 124702. https://doi.org/10.1016/j.biortech.2021.124702

Dianursanti, & Khalis, S.A (2018). The effect of compatibilizer addition on *Chlorella vulgaris* microalgae utilization as a mixture for bioplastic. *E3S Web of Conferences*, *67*, 03047. https://doi.org/10.1051/e3sconf/20186703047

Dróżdż, D., Malińska, K., Postawa, P., Stachowiak, T., & Nowak, D. (2022). End-of-life management of biodegradable plastic dog poop bags through composting of green waste. *Materials*, *15*(8), 2869. https://doi.org/10.3390/ma15082869

Elkaliny, N. E., Alzamel, N. M., Moussa, S. H., Elodamy, N. I., Madkor, E. A., Ibrahim, E. M., Elshobary, M. E., & Ismail, G. A. (2024). Macroalgae bioplastics: A sustainable shift to mitigate the ecological impact of petroleum-based plastics. *Polymers*, *16*(9). Multidisciplinary Digital Publishing Institute (MDPI). https://doi.org/10.3390/polym16091246

Friedrich, D. (2021). What makes bioplastics innovative for fashion retailers? An in-depth analysis according to the Triple Bottom Line Principle. *Journal of Cleaner Production*, *316*, 128257. https://doi.org/10.1016/j.jclepro.2021.128257

Gahlawat, G., Kumari, P., & Bhagat, N. R. (2020). Technological advances in the production of polyhydroxyalkanoate biopolymers. *Current Sustainable/Renewable Energy Reports*, *7*(3), 73–83. https://doi.org/10.1007/s40518-020-00154-4

Ghasemlou, M., Barrow, C. J., & Adhikari, B. (2024). The future of bioplastics in food packaging: An industrial perspective. *Food Packaging and Shelf Life*, *43*, 101279. https://doi.org/10.1016/j.fpsl.2024.101279

Gupta, P. L., Lee, S.-M., & Choi, H.-J. (2015). A mini review: Photobioreactors for large scale algal cultivation. *World Journal of Microbiology and Biotechnology*, *31*(9), 1409–1417. https://doi.org/10.1007/s11274-015-1892-4

Hamed, I. (2016). The evolution and versatility of microalgal biotechnology: A review. *Comprehensive Reviews in Food Science and Food Safety*, *15*(6), 1104–1123. https://doi.org/10.1111/1541-4337.12227

Hamin, S. H., Sayid Abdullah, S. H. Y., Lananan, F., Abdul Hamid, S. H., Kasan, N. A., Mohamed, N. N., & Endut, A. (2023). Effect of chemical treatment on the structural, thermal, and mechanical properties of sugarcane bagasse as filler for starch-based bioplastic. *Journal of Chemical Technology & Biotechnology*, *98*(3), 625–632. https://doi.org/10.1002/jctb.7218

Hannon, M., Gimpel, J., Tran, M., Rasala, B., & Mayfield, S. (2010). Biofuels from algae: Challenges and potential. *Biofuels*, *1*(5), 763–784. https://doi.org/10.4155/bfs.10.44

Huang, Q., Jiang, F., Wang, L., & Yang, C. (2017). Design of photobioreactors for mass cultivation of photosynthetic organisms. *Engineering*, *3*(3), 318–329. https://doi.org/10.1016/J.ENG.2017.03.020

Huesemann, M. H., Van Wagenen, J., Miller, T., Chavis, A., Hobbs, S., & Crowe, B. (2013). A screening model to predict microalgae biomass growth in photobioreactors and raceway ponds. *Biotechnology and Bioengineering*, *110*(6), 1583–1594. https://doi.org/10.1002/bit.24814

Huo, S., Wang, Z., Zhu, S., Shu, Q., Zhu, L., Qin, L., Zhou, W., Feng, P., Zhu, F., Yuan, Z., & Dong, R. (2018). Biomass accumulation of *Chlorella zofingiensis* G1 cultures grown outdoors in photobioreactors. *Frontiers in Energy Research*, *6*. https://doi.org/10.3389/fenrg.2018.00049

Ibrahim, N. I., Shahar, F. S., Sultan, M. T. H., Shah, A. U. M., Safri, S. N. A., & Mat Yazik, M. H. (2021). Overview of bioplastic introduction and its applications in product packaging. *Coatings*, *11*(11), 1423. https://doi.org/10.3390/coatings11111423

Iriani, E. S., Permana, A. W., Yuliani, S., Kailaku, S. I., & Sulaiman, A. A. (2019). The effect of agricultural waste nanocellulose on the properties of bioplastic for fresh fruit packaging. *IOP Conference Series: Earth and Environmental Science*, *309*(1), 012035. https://doi.org/10.1088/1755-1315/309/1/012035

Islam, M., Xayachak, T., Haque, N., Lau, D., Bhuiyan, M., & Pramanik, B. K. (2024). Impact of bioplastics on environment from its production to end-of-life. *Process Safety and Environmental Protection*, *188*, 151–166. https://doi.org/10.1016/j.psep.2024.05.113

IUCN. (2023, March 16). *From cleaner oceans to thriving communities: The economic impacts of plastic pollution – Story | IUCN*. www.iucn.org. www.iucn.org/story/202303/cleaner-oceans-thriving-communities-economic-impacts-plastic-pollution

Iyer, H., Grandgeorge, P., Jimenez, A. M., Campbell, I. R., Parker, M., Holden, M., Venkatesh, M., Nelsen, M., Nguyen, B., & Roumeli, E. (2023). Fabricating strong and stiff bioplastics from whole spirulina cells. *Advanced Functional Materials*, *33*(40). https://doi.org/10.1002/adfm.202302067

Jacob-Lopes, E., Scoparo, C. H. G., Lacerda, L. M. C. F., & Franco, T. T. (2009). Effect of light cycles (night/day) on CO_2 fixation and biomass production by microalgae in photobioreactors. *Chemical Engineering and Processing: Process Intensification*, *48*(1), 306–310. https://doi.org/10.1016/j.cep.2008.04.007

Jariyasakoolroj, P., Leelaphiwat, P., & Harnkarnsujarit, N. (2020). Advances in research and development of bioplastic for food packaging. *Journal of the Science of Food and Agriculture*, *100*(14), 5032–5045. https://doi.org/10.1002/jsfa.9497

Jegathese, S. J. P., & Farid, M. (2014). Microalgae as a renewable source of energy: A niche opportunity. *Journal of Renewable Energy*, *2014*, 1–10. https://doi.org/10.1155/2014/430203

Ji, Y., & Gao, K. (2021). Effects of climate change factors on marine macroalgae: A review, 91–136. https://doi.org/10.1016/bs.amb.2020.11.001

Jumaidin, R., Sapuan, S. M., Jawaid, M., Ishak, M. R., & Sahari, J. (2018). Seaweeds as renewable sources for biopolymers and its composites: A review. *Current Analytical Chemistry*, *14*(3), 249–267. https://doi.org/10.2174/1573411013666171009164355

Juneja, A., Ceballos, R., & Murthy, G. (2013). Effects of environmental factors and nutrient availability on the biochemical composition of algae for biofuels production: A review. *Energies*, *6*(9), 4607–4638. https://doi.org/10.3390/en6094607

Kai, D., Zhang, K., Liow, S. S., & Loh, X. J. (2019). New dual functional PHB-grafted lignin copolymer: Synthesis, mechanical properties, and biocompatibility studies. *ACS Applied Bio Materials*, *2*(1), 127–134. https://doi.org/10.1021/acsabm.8b00445

Khan, M. I., Shin, J. H., & Kim, J. D. (2018). The promising future of microalgae: current status, challenges, and optimization of a sustainable and renewable industry for biofuels, feed, and other products. *Microbial Cell Factories*, *17*(1), 36. https://doi.org/10.1186/s12934-018-0879-x

Khodaei, D., Álvarez, C., & Mullen, A. M. (2021). Biodegradable packaging materials from animal processing co-products and wastes: An overview. *Polymers*, *13*(15), 2561. https://doi.org/10.3390/polym13152561

Kibria, Md. Golam., Masuk, Nahid Imtiaz ., Safayet, Rafat., Nguyen, Huy Quoc ., & Mourshed, Monjur. (2023). Plastic waste: Challenges and opportunities to mitigate pollution and effective management. *International Journal of Environmental Research*, *17*(1), 20. https://doi.org/10.1007/s41742-023-00507-z

Kostas, E. T., Adams, J. M. M., Ruiz, H. A., Durán-Jiménez, G., & Lye, G. J. (2021). Macroalgal biorefinery concepts for the circular bioeconomy: A review on biotechnological developments and future perspectives. *Renewable and Sustainable Energy Reviews*, *151*, 111553. https://doi.org/10.1016/j.rser.2021.111553

Kumar, R., Verma, A., Shome, A., Sinha, R., Sinha, S., Jha, P. K., Kumar, R., Kumar, P., Shubham, Das, S., Sharma, P., & Vara Prasad, P. V. (2021). Impacts of plastic pollution on ecosystem services, sustainable development goals, and need to focus on circular economy and policy interventions. *Sustainability*, *13*(17), 9963. https://doi.org/10.3390/su13179963

Lam, M. K., & Lee, K. T. (2012). Potential of using organic fertilizer to cultivate *Chlorella vulgaris* for biodiesel production. *Applied Energy*, *94*, 303–308. https://doi.org/10.1016/j.apenergy.2012.01.075

Liu, F., & Zhu, J. (2014). Plant-Oil-Based Polymeric Materials and their Applications. In *Green Materials from Plant Oils* (pp. 93–126). The Royal Society of Chemistry. https://doi.org/10.1039/9781782621850-00093

Maroušek, J., Maroušková, A., Myšková, K., Váchal, J., Vochozka, M., & Žák, J. (2015). Techno-economic assessment of collagen casings waste management. *International Journal of Environmental Science and Technology*, *12*(10), 3385–3390. https://doi.org/10.1007/s13762-015-0840-z

Mata, T. M., Martins, A. A., & Caetano, N. S. (2010). Microalgae for biodiesel production and other applications: A review. *Renewable and Sustainable Energy Reviews*, *14*(1), 217–232. https://doi.org/10.1016/j.rser.2009.07.020

Mathiot, C., Ponge, P., Gallard, B., Sassi, J.-F., Delrue, F., & Le Moigne, N. (2019). Microalgae starch-based bioplastics: Screening of ten strains and plasticization of unfractionated microalgae by extrusion. *Carbohydrate Polymers*, *208*, 142–151. https://doi.org/10.1016/j.carbpol.2018.12.057

Mishra, A. K., Kaushik, M. S., & Tiwari, D. N. (2019). Nitrogenase and Hydrogenase: Enzymes for Nitrogen Fixation and Hydrogen Production in Cyanobacteria. In *Cyanobacteria* (pp. 173–191). Elsevier. https://doi.org/10.1016/B978-0-12-814667-5.00008-8

Mohan, S. V., Rohit, M. V., Subhash, G. V., Chandra, R., Devi, M. P., Butti, S. K., & Rajesh, K. (2019). Algal Oils as Biodiesel. In *Biofuels from Algae* (pp. 287–323). Elsevier. https://doi.org/10.1016/B978-0-444-64192-2.00012-3

Mojibayo, Ikusedun., Samson, Akinola O., (2020). A preliminary investigation of cassava starch potentials as natural polymer in bioplastic production. *The American Journal of Interdisciplinary Innovations and Research*, *02*(09), 31–39. https://doi.org/10.37547/tajiir/Volume02Issue09-05

Moreno-Bayona, D. A., Gómez-Méndez, L. D., Blanco-Vargas, A., Castillo-Toro, A., Herrera-Carlosama, L., Poutou-Piñales, R. A., Salcedo-Reyes, J. C., Díaz-Ariza, L. A., Castillo-Carvajal, L. C., Rojas-Higuera, N. S., & Pedroza-Rodríguez, A. M. (2019). Simultaneous bioconversion of lignocellulosic residues and oxodegradable polyethylene by *Pleurotus ostreatus* for biochar production, enriched with phosphate solubilizing bacteria for agricultural use. *PLoS One*, *14*(5), e0217100. https://doi.org/10.1371/journal.pone.0217100

Nanda, N., & Bharadvaja, N. (2022). Algal bioplastics: Current market trends and technical aspects. *Clean Technologies and Environmental Policy*, *24*(9), 2659–2679. https://doi.org/10.1007/s10098-022-02353-7

Nanda, S., Patra, B. R., Patel, R., Bakos, J., & Dalai, A. K. (2022). Innovations in applications and prospects of bioplastics and biopolymers: A review. *Environmental Chemistry Letters*, *20*(1), 379–395. https://doi.org/10.1007/s10311-021-01334-4

Nandakumar, A., Chuah, J.-A., & Sudesh, K. (2021). Bioplastics: A boon or bane? *Renewable and Sustainable Energy Reviews*, *147*, 111237. https://doi.org/10.1016/j.rser.2021.111237

Narancic, T., Cerrone, F., Beagan, N., & O'Connor, K. E. (2020). Recent advances in bioplastics: Application and biodegradation. *Polymers*, *12*(4), 920. https://doi.org/10.3390/polym12040920

Pathak, S., Sneha, C., & Mathew, B. B. (2014). Bioplastics: Its timeline based scenario & challenges. *Journal of Polymer and Biopolymer Physics Chemistry*, *2*(4), 84–90. https://doi.org/10.12691/jpbpc-2-4-5

Portugal-Nunes, D. J., Pawar, S. S., Lidén, G., & Gorwa-Grauslund, M. F. (2017). Effect of nitrogen availability on the poly-3-d-hydroxybutyrate accumulation by engineered *Saccharomyces cerevisiae*. *AMB Express*, *7*(1), 35. https://doi.org/10.1186/s13568-017-0335-z

Rafiqah, S. A., Khalina, A., Harmaen, A. S., Tawakkal, I. A., Zaman, K., Asim, M., Nurrazi, M. N., & Lee, C. H. (2021). A review on properties and application of bio-based poly(butylene succinate). *Polymers*, *13*(9), 1436. https://doi.org/10.3390/polym13091436

Rahman, A., & Miller, C. D. (2017). Microalgae as a Source of Bioplastics. In *Algal Green Chemistry* (pp. 121–138). Elsevier. https://doi.org/10.1016/B978-0-444-63784-0.00006-0

Rahman, M. H., & Bhoi, P. R. (2021). An overview of non-biodegradable bioplastics. *Journal of Cleaner Production*, *294*, 126218. https://doi.org/10.1016/j.jclepro.2021.126218

Rakhmi Sefrienda, A., Kurnianto, D., Jasmadi, J., & Frediansyah, A. (2024). Cassava and Microalgae Use in the Food Industry: Challenges and Prospects. In *Cassava – Recent Updates on Food, Feed, and Industry*. IntechOpen. https://doi.org/10.5772/intechopen.110518

Randrianarison, G., & Ashraf, M. A. (2017). Microalgae: A potential plant for energy production. *Geology, Ecology, and Landscapes*, *1*(2), 104–120. https://doi.org/10.1080/24749508.2017.1332853

Senko, O., Stepanov, N., Maslova, O., & Efremenko, E. (2023). Transformation of enzymatic hydrolysates of *Chlorella*–fungus mixed biomass into poly(hydroxyalkanoates). *Catalysts*, *13*(1), 118. https://doi.org/10.3390/catal13010118

Shah, M., Rajhans, S., Pandya, H. A., & Mankad, A. U. (2021). Bioplastic for future: A review then and now. *World Journal of Advanced Research and Reviews*, *9*(2), 056–067. https://doi.org/10.30574/wjarr.2021.9.2.0054

Sharif Hossain, A. B. M., Uddin, M. M., Veettil, V. N., & Fawzi, M. (2018). Nano-cellulose based nano-coating biomaterial dataset using corn leaf biomass: An innovative biodegradable plant biomaterial. *Data in Brief*, *17*, 162–168. https://doi.org/10.1016/j.dib.2017.12.046

Shen, Y., Yuan, W., Pei, Z. J., Wu, Q., & Mao, E. (2009). Microalgae mass production methods. *Transactions of the ASABE*, *52*(4), 1275–1287. https://doi.org/10.13031/2013.27771

Show, P., Tang, M., Nagarajan, D., Ling, T., Ooi, C.-W., & Chang, J.-S. (2017). A holistic approach to managing microalgae for biofuel applications. *International Journal of Molecular Sciences*, *18*(1), 215. https://doi.org/10.3390/ijms18010215

Simó-Cabrera, L., García-Chumillas, S., Hagagy, N., Saddiq, A., Tag, H., Selim, S., AbdElgawad, H., Arribas Agüero, A., Monzó Sánchez, F., Cánovas, V., Pire, C., & Martínez-Espinosa, R. M. (2021). Haloarchaea as cell factories to produce bioplastics. *Marine Drugs*, *19*(3), 159. https://doi.org/10.3390/md19030159

Singh, S. P., & Singh, P. (2015). Effect of temperature and light on the growth of algae species: A review. *Renewable and Sustainable Energy Reviews*, *50*, 431–444. https://doi.org/10.1016/j.rser.2015.05.024

Siracusa, V., & Blanco, I. (2020). Bio-polyethylene (Bio-PE), Bio-polypropylene (Bio-PP) and Bio-poly(ethylene terephthalate) (Bio-PET): Recent developments in bio-based polymers analogous to petroleum-derived ones for packaging and engineering applications. *Polymers*, *12*(8), 1641. https://doi.org/10.3390/polym12081641

Soto, L. R., Byrne, E., van Niel, E. W. J., Sayed, M., Villanueva, C. C., & Hatti-Kaul, R. (2019). Hydrogen and polyhydroxybutyrate production from wheat straw hydrolysate using *Caldicellulosiruptor* species and *Ralstonia eutropha* in a coupled process. *Bioresource Technology*, *272*, 259–266. https://doi.org/10.1016/j.biortech.2018.09.142

Sreenikethanam, A., & Bajhaiya, A. (2022). Algae Based Bio-Plastics: Future of Green Economy. In *Biorefineries – Selected Processes*. IntechOpen. https://doi.org/10.5772/intechopen.100981

Stabnikov, V., Ivanov, V., & Chu, J. (2015). Construction biotechnology: A new area of biotechnological research and applications. *World Journal of Microbiology and Biotechnology*, *31*(9), 1303–1314. https://doi.org/10.1007/s11274-015-1881-7

Stark, M., & O'Gara, I. (2012). An introduction to photosynthetic microalgae. *Disruptive Science and Technology*, *1*(2), 65–67. https://doi.org/10.1089/dst.2012.0017

Sudhakar, M. P., Maurya, R., Mehariya, S., Karthikeyan, O. P., Dharani, G., Arunkumar, K., Pereda, S. V., Hernández-González, M. C., Buschmann, A. H., & Pugazhendhi, A. (2024). Feasibility of bioplastic production using micro- and macroalgae – A review. *Environmental Research*, *240*, 117465. https://doi.org/10.1016/j.envres.2023.117465

Tan, J. S., Lee, S. Y., Chew, K. W., Lam, M. K., Lim, J. W., Ho, S.-H., & Show, P. L. (2020). A review on microalgae cultivation and harvesting, and their biomass extraction processing using ionic liquids. *Bioengineered*, *11*(1), 116–129. https://doi.org/10.1080/21655979.2020.1711626

Teixeira, Leonardo Vieira., Bomtempo, José Vitor Oroski, Fábio de Almeida ., & Coutinho, Paulo Luiz de Andrade. (2023). The diffusion of bioplastics: What can we learn from poly(lactic acid)? *Sustainability*, *15*(6), 4699. https://doi.org/10.3390/su15064699

Torres, S., Navia, R., Campbell Murdy, R., Cooke, P., Misra, M., & Mohanty, A. K. (2015). Green composites from residual microalgae biomass and poly(butylene adipate-*co*-terephthalate): Processing and plasticization. *ACS Sustainable Chemistry & Engineering*, *3*(4), 614–624. https://doi.org/10.1021/sc500753h

Torzillo, G., & Chini Zittelli, G. (2015). Tubular Photobioreactors. In *Algal Biorefineries* (pp. 187–212). Springer International Publishing. https://doi.org/10.1007/978-3-319-20200-6_5

Varghese, S., Dhanraj, N. D., Rebello, S., Sindhu, R., Binod, P., Pandey, A., Jisha, M. S., & Awasthi, M. K. (2022). Leads and hurdles to sustainable microbial bioplastic production. *Chemosphere*, *305*, 135390. https://doi.org/10.1016/j.chemosphere.2022.135390

Vasconcelos da Silva Pereira, G., Vasconcelos da Silva Pereira, G., Furtado de Araujo, E., Maria Paixão Xavier Neves, E., Regina Sarkis Peixoto Joele, M., & de Fátima Henriques Lourenço, L. (2019). Optimized process to produce biodegradable films with myofibrillar proteins from fish byproducts. *Food Packaging and Shelf Life*, *21*, 100364. https://doi.org/10.1016/j.fpsl.2019.100364

Verma, D., Dogra, V., Chaudhary, A. K., & Mordia, R. (2022). Advanced Biopolymer-Based Composites: Construction and Structural Applications. In *Sustainable Biopolymer Composites* (pp. 113–128). Elsevier. https://doi.org/10.1016/B978-0-12-822291-1.00010-5

Vo, H. N. P., Ngo, H. H., Guo, W., Nguyen, T. M. H., Liu, Y., Liu, Y., Nguyen, D. D., & Chang, S. W. (2019). A critical review on designs and applications of microalgae-based photobioreactors for pollutants treatment. *Science of the Total Environment*, *651*, 1549–1568. https://doi.org/10.1016/j.scitotenv.2018.09.282

Wang, S., & Hao, Z. (2008). Guest editorial—Environmental science and technology. *Asia-Pacific Journal of Chemical Engineering*, *3*(5), 471–472. https://doi.org/10.1002/apj.174

White, R. L., & Ryan, R. A. (2015). Long-term cultivation of algae in open-raceway ponds: Lessons from the field. *Industrial Biotechnology*, *11*(4), 213–220. https://doi.org/10.1089/ind.2015.0006

Yap, X. Y., Gew, L. T., Khalid, M., & Yow, Y.-Y. (2023). Algae-based bioplastic for packaging: A decade of development and challenges (2010–2020). *Journal of Polymers and the Environment*, *31*(3), 833–851. https://doi.org/10.1007/s10924-022-02620-0

Ye, C. P., Zhang, M. C., Yang, Y. F., & Thirumaran, G. (2012). Photosynthetic performance in aquatic and terrestrial colonies of *Nostoc flagelliforme* (Cyanophyceae) under aquatic and aerial conditions. *Journal of Arid Environments*, *85*, 56–61. https://doi.org/10.1016/j.jaridenv.2012.03.023

Zeng, X., Danquah, M. K., Chen, X. D., & Lu, Y. (2011). Microalgae bioengineering: From CO_2 fixation to biofuel production. *Renewable and Sustainable Energy Reviews*, *15*(6), 3252–3260. https://doi.org/10.1016/j.rser.2011.04.014

Zhao, X., Wang, Y., Chen, X., Yu, X., Li, W., Zhang, S., Meng, X., Zhao, Z.-M., Dong, T., Anderson, A., Aiyedun, A., Li, Y., Webb, E., Wu, Z., Kunc, V., Ragauskas, A., Ozcan, S., & Zhu, H. (2023). Sustainable bioplastics derived from renewable natural resources for food packaging. *Matter*, *6*(1), 97–127. https://doi.org/10.1016/j.matt.2022.11.006

Zhu, N., Ye, M., Shi, D., & Chen, M. (2017). Reactive compatibilization of biodegradable poly(butylene succinate)/*Spirulina* microalgae composites. *Macromolecular Research*, *25*(2), 165–171. https://doi.org/10.1007/s13233-017-5025-9

12 Algal Biorefinery in Waste Management and Recycling

Hirak Parikh and Archana Tiwari

12.1 INTRODUCTION

The world is currently grappling with an unprecedented waste crisis (Figure 12.1) marked by soaring pollution levels (Yang et al., 2018). Industrial sectors like textiles and paper processing contribute significantly to this issue, generating vast quantities of wastewater laden with organic pollutants and heavy metals (Shah et al., 2021). The textile industry alone is responsible for roughly 20% of global industrial water pollution. Moreover, a staggering 80% of municipal wastewater worldwide is discharged untreated, introducing pathogens and hazardous substances into ecosystems (Bhatia & Devraj, 2017). Compounding these issues is the annual production of over 380 million tons of plastic, much of which pollutes oceans and forms enormous garbage patches (Alabi et al., 2019). Therefore, effective waste management and recycling are critical as we confront this escalating crisis that threatens both environmental stability and public health. Though progress has been made in this realm, there remains a pressing need for green, eco-friendly, and cost-effective approaches to ensure sustainability (Giusti, 2009).

Algae, often considered nature's gift, present significant potential in addressing these challenges. Algal biorefineries offer a promising solution by harnessing algae's ability to convert various types of waste into valuable resources such as biofuels, pigments, and biochemicals (Liang et al., 2015; Wu et al., 2019). This helps mitigate pollution and promotes a circular economy model (Cheirsilp et al., 2023). By capitalizing on algae's natural capabilities, we can develop sustainable waste management practices that are both economically viable and environmentally sound (Singh et al., 2023a). Recent technological advancements in cultivation systems, genetic engineering, and harvesting methods have significantly enhanced the efficiency and scalability of algal biorefineries (Ahmad et al., 2022). Integrating these innovations into existing waste management frameworks addresses the urgent need for sustainable solutions and creates new economic opportunities. Algal biorefineries offer a cost-effective solution by converting waste materials into valuable products, which would otherwise incur expensive disposal fees, thereby lowering the overall production expenses for both industries and municipalities (Khoo et al., 2019). Despite challenges in scaling and commercialization, ongoing research and development show promising signs for algal biorefineries to significantly mitigate the waste crisis and advance environmental sustainability (Yadav et al., 2022). This underscores the importance of innovating and investing in these technologies to realize their full potential in shaping a cleaner and more resilient future.

DOI: 10.1201/9781003685814-12

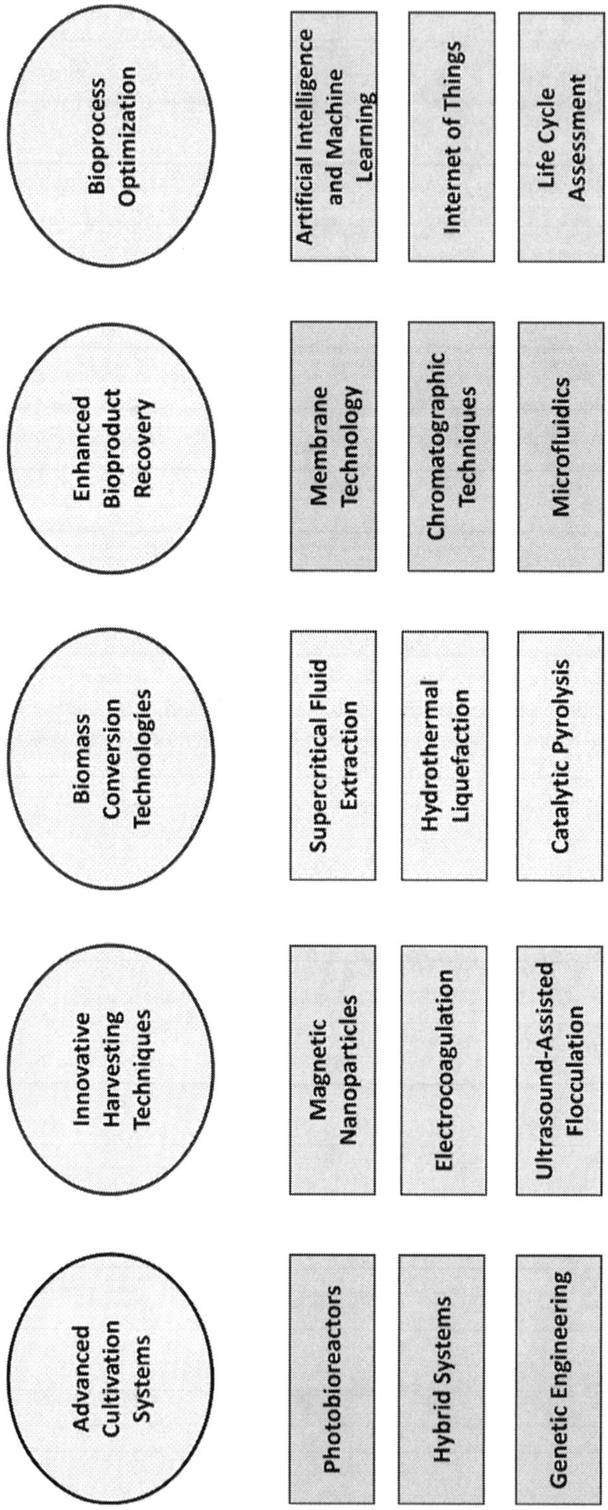

FIGURE 12.1 Schematic illustration of the common waste types.

12.2 FUNDAMENTALS OF ALGAL BIOREFINERY

The biorefinery concept, a versatile and innovative approach, integrates various biomass conversion processes to efficiently extract and utilize multiple valuable products from renewable biological resources (Baumberger & Scherrmann, 2024). It involves the sequential or simultaneous biomass processing into biofuels, biochemicals, biomaterials, and other sustainable products (Parikh, 2021a). Key components include biomass feedstock selection, pretreatment methods to enhance the accessibility of biomass components, enzymatic or chemical hydrolysis to break down complex molecules into sugars or other intermediates, fermentation or catalysis to convert sugars into desired products like biofuels or biochemicals, and separation and purification techniques to isolate and refine target compounds (Narayanan, 2024). This versatility is a testament to the potential of biorefineries in meeting diverse needs and challenges.

Algae represent a diverse group of photosynthetic organisms that thrive in many environments, ranging from oceans and lakes to soil and snow. It includes microscopic unicellular species, such as microalgae, to larger multicellular forms known as macroalgae or seaweeds (Hoek et al., 1995). Algae serve an essential ecological function as primary producers, transforming sunlight into chemical energy via photosynthesis (Chapman, 2013). Hence, it supports aquatic food chains and underscores their ecological importance. Their potential for diverse biotechnology and environmental sustainability applications, from biofuels to wastewater treatment, provides innovative solutions for improving waste management practices (Meeranayak et al., 2020).

Algal biorefinery is the integrated process of utilizing algae to produce a spectrum of valuable products such as biofuels, biochemicals, and bioproducts while minimizing waste and environmental impact (Figure 12.1). It encompasses a series of processes, including biomass production, harvesting, conversion, and downstream processing, ensuring a sustainable and efficient utilization of algal resources (Figure 12.3) (Laurens et al., 2017; Viswanathan et al., 2023).

Algal biorefineries are crucial for effective waste management and addressing global pollution. The unique ability of algae to thrive on various waste streams, such as agricultural runoff, municipal wastewater, and industrial effluents, enables them to convert these pollutants into valuable biomass (Goswami et al., 2023). This process not only reduces the environmental impact of waste but also recovers nutrients, playing a significant role in preventing eutrophication in water bodies. The biomass produced can be converted into valuable products for mankind, promoting a circular economy (Rao et al., 2019). Moreover, algae's capability to capture carbon dioxide from industrial emissions highlights their significant role in mitigating climate change and promoting sustainable waste management (Dahai et al., 2024).

Algal biorefineries employ two primary types of algae: microalgae and macroalgae. Microalgae are microscopic, unicellular organisms found in both freshwater and marine environments. They are characterized by their swift growth rates and high photosynthetic efficiency, rendering them excellent candidates for biomass production (Dolganyuk et al., 2020). Microalgae such as *Spirulina*, *Chlorella*, and *Nannochloropsis* are commonly utilized in biorefineries because of their high lipid content, essential for biofuel production (Chew et al., 2017). For example, *Nannochloropsis oceanica* can accumulate lipids up to 60% of its dry weight under nutrient-stress conditions, making it a prime candidate for biodiesel production. Similarly, diatoms, such as *Thalassiosira pseudonana*, are known for their rich biochemical profile. They produce a variety of bioactive compounds, including polyunsaturated fatty acids (PUFAs) like eicosapentaenoic acid (EPA) and docosahexaenoic acid (DHA), which have significant nutritional and pharmaceutical value (Schwarz et al., 2022). Microalgae such as *Nannochloropsis*, *Chlorella*, and *Spirulina* are widely used in biorefineries owing to their high lipid content (Russell et al., 2022; Parikh et al., 2024).

Macroalgae, or seaweeds, are multicellular, macroscopic algae predominantly found in marine environments. They are categorized into three primary groups: brown algae (Phaeophyceae), red algae (Rhodophyceae), and green algae (Chlorophyceae) (Leandro et al., 2019). Some common

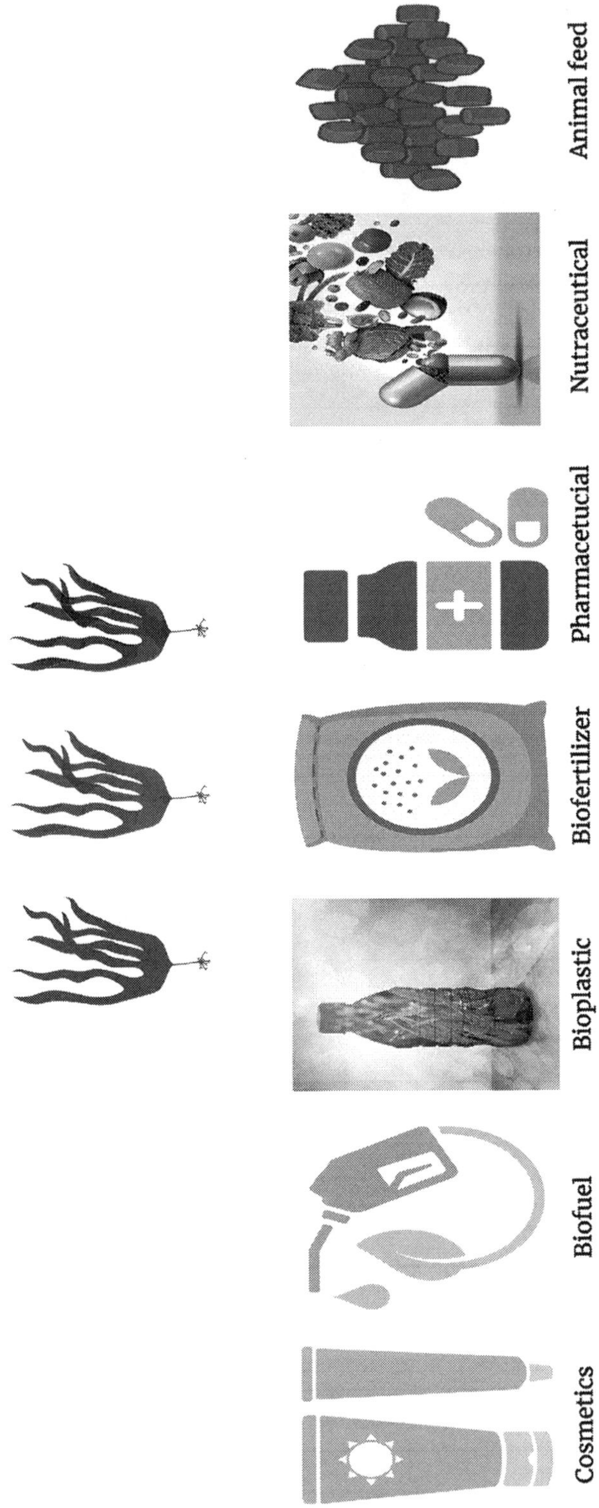

FIGURE 12.2 Comprehensive product spectrum from algal biorefineries.

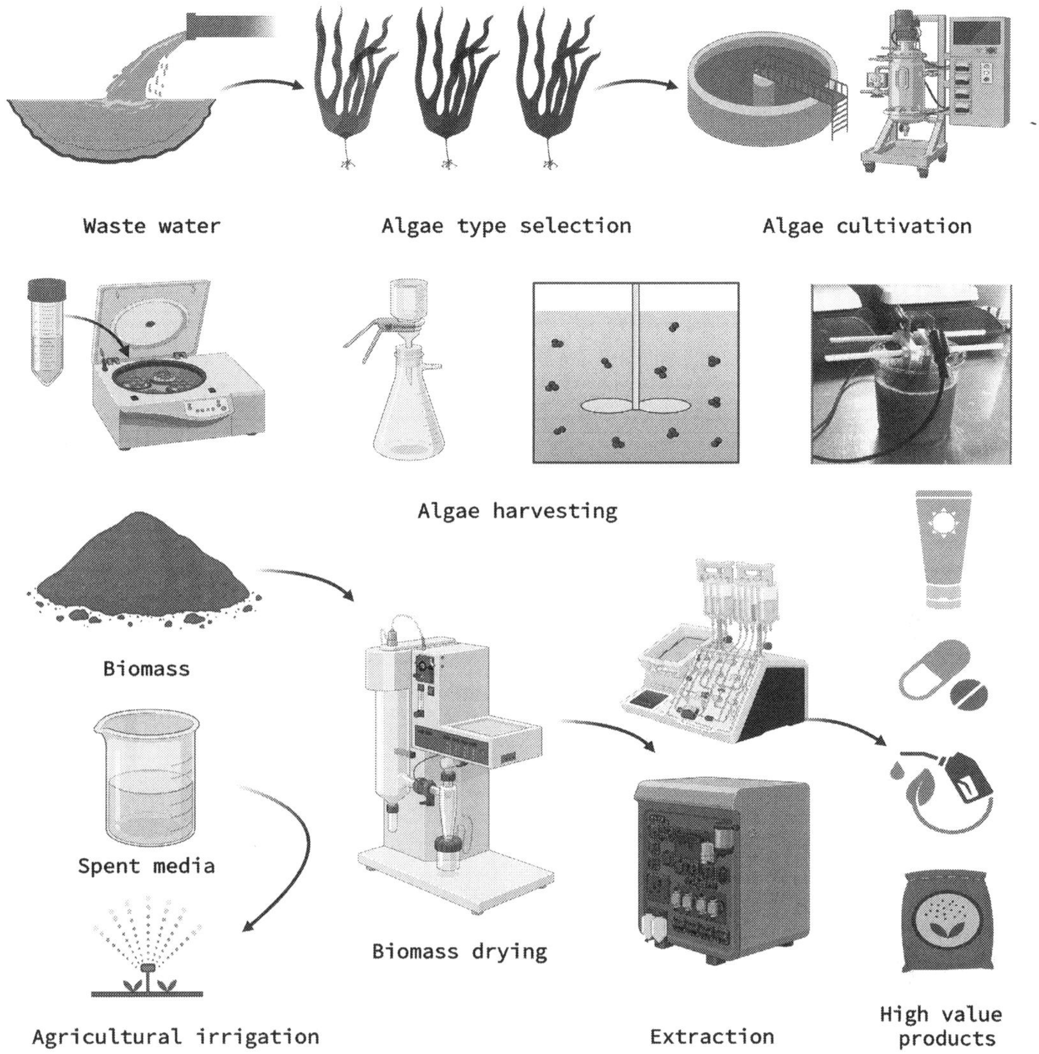

FIGURE 12.3 Schematic flow chart of algal biorefinery process.

examples that are utilized in biorefineries for their rich content of polysaccharides, such as alginate, agar, and carrageenan, are *Saccharina latissima*, *Gracilaria*, and *Ulva*. These polysaccharides are valuable for producing bioplastics, food additives, and pharmaceuticals (Filote et al., 2021). Moreover, microalgae can be cultivated in large-scale marine farms, providing an abundant and sustainable biomass source without competing for arable land and freshwater resources.

Together, microalgae and macroalgae in algal biorefineries exemplify sustainable waste management and recycling practices by converting organic waste streams into valuable biomass and bioproducts (Trivedi et al., 2015). Their dual roles in nutrient recycling, wastewater treatment, and biomass production contribute to environmental sustainability and advance the principles of the circular economy by enhancing resource efficiency and minimizing waste. Integrating these algae-based systems into industrial and municipal waste management strategies offers promising solutions for both environmental stewardship and economic viability in sustainable bioresource utilization (Laurens et al., 2017).

12.3 BIOMASS PRODUCTION AND HARVESTING TECHNIQUES

In recent years, the development and optimization of cultivation techniques for algae in biorefineries have garnered significant scientific interest due to their potential for high-yield biomass production and their role in waste management and recycling. Both micro- and macroalgae have demonstrated remarkable versatility and productivity under various cultivation conditions while effectively utilizing and recycling nutrients from waste streams. For instance, Peter et al. (2022) highlighted the efficiency of closed photobioreactors (PBRs) in maintaining optimal growth conditions for *Chlorella vulgaris*, leading to enhanced biomass yield and lipid content, and also effectively treating wastewater by removing nutrients and contaminants. Wastewater-fed open pond systems have been used for successfully cultivating *Scenedesmus obliquus*, yielding significant biomass suitable for biofuel production (Anuradha et al., 2023). *N. oceanica* has been effectively cultivated in tubular PBRs, where the configuration allows for high surface-area-to-volume ratios, optimizing light exposure and nutrient utilization from waste streams (Penloglou et al., 2024). Furthermore, hybrid systems combining different cultivation methods have shown promise in improving productivity and operational flexibility while enhancing waste nutrient recycling (Silva et al., 2024).

Macroalgae, including species like *S. latissima* (sugar kelp), *Gracilaria*, *Eucheuma*, *Laminaria*, *Undaria pinnatifida* (wakame), *Chondrus crispus* (Irish moss), *Kappaphycus alvarezii*, and *Macrocystis pyrifera* (giant kelp), are predominantly cultivated in marine farming systems, which can be integrated with coastal wastewater treatment plants (Arias et al., 2023). These systems leverage natural coastal environments to provide sustainable biomass sources without competing for arable land and freshwater resources. Leandro et al. (2019) discussed the successful cultivation of *S. latissima* in marine farms, emphasizing its potential for large-scale bioethanol production due to its high carbohydrate content and capacity to absorb excess nutrients from marine wastewater (Sæther et al., 2024). Moreover, red seaweed *Gracilaria edulis* has been extensively studied for its agar production, with marine farming techniques ensuring consistent yield and quality while also providing bioremediation benefits (Bhushan et al., 2023).

Innovative approaches such as heterotrophic and mixotrophic cultivation have also been explored to enhance algal biomass productivity and waste recycling efficiency (Tea et al., 2023). Heterotrophic cultivation, which involves growing algae like *C. vulgaris* without light using organic carbon sources, has yielded higher biomass densities and valuable bioproducts while efficiently recycling organic waste materials (Marques, 2023). Mixotrophic cultivation, combining both autotrophic and heterotrophic growth modes, has been reported to significantly boost growth rates and biomass production in species like *Skeletonema marinoi*, enhancing the recycling of both inorganic and organic waste nutrients (Villanova et al., 2024). Overall, the diverse array of cultivation techniques (Table 12.1) reflects the adaptability and potential of algae in biorefinery applications, particularly in waste management and recycling.

After successful and optimum biomass production, efficient harvesting techniques are crucial for separating algae biomass from culture media to maximize product yield and minimize energy consumption. Algae harvesting involves detaching or separating algae from its growth medium (Ananthi et al., 2021). The chosen harvesting method depends heavily on the characteristics of the target algae, including cell density and size, the specifications of the final product, and whether the culture medium can be reused (Kumar et al., 2023). The ideal harvesting process should be effective across various algal strains, achieving high biomass concentrations while maintaining moderate operations, energy, and maintenance costs. The value and properties of the end product also influence the selection of a suitable harvesting method (Pratama et al., 2024). When selecting a method, it's important to consider acceptable levels of moisture and salt and to minimize potential cell damage. The method should not introduce toxicity or contamination to the algal biomass and should allow for the recycling of the culture medium. In our current environmental and economic context, the need for cost-effective, sustainable methods is urgent (Singh & Patidar, 2018).

TABLE 12.1
Summary of Common Cultivation Techniques Used for Algae in Biorefineries: Advantages and Challenges

Cultivation Technique	Description	Advantages	Challenges
Closed photobioreactors	Sealed, controlled systems that provide optimized conditions for algae growth	Precise control over light, temperature, and nutrients; higher biomass yield and quality control	High initial investment; energy-intensive operation; limited scalability for large-scale production
Open pond systems	Outdoor ponds or raceways exposed to natural sunlight and ambient conditions	Scalable for large-scale production; lower infrastructure costs compared to closed systems	Less control over environmental factors; vulnerable to contamination and pests
Tubular photobioreactors	Vertical or horizontal tubes or panels are used to cultivate algae, providing increased surface area	Efficient use of space; good light exposure and mixing	Complex design and maintenance; limited scalability for large-scale production
Flat panel photobioreactors	Flat, horizontal panels or trays with circulating algae culture	Easy scalability; suitable for urban or constrained areas	Light penetration and shading issues; cleaning and fouling challenges
Bag or raceway photobioreactors	Large, shallow ponds or bags for algae cultivation, often used for large-scale operations	Low-cost production; suitable for large-scale operations	Contamination risk; nutrient and water management challenges
Heterotrophic cultivation	Algae grown in the absence of light, using organic carbon sources for growth	Higher biomass yield per unit area; reduced energy requirements	Requires additional organic carbon sources; higher production costs
Mixotrophic cultivation	Algae cultivated using both light and organic carbon sources for growth	Enhanced growth rates and biomass productivity; flexibility in nutrient utilization	Complex nutrient management; variability in growth performance
Marine farming	Cultivation of seaweeds in marine environments, such as coastal farms	Utilizes natural marine habitats; minimal land and freshwater requirements	Vulnerable to environmental factors; regulatory considerations for coastal farming

Screening, often used as a pre-processing step for algal cultures, plays a significant role in the overall harvesting process. Its efficiency is dependent on the spacing between screen openings and cell size. Commonly used methods include micro strainers and vibrating screens. The former involves a rotary drum covered with straining fabric that undergoes frequent backwashing, a process of reversing the flow to remove accumulated particles and maintain the efficiency of the straining fabric (Kumar et al., 2023).

Algae harvesting methods include mechanical, chemical, biological, and electrical techniques (Table 12.2) (Barros et al., 2015). For macroalgae, the harvesting process is relatively straightforward but labor-intensive. In contrast, microalgae require more complex methods, often involving chemical or mechanical processes. Mechanical methods are the most reliable and widely used for microalgae, but their efficiency can be significantly improved when combined with coagulation and flocculation steps (Deepa et al., 2023). This combination enhances harvesting efficiency and reduces operation and maintenance costs. Microalgal harvesting typically involves a two-step concentration process: thickening and dewatering. However, the process can be optimized to increase separation rates and lower costs. For instance, combining flocculation-sedimentation with centrifugation can

TABLE 12.2
Comprehensive Overview of Common Harvesting Techniques Used in Algae Biorefineries, Including Detailed Descriptions, Advantages, and Challenges Associated with Each Technique

Technique	Description	Advantages	Challenges
Centrifugation	Separates algae cells from the culture medium based on density differences using high-speed rotation	High recovery efficiency (>90), rapid processing, applicable to all microalgae, optimum for high-density cultures	Energy-intensive, equipment cost, potential cell damage, time-consuming expensive for large scale
Flocculation	Induces algae aggregation using chemicals (e.g., aluminum sulfate, chitosan) or polymers for easier separation from the culture medium	Effective for large-scale applications, reduces energy input for downstream processing, less cell damage, applied to a vast range of species	Chemical cost, residual contaminants in biomass, pH dependency, and culture medium recycling are limited
Filtration	Uses membranes (microfiltration, ultrafiltration) or screens to physically filter algae cells based on size	Scalable, suitable for various biomass concentrations, continuous operation is possible. No chemical required	Membrane fouling, clogging, high maintenance costs, not suitable for small algae
Sedimentation	Relies on gravity to settle algae cells in large settling tanks or ponds	Low energy consumption, simple operation, suitable for large-scale applications	Long settling times, large footprints, require large space for settling tanks
Electrocoagulation	Applies electrodes to destabilize algae cells, causing aggregation and easier separation from the culture medium	Effective for different alga types, reduces chemical usage	Electrode corrosion, process control complexity, high initial investment
Bio-flocculation	Utilizes genetically modified algae strains or bacteria to produce natural flocculants, promoting self-flocculation	Environmentally friendly, reduces chemical usage, potential for genetic engineering	Requires genetic modification, limited to specific strains, and scalability challenges
Foam fractionation	Generates foam from the culture medium, capturing algae cells, and then separating them from the foam	Efficient for low-density algae cultures, it can integrate with other processes like dewatering	Foam stability, foam generation efficiency, scalability issues

significantly reduce process costs (Kumar et al., 2024). Furthermore, emerging biological approaches hold the promise of further cutting operational expenses, demonstrating the potential for significant cost reduction in the harvesting process and a brighter future for algae harvesting (Gautam et al., 2024). Combining different techniques can optimize the process, balancing efficiency with cost and environmental considerations.

12.4 ADVANCEMENT IN ALGAL WASTEWATER TREATMENT

In the realm of sustainable waste management and recycling, algal biorefinery has been proven to effectively eliminate contaminants from diverse wastewater sources, including industrial, municipal, and agricultural wastewater (Mustafa et al., 2021). The typical components of wastewater, including oils and greases, heavy metals, pesticides, volatile organic compounds, and minerals, can be effectively and sustainably treated. This prevents soil and groundwater contamination during wastewater disposal or reuse and safeguards agricultural productivity and human health (Bhatt et al., 2022).

Algae, especially microalgae, with their rapid growth rates, high productivity even under nutrient-limited conditions, and exceptional ability to accumulate substantial amounts of fatty acids, play a crucial role in this process. They effectively remove nutrients like nitrogen, phosphorus, heavy metals, and organic compounds from wastewater via biosorption and biotransformation. Moreover, algal biorefineries have shown capabilities in reducing chemical oxygen demand and total dissolved solids while maintaining stable levels of dissolved oxygen.

Microalgae have been proven to reduce metal toxicity through several mechanisms: bonding metals to the cell surface, forming insoluble metal compounds, using influx pumps to move metals into the cytoplasm, altering the oxidation state to make metals less toxic, transforming metals into volatile forms for expulsion, and attaching metals to intracellular proteins or polysaccharides to limit toxicity (Monteiro et al., 2012; Leong & Chang, 2020). In bioaccumulation, pollutants are processed through various cellular routes, while in biosorption, pollutants are complexed within the cell membrane and then removed from the environment (Xiao et al., 2023). Hence, the natural uptake of harmful nutrients helps manage waste and significantly mitigates eutrophication when treated water is reintroduced into natural ecosystems (Yadav et al., 2021). Overall, algal biorefineries provide the dual benefit of treating wastewater in sustainable and environmentally friendly ways and producing valuable biomass that can be used for biofuels, animal feed, and other bioproducts (Abdelfattah et al., 2023). They also capture and utilize CO_2, contributing to greenhouse gas mitigation efforts. Recent studies summarizing the use of diverse algae for various types of wastewater treatment are compiled in Table 12.3.

However, challenges still need to be solved in scaling up these systems to industrial levels. Maintaining optimal growth conditions, managing biomass harvesting, and dealing with variable wastewater compositions can influence the efficiency and cost-effectiveness of algal wastewater treatment (Kumar et al., 2024). Despite these challenges, ongoing research and technological advancements play a vital role in enhancing the feasibility and performance of algal wastewater treatment systems, instilling hope for the future of this technology.

12.5 ALGAE IN SOLID WASTE MANAGEMENT

The rapid urban population growth and industrial activities have significantly increased solid waste production, presenting substantial challenges to waste management and environmental sustainability. Solid waste management (SWM) encompasses various techniques, such as segregation, composting, drainage, scrubbers or electrostatic precipitators, and incineration to handle solid wastes from collection to final disposal (Nanda & Berruti, 2021). However, traditional waste management practices, such as landfilling and incineration, have notable drawbacks, including greenhouse gas emissions, leachate production, and loss of valuable resources (Giusti, 2009). In contrast, algae offer a more sustainable approach to treating solid waste by converting organic materials into valuable biomass, reducing landfill dependency, and mitigating pollution (Satpati, 2021). This eco-friendly method enhances resource recovery and supports circular economy principles by closing resource loops, minimizing waste generation, and maximizing resource efficiency.

Organic solid waste, with significant environmental implications, demands effective and eco-friendly treatment. However, this waste contains high nutrient concentrations that algae can utilize for growth (Soobhany, 2019). SWM process begins with waste pretreatment to reduce particle size and enhance nutrient availability. Physical methods such as shredding and grinding, chemical methods like hydrolysis to break down complex organic molecules, and biological methods involving enzymes or microorganisms to further degrade organic matter and release essential nutrients are employed (Swar et al., 2023). After pretreatment, essential nutrients like nitrogen, phosphorus, and potassium, which are vital for algal growth, are extracted and supplied to algal cultivation systems. The choice of algal species depends on nutrient needs, growth conditions, cost, and desired end products (Ersahin et al., 2023). The harvested algal biomass can be processed

TABLE 12.3
Recent Research Outlining the Application of Diverse Algae in Treating Various Types of Wastewater

Algae Type	Wastewater Type	Outcomes	References
Chlorella vulgaris	Synthetic	Achieved 92% removal of nitrate and 98% removal of phosphate	Ms et al. (2023)
Chlamydomonas reinhardtii, *Monoraphidium braunii*, and *Scenedesmus obliquus*	Sewage	High efficiency to reduce nitrate, ammonia, phosphate, sulfate, heavy metals (Zn^{2+}, Cu^{2+}, Mn^{2+}, and Fe^{2+}), calcium, magnesium, sodium, and potassium	El-Sheekh et al. (2023)
Chlorella ellipsoidea	Biofloc	High growth kinetics, chlorophyll a, b, and β-carotene content High ammonia, nitrite, and phosphate removal efficiency	Afrin et al. (2024)
Spirulina platensis	Textile	High growth kinetics, lipid, and carbohydrate content. High COD, ammonia, and phosphate removal efficiency	Selvaraj and Arivazhagan (2024)
Chaetoceros sp. and *Thalassiosira* sp.	Municipal	*Thalassiosira* sp. – high phosphate and nitrate removal efficiency *Chaetoceros* sp. – COD reduction	Singh et al. (2024)
Chaetoceros gracilis and *Thalassiosira weissflogii*	Simulated dairy	High growth kinetics, lipid content and mono-polyunsaturated fatty acid (MUFA) content	Singh et al. (2024)
Dunaliella salina	Saline food industry	High cell density, achieved 84.82% removal of nitrate and 80.49% removal of phosphorus.	Balaji et al. (2024)
Scenedesmus sp. ASK22	Simulated dairy effluent	High biomass, high COD, nitrite and phosphate removal efficiency	Pandey et al. (2024)
Chlorella sp.	Synthetic	High lipid accumulation, achieved high ammonium removal efficiency	Wutthithien et al. (2024)
Gracilaria sp.	Electroplating	Achieved 96.70% for copper adsorption	Anwar (2024)
Oscillatoria sp. and *Phormidium* sp.	Textile	High efficiency in TDS removal, achieved high efficiency in the removal of nitrates, nitrites, phosphates, sulphates, chlorides, calcium, sodium, and magnesium	Mathimani et al. (2024)
Tetradesmus reginae, *Tetradesmus obliquus*, and *Chlorella sorokiniana*	Municipal	High biomass, achieved high efficiency in the removal of nitrates, nitrites, phosphates, and COD. Significantly reduced toxic heavy metals – As, Cd, Cr, Pb	Ahiahonu et al. (2024)
C. sorokiniana	Synthetic	Achieved high efficiency in the removal of copper and zinc	Oliomogbe et al. (2024)
Chloroidium ellipsoideum and *Desmodesmus subspicatus*	Agricultural	Achieved high efficiency in the removal of copper	Fawzy et al. (2024)

into various bio-products, contributing to a sustainable biorefinery model (Deepa et al., 2023). Several algal species, including *Ankistrodesmus*, *Anabaena*, *Chlorella pyrenoidosa*, *C. vulgaris*, *Closteriopsis longissima, Euglena* sp., *Galdieria sulphuraria*, *Nostoc*, *Spirulina* sp., *Synedra acus*, and *Trichocerca* sp., have demonstrated superior bioremediation potential for landfill leachate (Malyan et al., 2024).

To manage waste sustainably, anaerobic digestion is essential, as it efficiently converts organic waste into biogas and nutrient-rich digestate. Algae play a crucial role in this process by providing a balanced carbon-to-nitrogen ratio, enhancing microbial activity, and enhancing biogas production. Co-digestion of algae with other organic wastes has been shown to increase biogas yield and improve the stability of the digestion process (Barreiro-Vescovo et al., 2020). Additionally, the nutrient-rich digestate produced can be used as a biofertilizer, contributing to sustainable agriculture (Fernandes et al., 2022). Composting, an aerobic process involving the decomposition of organic matter by microorganisms, results in the production of compost. The addition of algae to compost piles not only accelerates the decomposition process but also significantly enhances the nutrient content of the final product. The high protein and carbohydrate content of algal biomass provides an excellent source of nutrients for composting microorganisms (Michalak & Chojnacka, 2013). Moreover, algae contribute to moisture balance and improve the physical structure of the compost pile (Anyaoha et al., 2024). Hence, the use of algae in composting not only speeds up the process but also results in composting with higher nutrient levels, making it more valuable for soil enrichment.

Studies suggest that the co-cultivation of algae and bacteria presents an effective green technology that can be integrated with landfill systems to remove toxic substances in leachate (Hu et al., 2021; Tang et al., 2023). This innovative approach harnesses the synergistic interactions between algae and bacteria to degrade contaminants, showcasing a promising solution for sustainable waste management and recycling within the framework of an algal biorefinery.

12.6 ALGAL CO-PRODUCTS: MAXIMIZING RESOURCE RECOVERY FROM WASTE

Algae are well-known for their adeptness at thriving in diverse environments and efficiently converting nutrients into biomass, underscoring their significant value as bioresources, particularly in transforming waste into beneficial compounds. Under stress conditions such as in wastewater environments, certain algae species have evolved to produce beneficial compounds that aid their survival and offer valuable applications in various industries. Extremophilic algae are known for their adaptability, rapid reproduction, resistance to environmental changes, and valuable storage products. For example, the presence of *G. sulphuraria* in wastewater is known for its ability to thrive in extreme conditions such as high acidity and temperature (Öztürk & Satpati, 2024).

Algal pigments like chlorophyll, carotenoids, and phycobiliproteins have extensive applications in food, cosmetics, nutraceuticals, and pharmaceuticals, serving as natural colorants and antioxidants (Choudhary et al., 2024). For example, astaxanthin, a potent antioxidant, is highly valued due to its superior ability to protect cells from oxidative damage. Sirotiya et al. (2023) found that *Haematococcus pluvialis* could convert industrial wastewater into a valuable source of astaxanthin, promoting both environmental remediation and the production of high-value bioproducts.

Phycobiliproteins are water-soluble pigments that aid photosynthesis by capturing light energy. They enhance photosynthetic efficiency and are crucial in biotechnological applications. In the study by Arashiro et al. (2020), *Nostoc* sp., *Arthrospira platensis*, and *Porphyridium purpureum* were cultivated in industrial wastewater. The researchers successfully extracted phycocyanin, allophycocyanin, and phycoerythrin from the resulting biomass.

The conversion of algal biomass into bioplastics, such as polyhydroxyalkanoates (PHAs), offers an eco-friendly alternative to conventional plastics, addressing plastic pollution and reducing reliance on fossil fuels. Bioplastics were developed from a microalgae consortium sourced from wastewater, composed of *Scenedesmus* sp., *Desmodesmus* sp., *Nannochloropsis gaditana*, and *A. platensis* (Rocha et al., 2020). These bioplastics demonstrated good thermal resistance and low water absorption, exhibiting promising mechanical properties.

Over the years, algae have been utilized to produce various biofuels such as biodiesel, bioethanol, and biogas from wastewater, providing renewable energy solutions and contributing to waste reduction (Parikh, 2021). A recent study from Singh et al. (2024b) demonstrated that diatoms *Chaetoceros gracilis* and *Thalassiosira weissflogii* cultivated in simulated dairy wastewater were not only notable for their high lipid content but also produced significant amounts of MUFAs, enhancing their potential for biodiesel applications. Similarly, *Dunaliella* sp., grown in fish processing wastewater, showed a remarkable four-fold increase in both MUFAs and PUFAs (Fakhri et al., 2024). In today's world, the management of fruit residue is particularly important due to the high organic load and nutrient content present in such waste. In light of this, the study by Condori et al. (2024) provided insightful details by demonstrating *Chlorella* sp.'s efficient removal of COD, nitrate, phosphate, and sulfate from fruit residue, along with its high growth kinetics and lipid content. Beyond these, algae can also produce animal feed, fertilizers, dyes, food supplements, and biochar, enhancing soil health and agricultural productivity.

12.7 TECHNOLOGICAL INNOVATIONS IN ALGAL BIOREFINERY

Recent advancements in algal biorefinery processes have introduced innovative technologies across various stages, enhancing efficiency, sustainability, and the economic viability of algal biomass conversion (Figure 12.4). For instance, by manipulating the genetic makeup of algae, scientists have developed strains capable of efficiently processing various types of waste and converting them into valuable bioproducts. A prominent example is the work conducted on utilizing CRISPR-Cas9 to bolster algae's capacity for carbon dioxide sequestration from industrial emissions, advancing the development of fourth-generation biofuels (Verdezoto-Prado et al., 2024). TALEN (transcription activator-like effector nuclease) technology allows precise genetic modifications in algae, improving traits such as biomass productivity and nutrient utilization from waste sources. For instance, researchers have used TALENs to engineer algae strains capable of more efficiently metabolizing organic waste into biodegradable polymers (Sharma et al., 2022). Metabolic engineering of

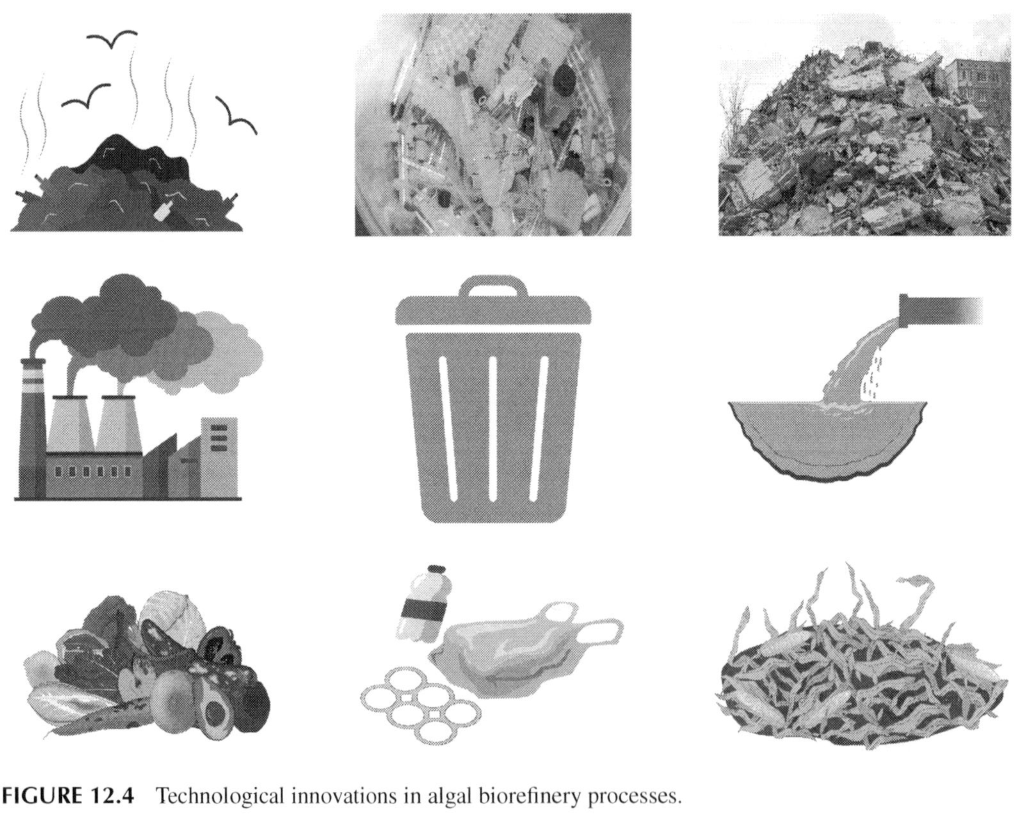

FIGURE 12.4 Technological innovations in algal biorefinery processes.

algae, particularly cyanobacteria, employs genetic modifications to strategically optimize metabolic pathways for the biosynthesis of a diverse range of high-value chemicals. This includes pharmaceutical intermediates, nutraceuticals, and specialized compounds like omega-3 fatty acids and antioxidants (Tiwari et al., 2024). Synthetic biology offers a powerful toolkit for advancing algae's capabilities in wastewater bioremediation and bioproduct synthesis. By designing and constructing novel biological systems, synthetic biology enables precise manipulation of algae's genetic and metabolic pathways to achieve desired traits. For example, synthetic biology techniques have been used to engineer algae with enhanced pollutant degradation capabilities and improved nutrient utilization efficiency in wastewater conditions (Webster et al., 2024).

Advancing technologies, especially genetic engineering and synthetic biology, are pivotal in enhancing algae's ability to efficiently convert waste into valuable products. By refining genetic manipulation tools and metabolic engineering approaches, researchers can optimize algae strains for higher yields of biofuels, bioplastics, and pharmaceuticals from various waste sources. These innovations not only improve resource efficiency but also promote sustainable solutions for waste management and environmental conservation. As these technologies continue to evolve, algae-based biorefinery systems hold promise in addressing global challenges through innovative biotechnological applications.

12.8 ECONOMIC AND ENVIRONMENTAL IMPACT ASSESSMENT

Integrating algal biorefineries into waste management systems offers substantial economic and environmental benefits, though it requires significant initial investments in infrastructure and technology. However, these investments are not just costs, but potential sources of revenue. Strategic planning and systematic implementation can mitigate these costs, generating revenue from high-value bioproducts. For example, wastewater treatment plants that integrate algal biorefineries can efficiently eliminate harmful nutrients such as nitrogen and phosphorus. Simultaneously, they can produce valuable biomass products like fertilizers, biofuels, cosmetics, and animal feed. These outputs create additional revenue streams and improve the economic viability and sustainability of the treatment processes.

The economic viability of algal biorefineries is not merely an aspiration but an achievable reality that can be expedited with the active involvement of start-up companies and government bodies. Start-ups are at the forefront of innovation, developing new biorefinery technologies and diversifying product portfolios. Their role is crucial in shaping the future of waste management. Simultaneously, government support through financial incentives, subsidies, and supportive policies can alleviate the initial financial burdens, making algal biorefineries more competitive with traditional fossil fuel-based technologies. This government assistance is not just beneficial but critical for promoting sustainable waste management solutions and facilitating the transition towards a circular economy.

Algal biorefineries are a beacon of hope for sustainability, significantly reducing carbon footprints, optimizing energy utilization, and enhancing resource efficiency. In urban areas with high industrial activity, pollution levels are notably elevated. However, incorporating algal bioreactors within industrial zones provides a viable solution for mitigating substantial carbon emissions. These biorefineries surpass traditional waste management methods, such as landfilling and incineration, by offering a more sustainable approach. Landfills are known to emit methane, a potent greenhouse gas, while incineration releases hazardous pollutants like particulate matter and toxic gases. In contrast, microalgae demonstrate a high carbon sequestration capacity per hectare, underscoring their potential in climate change mitigation. Furthermore, algae's ability to convert organic waste into high-value products enhances economic viability and reduces reliance on finite natural resources, aligning with the principles of a circular economy. Waste incineration presents significant environmental and health risks due to the emission of pollutants, such as particulate matter, heavy metals, and toxic gases. Algal biorefineries operate in controlled environments where emissions can be efficiently managed and by-products recycled. This controlled process reduces the overall environmental impact and offers a more sustainable alternative to traditional waste management practices.

The advancement and deployment of algal biorefineries are profoundly influenced by policies and regulations promoting sustainable waste management and recycling. Globally, governments increasingly acknowledge the potential of algae-based solutions in addressing environmental challenges and are enacting supportive frameworks to drive innovation and adoption. For example, the European Union's Renewable Energy Directive (RED II) sets ambitious targets for renewable energy production, including algae-derived biofuels. It provides incentives and funding for research and development in this sector (Maes et al., 2015). In the United States, the Department of Energy (DOE) has launched several initiatives to scale up algal biorefineries, focusing on cost reduction and environmental enhancement of algae-based technologies (Liang et al., 2015). Environmental regulations, such as the Clean Water Act and the Clean Air Act, further promote the implementation of algae-based solutions for wastewater treatment and carbon sequestration. These policies not only facilitate the use of algae in pollution mitigation but also support the development of a circular economy by transforming waste streams into valuable bioproducts (Talebi et al., 2022).

12.9 CHALLENGES AND FUTURE PROSPECTS

The algae biorefinery sector faces substantial technological and economic challenges. One primary technological limitation is the efficiency and cost-effectiveness of harvesting and extraction techniques. Current methods, including centrifugation, flocculation, and filtration, are often energy-intensive, time-consuming, and expensive (Bastiaens et al., 2017). Although innovations such as electrocoagulation and magnetic separation show promise, they require further development and scale-up to achieve industrial viability. Additionally, conversion technologies transforming specific algal biomass into valuable products need significant optimization (Kumar & Bharadvaja, 2020). Hydrothermal liquefaction, pyrolysis, and transesterification demand high energy inputs and specialized catalysts, which hinder economic viability. Furthermore, the design and operation of PBRs and open pond systems present challenges in light distribution, CO_2 supply, and contamination control. Improved reactor designs that maximize light penetration and nutrient distribution while minimizing contamination risks are essential for enhancing productivity (Beims et al., 2020).

Economic viability remains a major hurdle for algae biorefineries. High production costs are a significant barrier, as algae cultivation, harvesting, and processing are more expensive than traditional feedstocks. The costs associated with nutrient supply, energy consumption, and water usage contribute significantly to this issue. Innovations in nutrient recycling, energy-efficient technologies, and water reuse are critical to reducing these costs (Yadav et al., 2022). Moreover, algal biofuels and bioproducts must compete with established fossil fuels and other bio-based products. Achieving cost parity with these alternatives requires substantial production efficiency and scalability advancements. Market acceptance and regulatory support are also crucial for the commercial success of algal products. Scaling up from laboratory to industrial-scale production introduces additional challenges related to maintaining consistent quality, controlling contamination, and ensuring the sustainability of operations, all of which necessitate substantial capital investment and infrastructure development (Kumar & Bharadvaja, 2020).

Addressing research gaps is essential for advancing algae biorefinery technologies. One key area is strain improvement through genetic engineering and strain selection. Developing genetically modified algae strains with enhanced growth rates, higher lipid content, and resilience to environmental stresses can significantly improve productivity (Eppink et al., 2019). Advances in CRISPR/Cas9 and other genetic engineering techniques offer precise and targeted modification opportunities (Fayyaz et al., 2020). Identifying and breeding algae strains naturally suited for specific environmental conditions and desired product profiles is critical. High-throughput screening methods and bioprospecting can aid in discovering and optimizing such strains (Bhattacharya & Goswami, 2020).

Integrated biorefinery systems present a promising future direction. Developing systems that utilize all components of algal biomass can enhance economic viability. This includes the co-production of biofuels, bioplastics, nutraceuticals, and other high-value products. For example, the residual biomass after lipid extraction for biofuel can be used for animal feed or as a substrate for anaerobic digestion (Singh et al., 2023). Waste valorization is another critical aspect, as utilizing waste streams (e.g., wastewater, CO_2 emissions) as inputs for algal cultivation can reduce production costs and environmental impact (Moreira et al., 2023). Research into optimizing the use of industrial and municipal waste streams for algal growth is essential for sustainable development. Collaborative efforts among academia, industry, and government agencies are vital for advancing research, optimizing processes, and developing policies that support the growth of the algae biorefinery sector (Kostas et al., 2021).

In conclusion, while the algae biorefinery sector faces significant challenges, ongoing research and technological advancements hold promise for overcoming these barriers. By addressing technological limitations, improving economic viability, and fostering innovation, the algae biorefinery sector can play a crucial role in sustainable waste management and resource recovery, contributing to a circular economy and a more sustainable future. Integrating algal biorefineries into waste management systems offers a scientifically grounded and economically viable approach to environmental challenges. Combining innovative technologies, supportive policies, and strategic investments can drive the widespread adoption of algae-based solutions, enhancing waste management and resource recovery sustainability.

ACKNOWLEDGMENT

Financial support for this study had been provided by the AISTDF Secretariat, Science & Engineering Research Board (SERB), India [Grant number CRD/2022/000595].

DECLARATION

The authors have not used any generative Artificial Intelligence (AI) tools during the writing of this manuscript.

REFERENCES

Abdelfattah, A., Ali, S. S., Ramadan, H., El-Aswar, E. I., Eltawab, R., Ho, S. H., ..., & Sun, J. (2023). Microalgae-based wastewater treatment: Mechanisms, challenges, recent advances, and future prospects. *Environmental Science and Ecotechnology*, *13*, 100205.

Afrin, E., Akter, T., Baidya, A., Hossain, M. A., Islam, M. R., Das, M., ..., & Iqbal, M. A. (2024). Biofloc wastewater for microalgae (*Chlorella ellipsoidea*) production: An approach to algal biomass production and nutrient remediation. *Journal of Applied Aquaculture*, *37*, 86–106.

Ahiahonu, E. K., Anku, W. W., Roopnarain, A., Serepa-Dlamini, M. H., & Govender, P. P. (2024). Evaluation of green microalgal strains for simultaneous municipal wastewater treatment and heavy metal bioremediation. *Journal of Chemical Technology & Biotechnology*. *99*, 2035–2046. https://doi.org/10.1002/jctb.7705

Ahmad, A., Banat, F., Alsafar, H., & Hasan, S. W. (2022). Algae biotechnology for industrial wastewater treatment, bioenergy production, and high-value bioproducts. *Science of the Total Environment*, *806*, 150585.

Alabi, O. A., Ologbonjaye, K. I., Awosolu, O., & Alalade, O. E. (2019). Public and environmental health effects of plastic wastes disposal: a review. *Journal of Toxicology and Risk Assessment*, *5*(021), 1–13.

Ananthi, V., Balaji, P., Sindhu, R., Kim, S. H., Pugazhendhi, A., & Arun, A. (2021). A critical review on different harvesting techniques for algal based biodiesel production. *Science of the Total Environment*, *780*, 146467.

Anuradha, S., Thadikamala, S., Harish, B. S., Gayathri, G., Thulasidharan, D., & Uppuluri, K. B. (2023). Open system for the autotrophic cultivation of *Scenedesmus obliquus* NCIM 5586: Multiobjective optimization for the tradeoff between biomass and lipid. *Biomass Conversion and Biorefinery*, 2113–2123.

Anwar, C. (2024). Adsorption of copper metal from electroplating wastewater using activated charcoal of *Gracilaria* sp. *EnvironmentAsia, 17*(1), 95–102.

Anyaoha, K. E., Maletz, R., Rückert, A., & Dornack, C. (2024). Potentials for microalgae sequestration of carbon dioxide (CO_2) from composting off-gas; a review. *Resources, Conservation & Recycling Advances, 22*, 200213.

Arashiro, L. T., Boto-Ordóñez, M., Van Hulle, S. W., Ferrer, I., Garfí, M., & Rousseau, D. P. (2020). Natural pigments from microalgae grown in industrial wastewater. *Bioresource Technology, 303*, 122894.

Arias, A., Feijoo, G., & Moreira, M. T. (2023). Macroalgae biorefineries as a sustainable resource in the extraction of value-added compounds. *Algal Research, 69*, 102954.

Balaji, E., Swaminathan, D., & Deepa, V. S. (2024). Assessment of nutrient removal efficiency of marine microalgae *Dunaliella salina* in saline food industry wastewater. *Journal of Environmental Nanotechnology, 13*(2), 46–51.

Barreiro-Vescovo, S., Barbera, E., Bertucco, A., & Sforza, E. (2020). Integration of microalgae cultivation in a biogas production process from organic municipal solid waste: From laboratory to pilot scale. *ChemEngineering, 4*(2), 25.

Barros, A. I., Gonçalves, A. L., Simões, M., & Pires, J. C. (2015). Harvesting techniques applied to microalgae: A review. *Renewable and Sustainable Energy Reviews, 41*, 1489–1500.

Bastiaens, L., Van Roy, S., Thomassen, G., & Elst, K. (2017). Biorefinery of algae: Technical and economic considerations. In *Microalgae-based biofuels and bioproducts* (pp. 327–345). Woodhead Publishing.

Baumberger, S., & Scherrmann, M. C. (2024). Green Chemistry, Eco-Friendly Chemistry, Biorefinery. In *Green Chemistry and Agro-food Industry: Towards a Sustainable Bioeconomy* (pp. 3–22). Springer Nature Switzerland.

Beims, R. F., Hu, Y., Shui, H., & Xu, C. C. (2020). Hydrothermal liquefaction of biomass to fuels and value-added chemicals: Products applications and challenges to develop large-scale operations. *Biomass and Bioenergy, 135*, 105510.

Bhatia, S. C., & Devraj, S. (2017). *Pollution control in textile industry*. WPI Publishing.

Bhatt, P., Bhandari, G., Turco, R. F., Aminikhoei, Z., Bhatt, K., & Simsek, H. (2022). Algae in wastewater treatment, mechanism, and application of biomass for production of value-added product. *Environmental Pollution, 309*, 119688.

Bhattacharya, M., & Goswami, S. (2020). Microalgae – A green multi-product biorefinery for future industrial prospects. *Biocatalysis and Agricultural Biotechnology, 25*, 101580.

Bhushan, S., Veeragurunathan, V., Bhagiya, B. K., Krishnan, S. G., Ghosh, A., & Mantri, V. A. (2023). Biology, farming and applications of economically important red seaweed *Gracilaria edulis* (S. G. Gmelin) P. C. Silva: A concise review. *Journal of Applied Phycology, 35*(3), 983–996.

Chapman, R. L. (2013). Algae: The world's most important "plants"—An introduction. *Mitigation and Adaptation Strategies for Global Change, 18*, 5–12.

Cheirsilp, B., Maneechote, W., Srinuanpan, S., & Angelidaki, I. (2023). Microalgae as tools for bio-circular-green economy: Zero-waste approaches for sustainable production and biorefineries of microalgal biomass. *Bioresource Technology, 387*, 129620.

Chew, K. W., Yap, J. Y., Show, P. L., Suan, N. H., Juan, J. C., Ling, T. C., …, & Chang, J. S. (2017). Microalgae biorefinery: High value products perspectives. *Bioresource Technology, 229*, 53–62.

Choudhary, N., Tripathi, A., Singh, P. K., Parikh, H. S., & Tiwari, A. (2024). Application of algae for enhanced plant growth and food productivity. *Systems Microbiology and Biomanufacturing, 4*(2), 564–574.

Condori, M. A. M., Gutierrez, M. E. V., Oviedo, R. D. N., & Choix, F. J. (2024). Valorization of nutrients from fruit residues for the growth and lipid production of *Chlorella* sp.: A vision of the circular economy in Peru. *Journal of Applied Phycology, 36*(1), 101–111.

Dahai, H., Zhihong, Y., Lin, Q., Yuhong, L., Lei, T., Jiang, L., & Liandong, Z. (2024). The application of magical microalgae in carbon sequestration and emission reduction: Removal mechanisms and potential analysis. *Renewable and Sustainable Energy Reviews, 197*, 114417.

Deepa, P., Sowndhararajan, K., & Kim, S. (2023). A review of the harvesting techniques of microalgae. *Water, 15*(17), 3074.

Dolganyuk, V., Belova, D., Babich, O., Prosekov, A., Ivanova, S., Katserov, D., ..., & Sukhikh, S. (2020). Microalgae: A promising source of valuable bioproducts. *Biomolecules*, *10*(8), 1153.

El-Sheekh, M. M., Galal, H. R., Mousa, A. S. H., & Farghl, A. A. (2023). Coupling wastewater treatment, biomass, lipids, and biodiesel production of some green microalgae. *Environmental Science and Pollution Research*, *30*(12), 35492–35504.

Eppink, M. H., Olivieri, G., Reith, H., van den Berg, C., Barbosa, M. J., & Wijffels, R. H. (2019). From current algae products to future biorefinery practices: A review. *Biorefineries*, 166, 99–123.

Ersahin, M. E., Cicekalan, B., Cengiz, A. I., Zhang, X., & Ozgun, H. (2023). Nutrient recovery from municipal solid waste leachate in the scope of circular economy: Recent developments and future perspectives. *Journal of Environmental Management*, *335*, 117518.

Fakhri, M., Astryanti, S., Arifin, N. B., Putriani, O., Zalni, A., Riyani, E., ..., & Hariati, A. M. (2024). Cultivation of *Dunaliella* sp. using fish processing wastewater as a nutrient source: Effect on growth, biomass production, and biochemical profile. *Journal of microbiology, biotechnology and food sciences*, *13*(4), e10127–e10127.

Fawzy, M. A., Abdelsalam, S., Hafez, W. A., & Fathi, A. A. (2024). Bioremediation potential of microalgae for copper ion from wastewater and its impact on growth and biochemical contents: Equilibrium isotherm studies. *Materials Research Express*, *11*(5), 055404.

Fayyaz, MehmoodaM., Chew, Kit Wayne, Show, Pau Loke., Ling, Tau Chuan., Ng, I-Son., & Chang, Jo-Shu. (2020). Genetic engineering of microalgae for enhanced biorefinery capabilities. *Biotechnology Advances*, *43*, 107554.

Fernandes, F., Silkina, A., Gayo-Peláez, J. I., Kapoore, R. V., de La Broise, D., & Llewellyn, C. A. (2022). Microalgae cultivation on nutrient rich digestate: The importance of strain and digestate tailoring under PH control. *Applied Sciences*, *12*(11), 5429.

Filote, C., Santos, S. C., Popa, V. I., Botelho, C. M., & Volf, I. (2021). Biorefinery of marine macroalgae into high-tech bioproducts: A review. *Environmental Chemistry Letters*, *19*, 969–1000.

Gautam, R., Patial, S. K., & Singh, S. (2024). Algae biomass: Importance, harvesting techniques, extraction methods, and associated challenges. In *Value added products from bioalgae based biorefineries: Opportunities and challenges* (pp. 67–94). Singapore: Springer Nature Singapore.

Giusti, L. (2009). A review of waste management practices and their impact on human health. *Waste Management*, *29*(8), 2227–2239.

Goswami, R. K., Agrawal, K., & Verma, P. (2023). Multifaceted role of microalgae for municipal wastewater treatment: A futuristic outlook toward wastewater management. *CLEAN–Soil, Air, Water*, *51*(3), 2100286.

Hoek, C., Mann, D. G., & Jahns, H. M. (1995). *Algae: An introduction to phycology*. Cambridge University Press.

Hu, D., Zhang, J., Chu, R., Yin, Z., Hu, J., Nugroho, Y. K., ..., & Zhu, L. (2021). Microalgae *Chlorella vulgaris* and *Scenedesmus dimorphus* co-cultivation with landfill leachate for pollutant removal and lipid production. *Bioresource Technology*, *342*, 126003.

Khoo, C. G., Dasan, Y. K., Lam, M. K., & Lee, K. T. (2019). Algae biorefinery: Review on a broad spectrum of downstream processes and products. *Bioresource Technology*, *292*, 121964.

Kostas, E. T., Adams, J. M., Ruiz, H. A., Durán-Jiménez, G., & Lye, G. J. (2021). Macroalgal biorefinery concepts for the circular bioeconomy: A review on biotechnological developments and future perspectives. *Renewable and Sustainable Energy Reviews*, *151*, 111553.

Kumar, C., Sharma, M., Kaur, M., Arya, S. K., & Khatri, M. (2024). Cultivation of algae: Techniques and challenges. In *Value Added Products From Bioalgae Based Biorefineries: Opportunities and Challenges* (pp. 43–65). Singapore: Springer Nature Singapore.

Kumar, L., & Bharadvaja, N. (2020). A review on microalgae biofuel and biorefinery: Challenges and way forward. *Energy Sources, Part A: Recovery, Utilization, and Environmental Effects*, 46, 16599-16622.

Kumar, N., Banerjee, C., Negi, S., & Shukla, P. (2023). Microalgae harvesting techniques: Updates and recent technological interventions. *Critical Reviews in Biotechnology*, *43*(3), 342–368.

Laurens, L. M., Markham, J., Templeton, D. W., Christensen, E. D., Van Wychen, S., Vadelius, E. W., ..., & Pienkos, P. T. (2017). Development of algae biorefinery concepts for biofuels and bioproducts; a perspective on process-compatible products and their impact on cost-reduction. *Energy & Environmental Science*, *10*(8), 1716–1738.

Leandro, A., Pereira, L., & Gonçalves, A. M. (2019). Diverse applications of marine macroalgae. *Marine Drugs*, *18*(1), 17.

Leong, Y. K., & Chang, J. S. (2020). Bioremediation of heavy metals using microalgae: Recent advances and mechanisms. *Bioresource Technology*, *303*, 122886.

Liang, Y., Kashdan, T., Sterner, C., Dombrowski, L., Petrick, I., Kröger, M., & Höfer, R. (2015). Algal biorefineries. In *Industrial biorefineries & white biotechnology* (pp. 35–90). Elsevier.

Maes, D., Van Dael, M., Vanheusden, B., Goovaerts, L., Reumerman, P., Luzardo, N. M., & Van Passel, S. (2015). Assessment of the sustainability guidelines of EU Renewable Energy Directive: The case of biorefineries. *Journal of Cleaner Production*, *88*, 61–70.

Malyan, S. K., Kumar, S., Singh, R., Singh, S., Anand, G., Upadhyay, S., …, & Kumar, S. S. (2024). Algal intervention as nature-based solution for treatment of landfill leachate. In *Algae based bioelectrochemical systems for carbon sequestration, carbon storage, bioremediation and bioproduct generation* (pp. 21–32). Academic Press.

Marques, F. A. (2023). *Heterotrophic cultivation of Chlorella vulgaris using agri-food by-products* (Doctoral dissertation).

Mathimani, T., Alshiekheid, M. A., Sabour, A., Le, T. H. T., & Xia, C. (2024). Appraising the phycoremediation potential of cyanobacterial strains *Phormidium* and *Oscillatoria* for nutrient removal from textile wastewater (TWW) and synchronized biodiesel production from TWW-tolerant biomass. *Environmental Research*, *241*, 117628.

Meeranayak, U. F. J., Nadaf, R. D., Toragall, M. M., Nadaf, U., & Shivasharana, C. T. (2020). The role of algae in sustainable environment: A review. *Journal of Algal Biomass Utilization*, *11*(2), 28–34.

Michalak, I., & Chojnacka, K. (2013). Algal compost – Toward sustainable fertilization. *Reviews in Inorganic Chemistry*, *33*(4), 161–172.

Monteiro, C. M., Castro, P. M., & Malcata, F. X. (2012). Metal uptake by microalgae: Underlying mechanisms and practical applications. *Biotechnology Progress*, *28*(2), 299–311.

Moreira, J. B., Santos, T. D., Duarte, J. H., Bezerra, P. Q. M., de Morais, M. G., & Costa, J. A. V. (2023). Role of microalgae in circular bioeconomy: From waste treatment to biofuel production. *Clean Technologies and Environmental Policy*, *25*(2), 427–437.

Ms, K., Johnson, I., Ngo, H. H., Guo, W., & Kumar, M. (2023). Application of *Chlorella vulgaris* for nutrient removal from synthetic wastewater and MBR-treated bio-park secondary effluent: Growth kinetics, effects of carbon and phosphate concentrations. *Environmental Monitoring and Assessment*, *195*(3), 415.

Mustafa, S., Bhatti, H. N., Maqbool, M., & Iqbal, M. (2021). Microalgae biosorption, bioaccumulation and biodegradation efficiency for the remediation of wastewater and carbon dioxide mitigation: Prospects, challenges and opportunities. *Journal of Water Process Engineering*, *41*, 102009.

Nanda, S., & Berruti, F. (2021). Municipal solid waste management and landfilling technologies: A review. *Environmental Chemistry Letters*, *19*(2), 1433–1456.

Narayanan, M. (2024). Promising biorefinery products from marine macro and microalgal biomass: A review. *Renewable and Sustainable Energy Reviews*, *190*, 114081.

Oliomogbe, T. I., Nkollo, M. I., & Emegha, J. O. (2024). Assessing the suitability of green algae (*Chlorella sorokiniana*) as biosorbent for removal of Cu (II) and Zn (II) ions from synthetic wastewater in a batch system. *Journal of Applied Sciences and Environmental Management*, *28*(4), 1159–1164.

Öztürk, S., & Satpati, G. G. (2024). Extremophilic algae-based wastewater treatment, nutrient recovery, and animal feed production. *Trends in Biotechnology of Polyextremophiles*, 243–266.

Pandey, A., Srivastava, S., & Kumar, S. (2024). *Scenedesmus* sp. ASK22 cultivation using simulated dairy wastewater for nutrient sequestration and biofuel production: Insight into fuel properties and their blends. *Biomass Conversion and Biorefinery*, *14*(3), 3305–3317.

Parikh, H. (2021a). Algae is an Efficient Source of Biofuel. *International Research Journal of Engineering and Technology (IRJET)*, *8*(11).

Parikh, H. (2021b). Diatom biosilica as a source of nanomaterials. *International Journal of All Research Education and Scientific Methods*, *9*(11), 659.

Parikh, H. S., Singh, P. K. & Tiwari, A. (2024). Algal biorefinery: Focus on cosmeceuticals. *Systems Microbiology and Biomanufacturing*. *4*(4),1–23.

Penloglou, G., Pavlou, A., & Kiparissides, C. (2024). Recent advancements in photo-bioreactors for microalgae cultivation: A brief overview. *Processes*, *12*(6), 1104.

Peter, A. P., Tan, X., Lim, J. Y., Chew, K. W., Koyande, A. K., & Show, P. L. (2022). Environmental analysis of *Chlorella vulgaris* cultivation in large scale closed system under waste nutrient source. *Chemical Engineering Journal*, *433*, 134254.

Pratama, W. D., & Hadiyanto, H. (2024). Evaluation of different electrodes in electrocoagulation-flotation process for *Chlorella vulgaris* harvesting. *Case Studies in Chemical and Environmental Engineering, 10*, 100801.

Rao, P. H., Kumar, R. R., & Mohan, N. (2019). Phycoremediation: Role of algae in waste management. *Environmental Contaminants: Ecological Implications and Management*, 49–82.

Rocha, C. J. L., Álvarez-Castillo, E., Yáñez, M. R. E., Bengoechea, C., Guerrero, A., & Ledesma, M. T. O. (2020). Development of bioplastics from a microalgae consortium from wastewater. *Journal of Environmental Management, 263*, 110353.

Russell, C., Rodriguez, C., & Yaseen, M. (2022). High-value biochemical products & applications of freshwater eukaryotic microalgae. *Science of the Total Environment, 809*, 151111.

Sæther, M., Diehl, N., Monteiro, C., Li, H., Niedzwiedz, S., Burgunter-Delamare, B., …, & Forbord, S. (2024). The sugar kelp *Saccharina latissima* II: Recent advances in farming and applications. *Journal of Applied Phycology*, 36(4),1–33.

Satpati, G. G. (2021). Solid waste management by algae: Current applications and future perspectives. *Pollution Research, 40*(1), 260–265.

Schwarz, P., Herrfurth, C., Steinem, C., & Feussner, I. (2022). Lipidomics of *Thalassiosira pseudonana* as a function of valve SDV synthesis. *Journal of Applied Phycology, 34*(3), 1471–1481.

Selvaraj, D., & Arivazhagan, M. (2024). Synergistic effects of *Spirulina platensis* cultivation in textile wastewater towards nutrient removal and seed germination study. *Environmental Pollution, 357*, 124435.

Shah, S. N., Manzoor, S., & Asim, M. (2021). Impact of industrial pollution on our society. *Pakistan Journal of Science, 73*(1), 222.

Sharma, Y., Singh, A. D., Mehariya, S., Karthikeyan, O. P., Singh, G. P., Singh, C. P., & Molino, A. (2022). Genetic engineering in improving the output of algal biorefinery. In *Algal biorefineries and the circular bioeconomy* (pp. 1–60). CRC Press.

Silva, T. A., de Paula Pereira, A. S. A., Ferreira, J., Lorentz, J. F., de Assis, M. L., Assemany, P. P., …, & Calijuri, M. L. (2024). Enhancing microalgae biomass production: Exploring improved scraping frequency in a hybrid cultivation system. *Journal of Environmental Management, 355*, 120505.

Singh, G., & Patidar, S. K. (2018). Microalgae harvesting techniques: A review. *Journal of Environmental Management, 217*, 499–508.

Singh, P. K., Marella, T. K., Bhattacharjya, R., Tyagi, R., Plaha, N. S., Kaushik, N., & Tiwari, A. (2024a). Marine diatom algae cultivation in simulated dairy wastewater and biomass valorization. *Environmental Science and Pollution Research, 31*(46), 57466–57477.

Singh, P. K., Parikh, H., Saxena, A., Mishra, B., Tyagi, R., Awasthi, M. K., …, & Tiwari, A. (2024b). Managing municipal wastewater remediation employing alginate-immobilized marine diatoms and silver nanoparticles. *Energy & Environment, 0*(0). https://doi.org/10.1177/0958305X241246185 .

Singh, S., Singh, L., Kumar, V., Ali, W., Ramamurthy, P. C., Dhanjal, D. S., …, & Khan, N. A. (2023b). Algae-based approaches for Holistic wastewater management: A low-cost paradigm. *Chemosphere, 345*, 140470.

Sirotiya, V., Ahirwar, A., Mourya, M., Khan, M. J., Rai, A., Kwatra, R., ... & Vinayak, V. (2023). Astaxanthin bioaccumulation in microalgae under environmental stress simulated in industrial effluents highlighting prospects of *Haematococcus pluvialis*: Knowledge gaps and prospective approaches. *Phytochemistry Reviews, 22*(4), 1041–1066.

Soobhany, N. (2019). Insight into the recovery of nutrients from organic solid waste through biochemical conversion processes for fertilizer production: A review. *Journal of Cleaner Production, 241*, 118413.

Swar, S. S., Boonnorat, J., & Ghimire, A. (2023). Algae-based treatment of a landfill leachate pretreated by coagulation-flocculation. *Journal of Environmental Management, 342*, 118223.

Talebi, S., Edalatpour, A., & Tavakoli, O. (2022). Algal biorefinery: A potential solution to the food–energy–water–environment nexus. *Sustainable Energy & Fuels, 6*(11), 2623–2664.

Tang, C., Gao, X., Hu, D., Dai, D., Qv, M., Liu, D., & Zhu, L. (2023). Nutrient removal and lipid production by the co-cultivation of *Chlorella vulgaris* and *Scenedesmus dimorphus* in landfill leachate diluted with recycled harvesting water. *Bioresource Technology, 369*, 128496.

Tea, M., Francesco, S., Veronica, L., Concas, A., & Antonio, L. G. (2023). Dairy wastewater to promote mixotrophic and heterotrophic metabolism in *Chlorella Vulgaris*: Effect on growth and FAME profile. *Chemical Engineering Transactions, 98*, 81–86.

Tiwari, D., Kumar, N., Bongirwar, R., & Shukla, P. (2024). Nutraceutical prospects of genetically engineered cyanobacteria-technological updates and significance. *World Journal of Microbiology and Biotechnology*, *40*(9), 263.

Trivedi, J., Aila, M., Bangwal, D. P., Kaul, S., & Garg, M. O. (2015). Algae based biorefinery—How to make sense?. *Renewable and Sustainable Energy Reviews*, *47*, 295–307.

Verdezoto-Prado, J., Chicaiza-Ortiz, C., Mejía-Pérez, A. B., Freire-Torres, C., Viteri-Yánez, M., Deng, L., Barba-Ostria, C., & Guamán, L. P. (2024). Advances in environmental biotechnology with CRISPR/Cas9: Bibliometric review and cutting-edge applications. *7*(3).

Villanova, V., Engelbrektsson, J., Strömberg, N., Ekendahl, S., & Spetea, C. (2024). Unveiling the ecological resilience and industrial potential of *Skeletonema marinoi* through mixotrophic cultivation in Nordic winter condition. *Physiologia Plantarum*, *176*(3), e14308.

Viswanathan, K., Huang, J. M., Tsai, T. H., Chang, J. S., & Wu, W. (2023). Exploration of algal biorefinery frameworks: Optimization, quantification of environmental impacts and economics. *Algal Research*, *69*, 102903.

Webster, L. J., Villa-Gomez, D., Brown, R., Clarke, W., & Schenk, P. M. (2024). A synthetic biology approach for the treatment of pollutants with microalgae. *Frontiers in Bioengineering and Biotechnology*, *12*, 1379301.

Wu, W., & Chang, J. S. (2019). Integrated algal biorefineries from process systems engineering aspects: A review. *Bioresource Technology*, *291*, 121939.

Wutthithien, P., & Incharoensakdi, A. (2024). Nutrients removal from wastewater by microalga *Chlorella* sp. coupled to augmented lipid production with spent wastewater utilized by cyanobacterium *Fischerella muscicola* TISTR 8215 for hydrogen production. *Journal of Applied Phycology*, 1–12.

Xiao, X., Li, W., Jin, M., Zhang, L., Qin, L., & Geng, W. (2023). Responses and tolerance mechanisms of microalgae to heavy metal stress: A review. *Marine Environmental Research*, *183*, 105805.

Yadav, G., Shanmugam, S., Sivaramakrishnan, R., Kumar, D., Mathimani, T., Brindhadevi, K., …, & Rajendran, K. (2021). Mechanism and challenges behind algae as a wastewater treatment choice for bioenergy production and beyond. *Fuel*, *285*, 119093.

Yadav, K., Vasistha, S., Nawkarkar, P., Kumar, S., & Rai, M. P. (2022). Algal biorefinery culminating multiple value-added products: Recent advances, emerging trends, opportunities, and challenges. *3 Biotech*, *12*(10), 244.

Yang, H., Ma, M., Thompson, J. R., & Flower, R. J. (2018). Waste management, informal recycling, environmental pollution and public health. *Journal of Epidemiology and Community Health*, *72*(3), 237–243.

Index

Made in the USA
Las Vegas, NV
19 September 2026

6ed82df4-332f-449b-94a9-e5c4111ce0d4R01